# Natural Organic Matter and Disinfection By-Products

ACS SYMPOSIUM SERIES **761**

# Natural Organic Matter and Disinfection By-Products

## Characterization and Control in Drinking Water

**Sylvia E. Barrett,** EDITOR
*Metropolitan Water District of Southern California*

**Stuart W. Krasner,** EDITOR
*Metropolitan Water District of Southern California*

**Gary L. Amy,** EDITOR
*University of Colorado*

American Chemical Society, Washington, DC

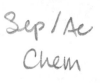

**Library of Congress Cataloging-in-Publication Data**

Natural organic matter and disinfection by-products characterization and control in drinking water / Sylvia E. Barrett, editor, Stuart W. Krasner, editor, Gary L. Amy, editor.

  p.    cm.–(ACS symposium series, ISSN 0097-6156 ; 761)

"Developed from a symposium sponsored by the Division of Environmental Chemistry at the 217th National Meeting of the American Chemical Society, Anaheim, CA, March 21-25, 1999.

Includes bibliographical references and index.

ISBN 0–8412–3676–3 (alk. paper)

  1. Organic water pollutants—Congresses. 2. Water—Purification—Disinfection—By-products—Congresses. 3. Drinking water—Purification—Congresses.

  I. Barrett, Sylvia E., 1943– . II. Krasner, Stuart W., 1949– . III. Amy, Gary L., 1947– . IV. American Chemical Society. Division of Environmental Chemistry. V. American Chemical Society. Meeting (217th : 1999 : Anaheim, Calif.) VI. Series.

628.1′662–dc21

                                                                                00-023307
                                                                                CIP

The paper used in this publication meets the minimum requirements of American National Standard for Information Sciences—Permanence of Paper for Printed Library Materials. ANSI Z39.48–1984.

# Foreword

THE ACS SYMPOSIUM SERIES was first published in 1974 to provide a mechanism for publishing symposia quickly in book form. The purpose of the series is to publish timely, comprehensive books developed from ACS sponsored symposia based on current scientific research. Occasionally, books are developed from symposia sponsored by other organizations when the topic is of keen interest to the chemistry audience.

Before agreeing to publish a book, the proposed table of contents is reviewed for appropriate and comprehensive coverage and for interest to the audience. Some papers may be excluded in order to better focus the book; others may be added to provide comprehensiveness. When appropriate, overview or introductory chapters are added. Drafts of chapters are peer-reviewed prior to final acceptance or rejection, and manuscripts are prepared in camera-ready format.

As a rule, only original research papers and original review papers are included in the volumes. Verbatim reproductions of previously published papers are not accepted.

ACS BOOKS DEPARTMENT

# Contents

## FORMATION OF CHLORINATION DISINFECTION BY-PRODUCTS

## CHEMISTRY OF ALTERNATIVE DISINFECTANTS

## DISINFECTION BY-PRODUCT METHODS DEVELOPMENT

## INDEXES

# Preface

An understanding of the relationship of natural organic matter (NOM) to disinfection by-product (DBP) formation in drinking water will make it possible to reach cost-effective decisions on source-water protection, treatment-plant process optimization, and distribution-system operations to control potentially harmful DBPs. The drinking water industry is required to minimize DBPs while ensuring adequate disinfection. This requirement is based on public health safety and is regulated in the United States by the U.S. Environmental Protection Agency and in other countries by various government agencies. Many complex issues arise when regulations are being established and enforced. Not only is a balance between disinfection efficiency and DBP control necessary, but also much work must still be done to determine the potential risk of compounds that have been identified in drinking water and to prioritize research on newly identified by-products.

The three general approaches to controlling DBPs are to control their formation during the treatment process, to remove them after formation, or to remove their precursors during the treatment process. NOM is a precursor to DBPs, as is the bromide ion. Because NOM varies from water supply to water supply, characterization of NOM is essential in understanding how best to remove NOM. In addition, an understanding of NOM's reactivity with disinfectants is critical in preventing the formation of DBPs and in determining which part of the NOM is most important to remove. NOM characterization and reactivity testing of specific fractions with various disinfectants help to optimize the water-treatment process. This book emphasizes characterization and reactivity of the previously less widely studied polar NOM. The procedure for comprehensive isolation of NOM from water for spectral characterization and reactivity testing that is presented in the book offers a unique approach. A number of different techniques for NOM isolation are available rather than a single uniform protocol. This means that selection of the proper method for NOM isolation could be quite challenging. We hope that this book will help to begin the process of standardization.

Although chlorination DBPs have been studied more extensively than any other group of DBPs, additional areas of investigation require more work. The chapter on modeling chlorine decay and DBP formation in the treated-water distribution system is timely, as more attention is currently directed toward providing water of high quality despite aging distribution systems.

As alternative disinfectants are used to meet increasingly stringent regulations, a better understanding of NOM–disinfectant reactions—such as ozone and hydroxyl radical chemistry with NOM—aids in decisions on treatment. In addition, alternative disinfectants may form different DBPs than are formed by chlorine. For example, chloramination has been reported to form significant levels of dihalogenated compounds compared to the levels of trihalogenated compounds formed by chlorination. Chapters in this book discuss similar findings and elaborate on better methods for the determination of nine haloacetic acids.

New analytical methods or modifications of existing methods are being developed for a full range of new DBPs to better characterize NOM and its reactivity with disinfectants and to determine polar and nonvolatile DBPs. This book emphasizes innovative new methods, such as a capillary electrophoresis method for haloacetic acids and a liquid chromatography–mass spectrometry method for the identification of polar drinking water DBPs.

We feel that this book will be of interest to a wide range of scientists, engineers, and consultants, including NOM and DBP experts, as well as water-utility chemists and engineers whose utilities will be expected to meet the Disinfectants/DBP Rule within the next few years. The series of books based on the chlorination conferences of the 1970s and 1980s and more recent books based on the American Chemical Society (ACS) symposia on NOM–DBPs in water treatment have helped advance our understanding of DBP formation and control. This book continues in that tradition.

A symposium entitled "Natural Organic Matter and Disinfection By-Products: Characterization and Control in Drinking Water," held at the 217th National Meeting of the ACS, sponsored by the ACS Division of Environmental Chemistry, in Anaheim, California, on March 21–23, 1999, provided the foundation for this book. We thank all of the people who participated in the symposium and those who provided chapters for this book. We also thank those who served as peer reviewers of the chapter manuscripts. We acknowledge and express our utmost appreciation to Metropolitan's technical editor, Peggy Kimball, whose dedicated effort greatly enhanced the quality of this book.

SYLVIA E. BARRETT
Metropolitan Water District
 of Southern California
700 Moreno Avenue
La Verne, CA 91750–3399

STUART W. KRASNER
Metropolitan Water District
 of Southern California
700 Moreno Avenue
La Verne, CA 91750–3399

GARY L. AMY
Department of Civil,
 Environmental, and
 Architectural Engineering
University of Colorado
Boulder, CO 80309–0421

# INTRODUCTION

Chapter 1

# Natural Organic Matter and Disinfection By-Products: Characterization and Control in Drinking Water—An Overview

Sylvia E. Barrett[1], Stuart W. Krasner[1], and Gary L. Amy[2]

[1]Metropolitan Water District of Southern California, 700 Moreno Avenue, La Verne, CA 91750–3399
[2]Department of Civil, Environmental, and Architectural Engineering, University of Colorado at Boulder, Campus Box 428, Boulder, CO 80309–0421

This chapter presents an overview of past research on natural organic matter (NOM) characterization and its relationship to disinfection by-product (DBP) formation and control in drinking water. The drinking water industry is required to minimize DBP formation while ensuring adequate disinfection. A large body of occurrence, treatment, and health effects data exists for certain DBPs that are present in drinking water, primarily for trihalomethanes and other chlorination by-products. This chapter provides background and a brief introduction to each section of the book, organized according to the book's five main topics: "Regulatory and Health Effects Background"; "NOM Characterization and Reactivity"; "Formation of Chlorination DBPs"; "Chemistry of Alternative Disinfectants"; and "DBP Methods Development."

When presenting an overview of past research on natural organic matter (NOM) characterization and its relationship to disinfection by-product (DBP) formation and control in drinking water, it is essential to refer to past conferences on the subject. The chlorination conferences of the 1970s and 1980s (1–7), more recent forums such as the American Chemical Society (ACS) symposia on NOM/DBPs in water treatment (8, 9), an American Water Works Association Research Foundation (AWWARF) workshop on NOM in drinking water (10), and an International Water Supply Association workshop on the influences of NOM characteristics on drinking water treatment and quality (11) provide an excellent background for this chapter and this book. The book's five main topics are "Regulatory and Health Effects Background," "NOM Characterization and Reactivity," "Formation of Chlorination

2

DBPs," "Chemistry of Alternative Disinfectants," and "DBP Methods Development." These topics will be discussed to provide background and a brief introduction to each section of the book.

## Regulatory and Health Effects Background

The drinking water industry is required to minimize DBP formation while ensuring adequate disinfection (12). Many complex issues arise when establishing DBP regulations (13). Not only is a balance between disinfection efficiency and DBP control necessary, but there may also be risk/risk tradeoff issues between by-products of chlorination and those of alternative disinfectants (14). However, much work remains to be done to determine the potential risk of compounds that have been identified in drinking water as well as to assign priorities to research on newly identified (and yet to be identified) by-products.

Initially, in the 1970s, Rook (15) in the Netherlands and researchers at the U.S. Environmental Protection Agency (USEPA) (16) demonstrated that trihalomethanes (THMs) were formed during the treatment process as a result of chlorination. Subsequently, the National Organics Reconnaissance Survey (17) and the National Organics Monitoring Survey (18) conducted by the USEPA found and confirmed, respectively, that THMs were present in the drinking water of utilities that used free chlorine in their treatment processes. After assessing the occurrence of THMs in drinking water, the USEPA began investigations to determine the health effects of THMs (19). Chloroform was shown to be a carcinogen in animal studies (20-21). Also, epidemiological studies showed an association between chlorinated surface water and cancer (22-24), and suggested a human health risk (25).

On November 29, 1979, because of the possibility of THMs being human carcinogens, the THM regulation was promulgated (26). In 1998, the USEPA promulgated the Stage 1 Disinfectants/DBP (D/DBP) Rule, which included a more stringent regulation for THMs and additional standards for other chlorination DBPs (i.e., haloacetic acids, or HAAs), a chlorine dioxide by-product (chlorite), an ozone by-product (bromate), and the disinfectants themselves, as well as requirements (i.e., enhanced coagulation) for the removal of DBP precursors (i.e., NOM) (12). During the rulemaking process in the 1990s, it was recognized that additional information was needed on how to balance the control of DBPs and the microbiological quality of the treated water. An Information Collection Rule (ICR) (27) was established that required extensive monitoring of water through the treatment process to provide the USEPA with occurrence data to evaluate which DBPs should be regulated under Stage 2 of the D/DBP Rule scheduled for promulgation by 2002.

Monitoring requirements in the ICR included four THMs, six to nine HAAs, four haloacetonitriles (HANs), two haloketones, chloral hydrate, chloropicrin, cyanogen chloride (CNCl), total organic halogen (TOX), seven aldehydes, chlorite, chlorate, and bromate. Analytical standards and methods for six HAAs have been available for routine monitoring (28). Analytical standards and methods for the remaining three HAAs have been recently introduced (29). Several laboratories were analyzing for nine HAAs at the time that ICR data were collected, so these data were submitted where possible. Data for the above compounds are currently being evaluated, and the rule-making process for Stage 2 began in December 1998.

Potential health effects associated with DBPs include carcinogenicity, adverse reproductive and developmental effects, and immunotoxic and neurotoxic effects (30). Recently, Boorman and colleagues (31) reviewed the existing cancer data on DBPs and outlined an approach for cancer hazard assessment for DBPs in drinking water. Because of the complexity of the problem, the USEPA, the National Institute of Environmental Health Sciences, and the U.S. Army are working together to develop a comprehensive biological and mechanistic DBP database (31). In addition, the USEPA convened an expert panel to evaluate the recent epidemiology data on adverse reproductive and developmental effects and DBPs (32).

Past toxicology studies have been useful in identifying potential health effects for various individual by-products, but little is known about the effects of mixtures in low concentrations such as those in drinking water to which a human being may be exposed. Chapters in the "Regulatory and Health Effects Background" section address the toxicology of DBPs and the need to understand risks at environmental exposure doses. In addition, another chapter shows how a better understanding of the occurrence of DBPs, their precursors, and the treatment process for controlling DBP formation provide for rule development based on sound science.

## NOM Characterization and Reactivity

During drinking-water treatment, NOM can be removed intact (i.e., separation) during coagulation (33, 34), granular activated carbon (GAC) adsorption (35, 36), membrane filtration (37, 38), or biological degradation (39, 40), or it can be transformed by oxidants such as ozone (40) or as part of an advanced oxidation process (41, 42). Characterization of NOM is essential in understanding how best to remove NOM (10). In addition, an understanding of NOM's reactivity with disinfectants is critical in minimizing the formation of DBPs and in identifying which fraction of the NOM is most important to remove. NOM characterization and reactivity testing of specific fractions with various disinfectants help optimize water treatment processes.

To control the bacteriological quality of drinking water while minimizing the amount of DBPs formed, some utilities currently use a minimal amount of disinfectant, or they move the disinfection to a point in the treatment process following the removal of a portion of the NOM. Some utilities in the U.S. are using alternative disinfectants (i.e., chloramines, ozone, and chlorine dioxide) (43). However, alternative disinfectants (e.g., ozone) also can react with NOM (44). Conventional coagulation removes a portion of the NOM. GAC, membranes, and enhanced coagulation can be used to improve the removal of NOM (45).

The NOM typically found in supplies used for drinking water has been classified as humic (nonpolar) and nonhumic (polar) material (46, 47). However, this definition of humic/nonhumic components of NOM has not been universally accepted. Humic and nonhumic NOM are functional definitions based on what is adsorbed (or not) on XAD resins (48). Indeed, hydrophilic ("nonhumic") fractions of NOM exhibit some of the properties typically observed for classic humic fractions. The humic NOM is most readily removed by conventional treatment, which consists of flocculation, sedimentation, and filtration (33, 46). The nonhumic NOM is more difficult to remove. In addition, inorganic constituents (e.g., bromide) can also affect DBP formation. Bromide passes conservatively through conventional treatment—coagulation or softening—and GAC adsorption and can lead to the formation of brominated DBPs (49, 50).

Traditionally, NOM characterization at most water utilities has focused on surrogate or general organic matter measurements, e.g., total organic carbon (TOC). Over the last two decades, the research community has begun to recognize the importance of the size, structure, and functionality of NOM. Elemental analysis helps characterize NOM (51), as does ultraviolet (UV) absorption analysis at 254 or 285 nanometers (nm) (47, 52). Moreover, the ratio of UV and dissolved organic carbon (DOC)—the specific UV absorbance (SUVA)—is a good indicator of the humic content of a water (53) and has been correlated with DBP formation (47). Reactivity with chlorine has been assessed through DBP formation potential (FP) tests (54, 55) and simulated distribution system (SDS) tests (56), and, more recently, uniform formation conditions (UFC) have been used (57). In addition, differential UV spectroscopy ($\Delta$UVA) at 272 nm has been used to monitor DBP formation (58). Moreover, there have been developments in other NOM characterization techniques such as carbon 13 nuclear magnetic resonance ($^{13}$C-NMR) (59, 60), pyrolysis-gas chromatography/mass spectrometry (Py-GC/MS) (61), fluorescence spectroscopy (62), size exclusion chromatography for molecular weight and size determinations (59, 63), and on-line TOC measurements (64).

The laboratory isolation and fractionation protocol for the hydrophobic fraction (humic fraction) of NOM is well established (48). The isolation of the hydrophilic (nonhumic) fraction is being developed (65, 66). The fractionation techniques that have traditionally been used to separate humic and nonhumic NOM are XAD resins (48). XAD-8 resins do not adsorb polar NOM (67), whereas XAD-4 resins isolate some of, but not all, the more polar NOM from inorganic matter (68-70).

Newer techniques employ ultrafiltration membranes to recover and desalt collodial NOM (71). Nanofiltration and reverse osmosis recover most DOC (72, 73), but desalting the inorganic matrix remains a problem. Resin combinations (74) and selective precipitation are being used to isolate fractions of polar NOM (65). The "NOM Characterization and Reactivity" section in the book elaborates on ways to characterize NOM, including new comprehensive isolation techniques. Some of these methods need further sophistication, improvements in efficiency, and eventually more standardization. A number of different techniques for NOM isolation are available rather than a single uniform protocol. This means that the selection of the proper method for NOM isolation might be quite challenging.

Detection methods that have been used in conjunction with the above separation techniques include organic carbon analysis, UV and fluorescence spectroscopy (47, 70), elemental analysis (51), molecular size distributions (63), pyrolysis–GC/MS (61), Fourier transform infrared spectroscopy (FTIR) (66), and $^{13}$C-NMR (60). Using a combination of these detection methods provides significantly more structural and functional group information than is provided with the use of a single method (47).

## Formation of Chlorination DBPs

DBPs found during the chlorination process have been studied more than any other group of DBPs (i.e., those formed during the use of alternative disinfectants). The Jolley water chlorination conferences, previous ACS symposia, and other work include postulations on the formation mechanisms of THMs, HAAs, and other DBPs. The detailed literature reviews by Rice and Gomez-Taylor (75) and Richardson (76) describe organic and inorganic by-products that have been identified in water with chlorine, chlorine dioxide, chloramines, and ozone. In the presence of bromide, higher concentrations of brominated compounds are formed in comparison to the formation of chlorinated compounds (77, 78). However, less work has been done on identification of the brominated compounds and their health effects. The "Formation of Chlorination DBPs" section includes modeling chlorine decay and DBP formation, particularly in the treated-water distribution system.

## Chemistry of Alternative Disinfectants

As alternative disinfectants are used to meet increasingly stringent regulations, a better understanding of NOM/disinfectant reactions—such as ozone and hydroxyl radical chemistry with NOM—aids in treatment decisions. Richardson (76) has published a comprehensive list of specific DBPs produced when humic material reacts with chlorine, ozone, chlorine dioxide, and chloramines or when these disinfectants are used in actual drinking-water treatment plants. To date, it is known that ozonation forms bromate (79), aldehydes (80), ketones, and carboxylic acids (81); chloramination forms HAAs plus CNCl (82-84) and cyanogen bromide (CNBr); and chlorine dioxide reactions in water result in the formation of chlorite and chlorate (85). In addition, chlorine dioxide has been shown to form organic by-products (86). Heller-Grossman and colleagues (87) found that chloramination yielded less CNBr, THMs, and HANs than chlorination of bromide-rich lake water, and chlorine dioxide yielded even less of these specific DBPs. However, Symons and colleagues (83) and Smith and colleagues (84) have shown that chloramination may form significant levels of dihalogenated compounds (e.g., dichloroacetic acid) compared to trihalogenated compounds (e.g., THMs, trichloroacetic acid). AWWARF and USEPA have sponsored research on by-products of ozonation (88), chlorine dioxide (86), chloramines (83), and UV disinfection (89). The "Chemistry of Alternative Disinfectants" section will discuss the application of radiation chemistry techniques to the study of DBP formation. In addition, one chapter compares DBP formation for four different disinfectants..

## DBP Methods Development

New analytical methods or modifications of existing methods are being developed for a full range of new DBPs to better characterize NOM and its reactivity with disinfectants. Liquid/liquid extraction is the classical solvent extraction approach, coupled with GC/electron capture (EC) or GC/MS to analyze for neutral chlorination by-products such as THMs and HANs (90). THMs can also be analyzed by purge–and–trap extraction coupled with GC (16). Derivatization methods are used in conjunction with GC where direct extraction is not feasible or chromatography and/or detection of the underivatized compounds is poor. For example, aldehydes cannot be easily extracted from water, but with the formation of a fluorinated oxime they can be extracted, separated in a GC, and detected with an EC detector (91). HAAs are derivatized (e.g., with diazomethane) to form methyl esters that are more readily chromatographable (55). The USEPA sponsored some of this methods development as part of a 1988-89 study of 21 DBPs (92). Derivatization GC/MS and FTIR spectroscopy have been used to study organic by-products of chlorine dioxide (86). Capillary electrophoresis has been used to study polar carbonyl (93) by-products of ozonation. Ion chromatography has been used to study bromate (*79, 88*) and carboxylic acids (*81*), by-products of ozonation, as well as chlorite and chlorate, by-products of chlorine dioxide treatment. Flow injection analysis has been used to analyze for chlorite and chlorate (*94*).

At a workshop of experts from diverse fields in 1998 (*95*), various needs related to DBP methods development were recognized, including a need for additional extraction and derivatization techniques, a need for higher concentration factors or improved instrumental sensitivity, and a need to explore other instrumental techniques. For example, there are methods of identifying nonhalogenated, nonvolatile DBPs by using liquid chromatography (LC)/MS or MS/MS ("tandem MS"). It was also recognized that new biological and toxicological screening tools are needed to help guide DBP methods research and regulatory development. For example, since the workshop, the USEPA has sponsored a study of 49 DBPs (*95, 96*). These DBPs, which had been tentatively or qualitatively identified in disinfected drinking water (*76*) were, in part, prioritized on the basis of structural activity relationships.

The "DBP Methods Development" section discusses current information on LC/MS analysis for polar, nonvolatile organic compounds and other new techniques. It also discusses improvements to one of the derivatization methods for HAA analysis for all nine species.

## Conclusion

The characterization and control of NOM and DBPs in drinking water present a complex problem. A multidisciplinary effort is needed that brings together the expertise and collaboration of chemists, engineers, toxicologists, and representatives of other disciplines to continue to move forward in an effective manner.

## References

1. Jolley, R. L., Ed. *Water Chlorination: Chemistry, Environmental Impact and Health Effects*; Ann Arbor Science Publishers: Ann Arbor, MI, 1978; Vol. 1.
2. Jolley, R. L., Gorchev, H., Hamilton, D. H., Eds. *Water Chlorination: Chemistry, Environmental Impact and Health Effects*; Ann Arbor Science Publishers: Ann Arbor, MI, 1978; Vol. 2.
3. Jolley, R. L., Brungs, W. A., Cumming, R. B., Jacobs, V. A., Eds. *Water Chlorination: Chemistry, Environmental Impact and Health Effects*; Ann Arbor Science Publishers: Ann Arbor, MI, 1980; Vol. 3.
4. Jolley, R. L., Brungs, W. A., Cotruvo, J. A., Cumming, R. B., Mattice, J. S., Jacobs, V. A., Eds. *Water Chlorination: Chemistry, Environmental Impact and*

*Health Effects—Chemistry and Water Treatment*; Ann Arbor Science Publishers: Ann Arbor, MI, 1983; Vol. 4 (Part 1).

5. Jolley, R. L., Brungs, W. A., Cotruvo, J. A., Cumming, R. B., Mattice, J. S., Jacobs, V. A., Eds. *Water Chlorination: Chemistry, Environmental Impact and Health Effects—Environment, Health, and Risk*; Ann Arbor Science Publishers: Ann Arbor, MI, 1983; Vol. 4 (Part 2).

6. Jolley, R. L., Bull, R. J., Davis, W. P., Katz, S., Roberts, M. H., Jacobs, V. A., Eds. *Water Chlorination: Chemistry, Environmental Impact and Health Effects*, Lewis Publishers: Chelsea, MI, 1985; Vol. 5.

7. Jolley, R. L., Condie, L. W., Johnson, J. D., Katz, S., Minear, R. A., Mattice, J. S., Jacobs, V. A., Eds. *Water Chlorination: Chemistry, Environmental Impact and Health Effects—Environment, Health, and Risk*; Lewis Publishers: Chelsea, MI, 1990; Vol. 6.

8. Minear, R. A., Amy, G. L., Eds. *Disinfection By-Products in Water Treatment: The Chemistry of Their Formation and Control*; CRC Press: Boca Raton, FL, 1996.

9. Minear, R. A., Amy, G. L. *Water Disinfection and Natural Organic Matter: Characterization and Control*; American Chemical Society: Washington, DC, 1996.

10. AWWARF. *Natural Organic Matter in Drinking Water: Origin, Characterization, and Removal*, Workshop Proceedings, Chamonix, France, Sept 19-22, 1993; AWWARF and AWWA: Denver, CO, 1994.

11. International Water Supply Association, *Influence of Natural Organic Matter Characteristics on Drinking Water Treatment and Quality*, Workshop Proceedings, Université de Poitiers, Poitiers, France, Sept 18-19, 1996.

12. U.S. Environmental Protection Agency. *Fed. Reg.* **1998,** *63* (241), 69390.

13. Means, E. G., III; Krasner, S. W. *J.—Am. Water Works Assoc.* **1993,** *85* (2), 68.

14. U.S. Environmental Protection Agency. *Fed. Reg.* 1994, 59 (145), 38668.

15. Rook, J. J. *Wat. Treat. Exam.* **1974,** *23* (Part 2), 234.

16. Bellar, T. A.; Lichtenberg, J. J.; Kroner, R. C. *J.—Am. Water Works Assoc.* **1974,** *66* (12), 703.

17. Symons, J. M.; Bellar, T. A.; Carswell, J. K.; DeMarco, J.; Kropp, K. L.; Robeck, G. G.; Seeger, D. R.; Slocum, C. J.; Smith, B. L.; Stevens, A. A.. *J.— Am. Water Works Assoc.* **1975,** *67,* 634.

18. Brass, H. J.; Freige, M. A.; Halloran, T.; Mello, J. W.; Munch, D.; Thomas, R. F. The National Organic Monitoring Survey: Sampling and Analysis for Purgeable Organic Compounds. In *Drinking Water Quality Enhancement Through Source Protection*; Pojasek, R. B., Ed.; Ann Arbor Science Publishers: Ann Arbor, MI, 1977.

19. National Research Council. *Drinking Water and Health: Disinfectants and Disinfection By-Products*; National Academy Press: Washington, DC, 1987; Vol. 7.

10

20. National Cancer Institute (NCI). *Report on Carcinogenesis Bioassay of Chloroform;* NTIS PB-264018. U.S. Government Printing Office: Washington, DC, 1976.
21. Jorgenson, T. A.; Meierhenry, E. F.; Rushbrook, C. J.; Bull, R. J.; Robinson, M. *Fundam. Appl. Toxicol.* **1985,** *5,* 760.
22. Wilkins, J. R.; Comstock, G. W. *Am. J. Epidemiol.* **1981,** *114,* 178.
23. Cantor, K. P.; Hoover, R.; Hartge, P.; Mason, T. J.; Silverman, D. Bladder Cancer, Tap Water Consumption, and Drinking Water Source. In *Water Chlorination: Chemistry, Environmental Impact and Health Effects— Environment, Health, and Risk;* Jolley, R. L., Condie, L. W., Johnson, J. D., Katz, S., Minear, R. A., Mattice, J. S., Jacobs, V. A., Eds.; Lewis Publishers: Chelsea, MI, 1990; Vol. 6.
24. Morris, R. D.; Audet, A.; Angelillo, I. O. *Am. J. Public Health* **1992,** *82,* 955.
25. U.S. Environmental Protection Agency. *Treatment Techniques for Controlling Trihalomethanes in Drinking Water;* Symons, J. M., Stevens, A. A., Clark, R. M., Geldreich, E. E., Love , O. T. Jr., DeMarco, J., Eds.; EPA-600/2-81-156. Drinking Water Research Division, Municipal Environmental Research Laboratory, Office of Research and Development: Cincinnati, OH, 1981.
26. U.S. Environmental Protection Agency. *Fed. Reg.* **1979,** *44* (231), 68624.
27. U.S. Environmental Protection Agency. *Fed. Reg.* **1996,** *61* (94), 24354.
28. U.S. Environmental Protection Agency. *Determination of Haloacetic Acids and Dalapon in Drinking Water by Ion-Exchange Liquid-Solid Extraction and Gas Chromatography with an Electron Capture Detector;* Hodgeson, J. W., Becker, D., Eds.; Environmental Monitoring Systems Laboratory, Office of Research and Development: Cincinnati, OH, 1992.
29. Brophy, K. S.; Weinberg, H. S.; Singer, P. C. *Natl. Meet.—Am. Chem. Soc., Div. Environ. Chem.* **1999,** *39* (1), 256.
30. International Life Sciences Institute (ILSI). *Disinfection By-Products in Drinkng Water: Critical Issues in Health Effects Research,* Workshop Report; ILSI Health and Environmental Sciences Institute: Washington, DC, 1995.
31. Boorman, G. A.; Dellarco, V.; Dunnick, J. K.; Chapin, R. E.; Hunter, S.; Hauchman, F.; Gardner, H.; Cox, M.; Sills, R. C. *Environ. Health Persp.* **1999,** *107* (Suppl. 1).
32. U.S. Environmental Protection Agency. *Recommendations for Conducting Epidemiological Research on Possible Reproductive and Developmental Effects of Exposure to Disinfected Drinking Water,* Panel Report; National Health and Environmental Effects Research Laboratory: Research Triangle Park, NC, 1998.
33. White, M. C.; Thompson, J. D.; Harrington, G. W.; Singer, P. C. *J.—Am. Water Works Assoc.* **1997,** *89* (5), 64.
34. Vilgé-Ritter, A.; Masiou, A.; Boulange, T.; Rybacki, D.; Bottero, J.-Y. *Environ. Sci. Technol.* **1999,** *33* (17), 3027.
35. Lykins, B. W. Jr.; Clark, R. M.; Adams, J. Q. *J.—Am. Water Works Assoc.* **1988,** *80* (5), 85.

36. McGuire, M. J.; Davis, M. K.; Tate, C. H.; Aeita, E. M.; Howe, E. W.; Crittenden, J. C. *J.—Am. Water Works Assoc.* **1991,** *83* (1), 38.
37. Taylor, J. S.; Thompson, D. M.; Carswell, J. K. *J.—Am. Water Works Assoc.* **1987,** *79* (8), 72.
38. Fu, P.; Ruiz, H.; Thompson, K.; Spangenberg, C. *J.—Am. Water Works Assoc.* **1994,** *86* (12), 55.
39. Hozalski, R. M.; Bouwer, E. J. Biofiltration for Removal of Natural Organic Matter. In *Formation and Control of Disinfection By-Products in Drinking Water;* Singer, P. C., Ed.; American Water Works Association: Denver, CO, 1999.
40. Miltner, R. J.; Shukairy, H. M.; Summers, R. S. *J.—Am. Water Works Assoc.* **1992,** *84* (11), 53.
41. Symons, J. M.; Worley, K. L. *J.—Am. Water Works Assoc.* **1995,** *87* (11), 66.
42. Andrews, S. A.; Huck, P. M.; Chute, A. J.; Bolton, J. R.; Anderson, W. A. UV Oxidation for Drinking Water: Feasibility Studies for Addressing Specific Water Quality Issues. In *Proceedings,* AWWA Water Quality Technology Conference, New Orleans, LA, Nov 12-16, 1995; American Water Works Association: Denver, CO, 1996.
43. AWWA; AWWARF. *WATER:\STATS, The Water Utility Database: 1996 Survey;* Denver, CO, 1998. [http://www.awwa.org/h20stats/h20stats.htm]
44. Najm, I. N.; Krasner, S. W. *J.—Am. Water Works Assoc.* **1995,** *87* (1), 106.
45. Jacangelo, J. G.; DeMarco, J.; Owen, D. M.; Randtke, S. J. *J.—Am. Water Works Assoc.* **1995,** *87* (1), 64.
46. Owen, D. M.; Amy, G. L.; Chowdhury, Z. K. *Characterization of Natural Organic Matter and Its Relationship to Treatability;* AWWARF and AWWA: Denver, CO, 1993.
47. Krasner, S. W.; Croué, J.-P.; Buffle, J.; Perdue, E. M. *J.—Am. Water Works Assoc.* **1996,** *88* (6), 66.
48. Malcolm, R. L.; MacCarthy, P. *Envir. International* **1992,** *18*, 597.
49. Amy, G.; Tan, L.; Davis, M. K. *Wat. Res.* **1991,** *25* (2), 191.
50. Summers, R. S.; Benz, M. A.; Shukairy, H. M.; Cummings, L. *J.—Am. Water Works Assoc.* **1993,** *85* (1), 88.
51. Huffman, E. W. D. Jr.; Stuber, H. A. *Humic Substances in Soil, Sediment, and Water-Geochemistry, Isolation, and Characterization;* Aiken, G. R., Ed.. Wiley-Interscience: New York, 1985.
52. Korshin, G. V.; Li, C.-W.; Benjamin, M. M. The Use of UV Spectroscopy to Study Chlorination of Natural Organic Matter. In *Water Disinfection and Natural Organic Matter: Characterization and Control;* Minear, R. A., Amy, G. L., Eds.; American Chemical Society: Washington, DC, 1996.
53. Edzwald, J. K.; Van Benschoten, J. E. Aluminum Coagulation of Natural Organic Matter. In *Proceedings of Fourth International Gothenburg Symposium on Chemical Treatment,* Madrid, Spain, 1990.
54. Reckhow, D. A.; Singer, P. C. *J.—Am. Water Works Assoc.* **1990,** *82* (4), 173.

55. Eaton, A. D.; Clesceri, L. S.; Greenberg, A. E., Eds. *Standard Methods for the Examination of Water and Wastewater*. American Public Health Association, American Water Works Association, Water Environment Federation: Washington, DC, 1995; 19th Ed.

56. Koch, B.; Krasner, S. W.; Sclimenti, M. J.; Schimpff, W. S. *J.—Am. Water Works Assoc.* **1991**, *83* (10), 62.

57. Summers, R. S.; Hooper, S. M.; Shukairy, H. M.; Solarik, G.; Owen, D. M. *J.—Am. Water Works Assoc.* **1996**, *88* (6), 80.

58. Li, C.-W.; Korshin, G. V.; Benjamin, M. M. *J.—Am. Water Works Assoc.* **1998**, *90* (8), 88.

59. Croué, J.-P.; DeBroux, J.-F.; Amy, G. L.; Aiken, G. R.; Leenheer, J. A. Natural Organic Matter: Structural Characteristics and Reactive Properties. In *Formation and Control of Disinfection By-products in Drinking Water*; Singer, P. C., Ed.; American Water Works Association: Denver, CO, 1999.

60. Harrington, G. W.; Bruchet, A.; Rybacki, D.; Singer, P. Characterization of Natural Organic Matter and Its Reactivity with Chlorine. In *Water Disinfection and Natural Organic Matter: Characterization and Control*; Minear, R. A., Amy, G. L., Eds.; American Chemical Society: Washington, DC, 1996.

61. Bruchet, A.; Rousseau, C.; Mallevialle, J. *J.—Am. Water Works Assoc.* **1990**, *82* (9), 66.

62. Ewald, M, Belin, C.; Weber, J. H.; *Environ. Sci. Technol.* **1983**, *17* (8), 501.

63. Chin, Y.-P.; Aiken, G.; O'Loughlin, E. *Environ. Sci. Technol.* **1994**, *26*, 11.

64. Frimmel, F. H.; Hasse, S.; Kleiser, S. *Natl. Meet.—Am. Chem. Soc., Div. Environ. Chem.* **1999**, *39* (1), 206.

65. Benjamin, M. M.; Croué, J.-P.; Leenheer, J. A. *Novel and Advanced Approaches for Concentrating, Isolating and Characterizing Natural Organic Matter*; AWWARF and AWWA: Denver, CO, 1999.

66. Leenheer, J. A.; Croué, J.-P.; Benjamin, M.; Korshin, G. V.; Hwang, C. J. *Natl. Meet.—Am. Chem. Soc., Div. Environ. Chem.* **1999**, *39* (1), 220.

67. Thurman, E. M. *Developments in Biochemistry: Organic Geochemistry of Natural Waters*; Nijhoff & Junk: Dordecht, Germany, 1985.

68. Leenheer, J. A. *Environ. Sci. Technol.* **1981**, *15*, 578.

69. Aiken, G. R.; Leenheer, J. A. *Chem. Ecol.* **1993**, *8*, 135.

70. Croué, J.-P.; Martin, B.; Deguin, A.; Legube, B. Isolation and Characterization of Dissolved Hydrophobic and Hydrophilic Organic Substances of a Reservoir Water. In *Natural Organic Matter in Drinking Water: Origin, Characterization and Removal*, Workshop Proceedings, Chamonix, France, Sept 19-22, 1993; AWWARF and AWWA: Denver, CO, 1994.

71. Rostad, C. E.; Katz, B. G.; Leenheer, J. A.; Martin, B. S.; Noyes, T. I. *Natl. Meet.—Am. Chem. Soc., Div. Environ. Chem.* **1999**, *39* (1), 222.

72. Serkiz, S. M.; Perdue, E. M. *Wat. Res.* **1990**, *24* (7), 911.

73. Croué, J.-P.; Labouyrie-Rouiller, L.; Belin, C.; Lamotte, M.; Lefebvre, E.; Legube, B. Isolation of Aquatic Natural Organic Matter Using Membrane Filtration and XAD Resin Adsorption: A Comparative Study. Presented at the

8th Meeting of the International Humic Substances Society, Wroclaw, Poland, Sept 9-14, 1996.

74. Leenheer, J. A. Fractionation, Isolation and Characterization of Hydrophilic Constituents of Dissolved Organic Matter in Water. In *Influence of Natural Organic Matter Characteristics on Drinking Water Treatment and Quality*, Workshop Proceedings, International Water Supply Association, Université de Poitiers, Poitiers, France, Sept 18-19, 1996.

75. Rice, R. G.; Gomez-Taylor, M. *Environ. Health Persp.* **1986,** *69,* 31.

76. Richardson, S. D. Drinking Water Disinfection By-Products. In *Encyclopedia of Environmental Analysis and Remediation*; Meyers, R. A., Ed.; Wiley & Sons: New York, 1998.

77. Doré, M.; Merlet, N.; Legube, B.; Croué, J.-P. *Ozone Sci. Eng.* **1988,** *10* (Spring), 153.

78. Symons, J. M.; Krasner, S. W.; Simms, L. A.; Sclimenti, M. J. *J.—Am. Water Works Assoc.* **1993,** *85* (1), 51.

79. Krasner, S. W.; Glaze, W. H.; Weinberg, H. S.; Daniel, P. A.; Najm, I. N. *J.—Am. Water Works Assoc.* **1993,** *85* (1), 73.

80. Weinberg, H. S.; Glaze, W. H.; Krasner, S. W.; Sclimenti, M. J. *J.—Am. Water Works Assoc.* **1993,** *85* (5), 72.

81. Kuo, C.-Y.; Wang, H.-C.; Krasner, S. W.; Davis, M. K. Ion-Chromatographic Determination of Three Short-Chain Carboxylic Acids in Ozonated Drinking Water. In *Water Disinfection and Natural Organic Matter: Characterization and Control*; Minear, R. A., Amy, G. L., Eds.; American Chemical Society: Washington, DC, 1996.

82. Krasner, S. W.; McGuire, M. J.; Jacangelo, J. G.; Patania, N. L.; Reagan, K. M.; Aieta, E. M. *J.—Am. Water Works Assoc.* **1989,** *81* (8), 41.

83. Symons, J. M.; Xia, R.; Speitel, G. E., Jr.; Diehl, A. C.; Hwang, C. J.; Krasner, S. W.; Barrett, S. E. *Factors Affecting Disinfection By-Product Formation During Chloramination*; AWWARF and AWWA: Denver, CO, 1998.

84. Smith, M. E.; Cowman, G. A.; Singer, P. C. The Impact of Ozonation and Coagulation on Disinfection By-Product Formation. In *Proceedings: Water Research*, AWWA Annual Conference, San Antonio, TX, June 6-10, 1993; American Water Works Association: Denver, CO, 1994.

85. Aieta, E. M.; Berg, J. D. *J.—Am. Water Works Assoc.* **1986,** 78 (6), 62.

86. Richardson, S. D.; Thurston, A. D., Jr.; Collette, T. W.; Patterson, K. S. *Environ. Sci. Technol.* **1994,** *28* (4), 592.

87. Heller-Grossman, L.; Idin, A.; Limoni-Relis, B.; Rebhun, M. *Environ. Sci. Technol.* **1999,** *33* (6), 932.

88. Glaze, W. H., Weinberg, H. S. *Identification and Occurrence of Ozonation By-products in Drinking Water*; AWWARF and AWWA: Denver, CO, 1993.

89. Malley, J. P.; Shaw, J. P.; Ropp, J. R. *Evaluation of By-Products Produced by Treatment of Groundwaters with Ultraviolet Irradiation*. AWWA Research Foundation: Denver, CO, 1995.

**14**

90. Koch, B.; Crofts, E. W.; Davis, M. K.; Schimpff, W. K. Analysis of Halogenated Disinfection By-Products by Capillary Chromatography. In *Disinfection By-Products: Current Perspectives*; American Water Works Association: Denver, CO, 1989.
91. Sclimenti, M. J.; Krasner, S. W.; Glaze, W. H.; Weinberg, H. S. Ozone Disinfection By-Products: Optimization of the PFBHA Derivatization Method for the Analysis of Aldehydes. In *Proceedings*, AWWA Water Quality Technology Conference, San Diego, CA, Nov 11-15, 1990; American Water Works Association: Denver, CO, 1991.
92. Krasner, S. W.; Sclimenti, M. J.; Hwang, C. J. Experiences with Implementing a Laboratory Program to Sample and Analyze for Disinfection By-Products in a National Study. In *Disinfection By-Products: Current Perspectives*; AWWA, Denver, CO, 1989.
93. Feinberg, T. N., Charles, M. J. Analysis of Polar Carbonyls Using Capillary Electrophoresis/Electrospray Mass Spectrometry. In *Disinfection By-Products in Drinking Water: Critical Issues in Health Effects Research*; International Life Sciences Institute: Washington, DC, 1995.
94. Miller, K. G.; Pacey, G. E.; Gordon, G. *Anal. Chem.* **1985**, 57, 734.
95. International Life Sciences Institute (ILSI). *Identification of New and Uncharacterized Disinfection By-Products in Drinking Water*, Workshop Report; ILSI Health and Environmental Sciences Institute: Washington, DC, 1999.
96. Weinberg, H. S.; Krasner, S. W.; Richardson, S. D.; Sargaiah, R.; Singer, P. C. *Natl. Meet.—Am. Chem. Soc., Div. Environ. Chem.* **1999**, *39* (1), 265.

# REGULATORY AND HEALTH EFFECTS BACKGROUND

Chapter 2

# A New Assessment of the Cytotoxicity and Genotoxicity of Drinking Water Disinfection By-Products

Yahya Kargalioglu[1], Brian J. McMillan[1], Roger A. Minear[2], and Michael J. Plewa[1,3]

Departments of [1]Crop Sciences and [2]Civil Engineering, University of Illinois at Urbana-Champaign, Urbana, IL 61801
[3]Corresponding author.

The disinfection of drinking water generates cytotoxic and mutagenic compounds. The cytotoxic and mutagenic properties of known disinfection by-products (DBPs) were quantitatively compared. Using *Salmonella typhimurium* strain TA100 a rapid, semi-automated, microplate cytotoxicity assay was developed. The assay can accommodate a concentration range of six log orders of magnitude with 6 replicates per concentration and requires approximately 5 h. Data were automatically transferred from a microplate reader to a computer spreadsheet. The DBP concentration that induced 50% repression of growth in the cytotoxicity assay was used as the highest concentration for the *S. typhimurium* mutagenicity assay. Selected DBPs were assayed in *S. typhimurium* strains TA98, TA100 and RSJ100 under suspension test conditions. The mutagenic potency of the DBPs were calculated and compared with the cytotoxicity data.

## Introduction

The use of oxidants and disinfectants in drinking water treatment leads to the formation of mutagenic and potentially carcinogenic oxidation and disinfection by-products (DBPs). Natural organic matter (NOM) and bromide ion (Br$^-$) serve as the organic and inorganic precursors, respectively, to DBP formation associated with oxidants/disinfectants such as ozone (O$_3$) and chlorine (Cl$_2$) (*1, 2, 3, 4, 5, 6*). The use of O$_3$ as a pre-oxidant/pre-disinfectant has become increasingly common, while Cl$_2$ is still widely used as a post-disinfectant. Studies conducted since the 1970s on the safety of drinking water after disinfection have primarily focused on the biological effects of chlorination of water containing NOM (*1*). In a review on the genotoxic activity of organic chemicals in drinking water, Meier (*7*) stated that the overwhelming majority of genotoxic agents in drinking water was generated during the chlorination stage of water treatment. In general, epidemiological studies suggest that individuals who consume chlorinated drinking water have a somewhat elevated risk of rectal, kidney and bladder cancer as compared to those who do not drink chlorinated water (*8, 9*). Using the estimated mutagenic potency of drinking water based on *S. typhimurium* TA100, Koivusalo et al. (*10*) determined that there was a significant exposure-dependent response with the consumption of mutagenic water and the development of kidney and bladder cancer (*9*). Considering the assembled information on the genotoxic and carcinogenic potency of chlorination by-products it has been suggested that there should be a more judicious use of chlorine in the disinfection of drinking water (*7, 11*). In the presence of bromide, brominated trihalomethanes are preferentially produced during chlorine disinfection (*4*). Water with elevated bromide (1 mg/L) and treated with ozone followed by secondary chlorine or chloramine induced a shift in DBP formation from the chlorinated to the brominated species (*6*). There is a general opinion that brominated DBPs are more genotoxic than chlorinated DBPs and the toxicity of these brominated DBP products is now attracting increased attention (*12*). The most prevalent DBPs in drinking water in the United States are the trihalomethanes and haloacetic acids. Although epidemiological studies demonstrated a relationship of DBPs to specific cancers, it is unclear which DBPs pose the greatest risk.

The work presented here is part of a larger on-going study to quantitatively compare the cytotoxicity and genotoxicity of DBPs using bacterial and mammalian cell assays. Genetic assays based on mammalian cells are more relevant to the assessment of human risks. The goal is to quantitatively compare and rank order the cytotoxicity and genotoxicity in *S. typhimurium* and mammalian cells of DBP standards and DBP mixtures from disinfected waters. In this paper a method is described that allows for the quantitative comparison of the cytotoxicity and mutagenicity of DBPs in *S. typhimurium*. In the future this general approach will be expanded to include rapid, quantitative cytotoxicity and genotoxicity analysis of DBPs with mammalian cells.

## *Salmonella* Microplate Cytotoxicity Assay

While the *S. typhimurium* plate incorporation test is an excellent rapid qualitative mutagenicity assay, it cannot quantitatively determine the cytotoxicity of test agents (*13*). Cytotoxicity is usually ignored throughout the concentration range of a test agent in the standard *Salmonella* plate incorporation assay. A cytotoxicity concentration-response curve for a DBP is an indication of its biological response. It is also useful in establishing the concentration range for mutagenicity tests. However, standard methods to measure toxicity are laborious and time consuming. To address these issues a rapid, semi-automatic cytotoxicity assay to quantitatively compare selected DBPs was developed. Log-phase *S. typhimurium* TA100 cells previously frozen (−80°C) in Luria Broth (LB) plus 10% dimethylsulfoxide (DMSO) were thawed and grown in 5 mL LB at 37°C for 2 h while shaking. The cell titer was adjusted to an optical density (OD) at 595 nm of 0.030. Treatment with each DBP standard was conducted in 2-mL glass vials sealed with Teflon in a total volume of 1 mL. Each treatment vial contained 300 μL of the titered TA100 cells, a known concentration of the DBP (<10 μL) and 100 mM potassium phosphate buffer, pH 7.4. The cells were treated for 1 h at 37°C at 200 rpm shaking. At the end of the treatment time, 100 μL aliquots (~1.2×10⁶ cells) from each treatment vial were transferred into a series of six microplate wells per concentration group. An aliquot (100 μL) of 2× LB medium was then added to each microplate well. At time 0, the initial OD of each well was measured using a Bio-Rad microplate reader at 595 nm (Figure 1). This initial reading generated the OD value as a blank for each microplate well. The data were transferred to a computer spreadsheet program.

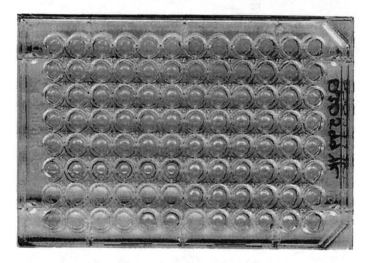

*Figure 1. Salmonella microplate cytotoxicity assay at time = 0 min.*

The microplate was placed in a padded plastic container and incubated at 37°C and shaken at 200 rpm for 210 min. The final OD of each well was analyzed at 595 nm with a microplate reader and the data stored on a spreadsheet file (Figure 2). For each well, the blank OD value (time 0 reading) was subtracted from the OD reading of that specific well after 210 min of incubation. The blank corrected data for each DBP concentration were averaged. The negative control consisted of a reaction tube without DBP and the negative control OD value was set at 100%. The OD reading for each concentration of a DBP was converted into a percentage of its corresponding negative control. These data were plotted as a function of the DBP concentration versus the percentage of the negative control. The DBPs that were analyzed were 3-chloro-4-(dichloromethyl)-5-

hydroxy-2-(*5H*)-furanone (MX), bromoacetic acid (BA), bromoform (BF), dibromo-acetic acid (DBA), tribromoacetic acid (TBA), chloroform (CF), and the solvents, ethanol (EtOH), and dimethylsulfoxide (DMSO).

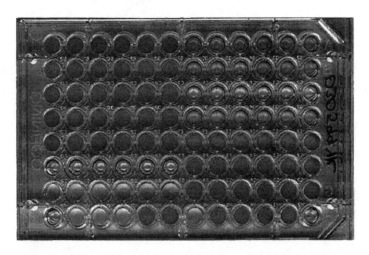

*Figure 2. Salmonella microplate cytotoxicity assay at time = 210 min.*

The concentration-response curves of the DBP standards and solvents demonstrated a range of cytotoxic potencies in *S. typhimurium* that encompassed 8 orders of magnitude (Figure 3). To rank order these agents we calculated their %C½ values (Table I). The DBP concentration that corresponded to 50% of the growth of the negative control was designated as the %C½ value.

**Table I. Rank Order of the Cytotoxicity for DBPs and Sample #1014**

| Test Agent | %C½ Value (M) | %C½ Value (µg/mL) |
|---|---|---|
| MX | $3.80 \times 10^{-6}$ | 0.8 |
| Bromoacetic Acid | $5.22 \times 10^{-4}$ | 72.5 |
| Bromoform | $1.24 \times 10^{-2}$ | 3,130.0 |
| Dibromoacetic Acid | $1.54 \times 10^{-2}$ | 3,355.0 |
| Tribromoacetic Acid | $2.02 \times 10^{-2}$ | 5,990.0 |
| Chloroform | $2.82 \times 10^{-2}$ | 3,337.0 |
| Dimethylsulfoxide | 1.07 | 83,599.0 |
| SRFA Sample #1014 | — | 35.2 |

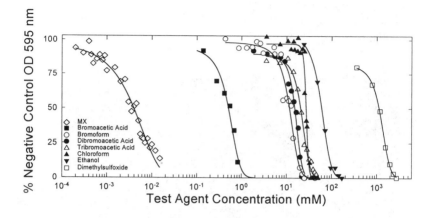

*Figure 3. Comparative cytotoxicity of water DBPs as measured as a repression of the growth rate as compared to the concurrent negative control in S. typhimurium strain TA100.*

### Cytotoxicity of Suwannee River Fulvic Acid Chlorination Sample

The cytotoxicity assay was used to evaluate a chlorinated water sample. A chlorination experiment with a model water was conducted with the following conditions: DOC source, Suwannee River Fulvic Acid; initial DOC concentration, 108 mg/L as C; chlorine dose, 500 mg/L; initial bromide ion concentration, 7 mg/L; pH, 7.8; NaHCO$_3$ concentration, 25 mM; incubation time, 5.5 days; incubation temperature, 25°C. The sample solution was prepared in a 500-mL, glass-stopped bottle. After mixing the sample thoroughly, the bottle was stored headspace-free in the dark. At the end of the incubation period the residual chlorine was quenched by sodium thiosulfate. The use of sodium thiosulfate may decompose some fraction of the DBPs, however, for methods development this was not a concern. Since this study involves the analysis of cytotoxicity it would have been unwise to allow the presence of free chlorine in the sample. The chlorination of the Suwannee River Fulvic Acid water sample (#1014) resulted in a 70-mL sample with a total organic carbon (TOC) concentration of 108 mg/L. The sample was lyophilized to dryness and the organic material was recovered by three extractions with nanograde ethyl acetate. This material was taken to dryness under a rotary vacuum and dissolved in 200 μL of dimethylsulfoxide resulting in a 350× concentration.

The sample was analyzed using the *Salmonella* microplate cytotoxicity assay (Figure 4). In order to compare the cytotoxic potency of sample #1014 with the DBP standards the concentrations of the test agents were determined in μg/mL units (Table I). The %C½ value for the Suwannee River Fulvic Acid sample #1014 was calculated as 35.23

µg/mL. Based on the comparison with the DBP standards, the Suwannee River Fulvic Acid sample #1014 was approximately twice as cytotoxic as bromoacetic acid (Table I).

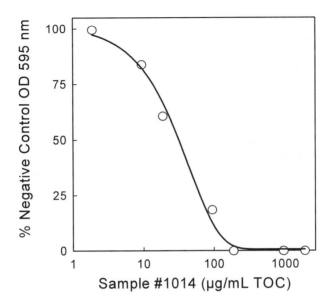

*Figure 4. Cytotoxicity of the concentrated Suwannee River Fulvic Acid chlorination sample #1014.*

## *Salmonella* Mutagenicity Assay

For each mutagenicity experiment a culture of *S. typhimurium* was grown overnight from a single colony isolate in 100 mL of LB medium supplemented with ampicillin (50 µg/mL) at 37°C with shaking (200 rpm). The cells were harvested by centrifugation, washed in phosphate buffer and the titer was adjusted to $2\times10^{10}$ cells/mL. In general, the %C½ values for each DBP standard was used as the highest concentration of the DBP in the mutagenicity experiments. Each 2-mL reaction vial was composed of known µL volumes of the DBP, $2\times10^9$ bacterial cells and phosphate buffer in a total volume of 1 mL. Experiments with hepatic S9 activation included 500 µL of S9 mix (5 mL of S9 mix contains 50 mM sodium phosphate, pH:7. 4, 10mM $MgCl_2$, 30 mM KCl, 5 mM glucose-6-phosphate, 4 mM NADP, 10% S9 microsomal fraction) in the reaction vials. Each reaction tube was sealed and incubated for 1 h at 37°C while shaking. After the treatment period, triplicate 250 µL aliquots of the reaction vials (containing $5\times10^8$ cells) were added to 2 mL of molten top agar supplemented with a final concentration of 55μM histidine plus biotin. Each top agar tube was mixed and poured onto selective Vogel Bonner (VB) medium plates and incubated at 37°C for 72 h. A New Brunswick Biotran III automatic colony counter was used to determine the number of histidine revertant colonies present on each VB plate. The data were saved on an Excel 7 spreadsheet and plotted using the graphical and statistical functions of SigmaPlot 4 (SPSS). The *S. typhimurium* concentration-response curves for BA employing strains TA98, TA100 and RSJ100 are illustrated in Figure 5.

**Calculation of the Mutagenic Potency of DBPs**

To make comparisons among the *S. typhimurium* mutagenicity data, the mutagenic potency of each agent was used as the measure of mutagenic strength. The detailed procedure to calculate the mutagenic potency of bromoacetic acid is presented here as an example. For each DBP the linear region of the concentration-response curve was determined by running a linear regression for specific regions of the data set. The linear portion of the bromoacetic acid concentration-response curve with TA100 −S9 is presented in Figure 6. The linear region for mutation induction in strain TA100 (−S9) was from 222 μM to 522 μM. This range was used for determining the mutagenic potency of bromoacetic acid (Table II). The induced revertants per plate were calculated by subtracting the average negative control revertant frequency from the treatment revertant frequency. The induced revertants per reaction tube was a calculation of the revertants that were generated while the cells were in suspension during the treatment time. Finally the induced revertants per μmole were calculated for each concentration of bromoacetic acid used in the treatment groups. The average (± standard error) revertants per μmole value was 5,465 ± 393 which is the mutagenic potency for bromo-acetic acid using *S. typhimurium* strain TA100 without S9 activation.

**Table II. *S. typhimurium* Mutagenic Potency for Bromoacetic Acid**

| BA μM | Revertants per Plate | Induced Revertants per Tube | Mole per Reaction Tube | Revertants per μMole |
|-------|------------------|---------------------------|----------------------|---------------------|
| 222 | 363 | 828 | $222 \times 10^{-7}$ | 3,730 |
| 272 | 544 | 1,196 | $272 \times 10^{-7}$ | 4,397 |
| 322 | 605 | 1,796 | $322 \times 10^{-7}$ | 5,578 |
| 372 | 700 | 2,176 | $372 \times 10^{-7}$ | 5,849 |
| 422 | 789 | 2,532 | $422 \times 10^{-7}$ | 6,000 |
| 472 | 857 | 2,804 | $472 \times 10^{-7}$ | 5,941 |
| 522 | 1,038 | 3,528 | $522 \times 10^{-7}$ | 6,759 |

The reversion frequency for the negative control was 156 revertants/$5 \times 10^{8}$ cells plated.
The mean (±SE) revertants per μmole BA was 5,465 ±393.

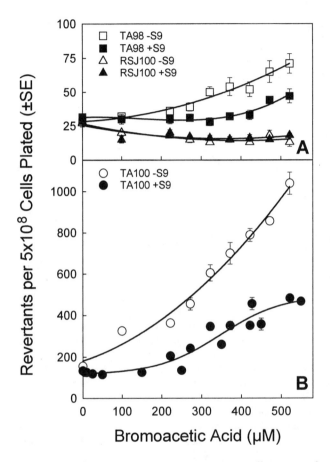

*Figure 5. The concentration-response curves illustrating the mutagenic activity of bromoacetic acid. (A) S. typhimurium strains TA98 and RSJ100 ±S9. (B) S. typhimurium strain TA100 ±S9. These data represent a minimum of 3 experiments per strain with triplicate VB plates per treatment group.*

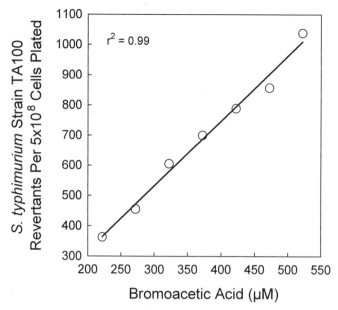

*Figure 6. Linear region of the concentration-response curve for bromoacetic acid.*

Using the procedure outlined above, the mutagenic potency was calculated for MX, bromoacetic acid, bromoform, dibromoacetic acid and tribromoacetic acid (Table III). Although the suspension assay requires more labor than the standard *S. typhimurium* plate incorporation assay, it provides greater control of the experimental variables such as known cell titer, precise dose and consistent exposure time. From the data presented in Table III the DBPs analyzed preferentially induced base pair substitutions (strains TA100 and RSJ100) over frameshift mutations (strain TA98). In general cytochrome P-450-mediated metabolic activation reduced the mutagenic potencies of the DBPs. Finally, the glutathione-S-transferase expressing *S. typhimurium* strain, RSJ100, was effective in identifying the mutagenic activity of bromoform. These data confirmed the findings of DeMarini and colleagues (*14*).

**Integration of DBP Cytotoxicity and Mutagenicity Data**

Since the *S. typhimurium* cytotoxicity and the mutagenicity tests were conducted under identical exposure conditions the data may be combined. An estimated reversion frequency as well as an estimated mutagenic potency may be calculated employing data that were normalized according to the cytotoxicity at each concentration of the test agent. The data for bromoacetic acid was used as an example. The reversion frequency at each bromoacetic acid concentration was divided by the cytotoxicity factor (the percentage

**Table III. *S. typhimurium* Mutagenic Potency of DBP Standards**

| DBP Standard | Range | Strain | Average Mutagenic Potency ±SE (Revertants/μmole) |
|---|---|---|---|
| MX | 0. 2 - 4 μM | TA100 -S9 | 806,819 ± 28,794 |
| MX | 0. 8 - 9 μM | TA98 -S9 | 89,160 ± 6,330 |
| MX | 1. 25 - 2 μM | TA98 +S9 | 62,178 ± 15,825 |
| Bromoacetic Acid | 222 - 522 μM | TA100 -S9 | 5,465 ± 393 |
| Bromoacetic Acid | 222 - 522 μM | TA100 | 2,088 ± 184 |
| Bromoacetic Acid | 272 - 522 μM | TA98 -S9 | 264 ± 25 |
| Bromoacetic Acid | 472 - 522 μM | TA98 +S9 | 116 ± 6 |
| Bromoform | 0. 5 - 3 mM | RSJ100 - | 183 ± 16 |
| Dibromoacetic Acid | 1 - 16 mM | TA100 -S9 | 148 ± 25 |
| Dibromoacetic Acid | 1. 2 - 15 mM | TA100 | 136 ± 19 |
| Dibromoacetic Acid | 8 - 16 mM | TA98 -S9 | 13 ± 1 |
| Dibromoacetic Acid | 2. 5 - 15. 5 mM | TA98 +S9 | 9 ± 1 |
| Tribromoacetic Acid | — | TA100 -S9 | NS |

NS = not significant

of growth in a treatment group as compared to its negative control) determined from the microplate assay. The resulting value was an estimated reversion frequency after normalizing for cytotoxicity. The impact of including the cytotoxicity of bromoacetic acid in the interpretation of the mutation concentration response curve is illustrated in Figure 7. The estimated TA100 mutagenic potency for bromoacetic acid normalized for cytotoxicity was 10,264 revertants per μmole as compared to a mutagenic potency of 5,465 revertants per μmole calculated without including the impact of cytotoxicity (Table III).

## Conclusions

The *Salmonella* microplate cytotoxicity assay complements the suspension mutation assay. The rank order of decreasing cytotoxicity of the DBP standards and solvents were MX >> bromoacetic acid >> bromoform > dibromoacetic acid >> tribromoacetic acid > chloroform >> dimethylsulfoxide. The rank order of the mutagenic potency of the DBPs from highest to lowest were MX >>> bromoacetic acid >> bromoform > dibromoacetic acid >> tribromoacetic acid. These two quantitative biological assays will be used to rank order DBP mixtures isolated from disinfected waters. In the future the cytotoxicity and genotoxicity of DBPs will be evaluated in mammalian cells and compared with the results from *Salmonella*.

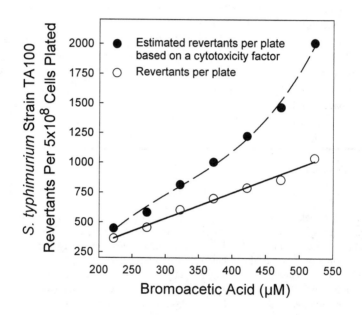

*Figure 7. Comparison of the linear region of the concentration-response curve for bromoacetic acid using S. typhimurium TA100 before (O) and after (●) the data were normalized for cytotoxicity.*

# Acknowledgments

This research was funded by AWWARF grant 554, U. S. Environmental Protection Agency grant R825956-01 and a UIUC Campus Honors Program Summer Research Grant.

# References

1. Rook, J. J. *Environ. Sci. Technol.* **1977**, *11*, 478-482.
2. Doré, M.; Merlet, N.; Legube, B.; Croue, J. Ph. *Ozone Sci. Engrg.* **1988**, *10*, 153-172.
3. Krasner, S. W.; McGuire, M. J.; Jacangelo, J. G.; Patania, N. L.; Reagan, K. M.; Aieta, E. M. *J. Am. Water Works Assoc.* **1989**, *81*, 41-53.
4. Symons, J. M.; Krasner, S. W.; Simms, L. A.; Sclimenti, M. *J. Am. Water Works Assoc.* **1993**, *85*, 41-53.
5. Richardson, S. D.; Thruston, A. D. Jr.; Caughran, T. V.; Chen, P. H.; Collette, T. W.; Floyd, T. L.; Schenck, K. M.; Lykins, B. W.; Sun, G. R.; Majetich G. *Environ. Sci. Technol.* **1999**, *in press*.
6. Richardson, S. D.; Thruston, A. D. Jr.; Caughran, T. V.; Chen, P. H.; Collette, T. W.; Floyd, T. L.; Schenck, K. M.; Lykins, B. W.; Sun, G. R.; Majetich G. *Environ. Sci. Technol.* **1999**, *in press*.
7. Meier, J. R. *Mutation Res.* **1988**, *196*, 211-245.
8. Morris, R. D.; Audet, A. M.; Angelillo, I. F.; Chalmers, T. C.; Mosteller, F. *Am. J. Public Health* **1992**, *82*, 955-963.
9. Koivusalo, M.; Jaakkola, J. J.; Vartianen, T.; Hakulinen, T.; Karjalainen, S. ; Pukkala E.; Tuomisto, J. *Am. J. Public Health* **1994**, *84*, 1223-1228.
10. Koivusalo, M.T.; Jaakkola, J. J.; Vartianen, T. *Environ. Res.* **1994**, *64*, 90-101.
11. Wilcox, P.; Williamson, S. *Environ. Health Perspect.* **1986**, *69*, 141-149.
12. Bull, R. J.; Birnbaum, L. S.; Cantor, K. P.; Rose, J. B.; Butterworth, B. E.; Pegram, R.; Tuomisto, J. *Fund. Applied Toxicol.* **1995**, *28*, 155-166.
13. Maron, D. M.; Ames, B. N. *Mutation Res.* **1983**, *113*, 173–215.
14. DeMarini, D. M.; Shelton, M. L.; Warren, S. H.; Ross, T. M.; Shim, J. Y.; Richard, A. M.; Pegram, R. A. *Environ. Mol. Mutagen.* **1997**, *30*, 440-447.

Chapter 3

# An Illustration of the Use of Scientific Information To Support Disinfection By-Product Rule Development

**Hiba M. Shukairy[1] and Alison M. Gusses[2]**

[1]U.S. Environmental Protection Agency and [2]Oak Ridge Institute for Science and Education, USEPA Technical Support Center, MS 140, 26 West M. L. King Drive, Cincinnati, OH 45268

The United States Environmental Protection Agency (USEPA) is addressing concerns to balance risks between microbial pathogens and disinfection by-products (DBPs). The Microbial-DBP Rule cluster is being developed in stages using a collaborative effort between the USEPA and stakeholders. To support rule development, historical and up-to-date information, including health effects research, occurrence, treatment technologies, and costs are used. Databases on contaminant occurrence, operation, design and treatment information from systems serving $\geq$ 100,000 persons are created to house data collected under the Information Collection Rule (ICR). This paper presents examples of technical information that resulted in changes in DBP-related rules between the proposal and the final Stage 1 Disinfectants/Disinfection Byproduct Rule. Information on the status of using ICR data to support Stage 2 Rule development is summarized.

## Introduction

Disinfectants used in drinking water treatment to provide microbiologically safe water to the consumers react with a portion of the natural organic matter (NOM) and inorganic compounds, such as bromide, to form disinfection by-products (DBPs). In

1974, trihalomethanes (THMs) were identified as by-products of the disinfection of drinking water with free chlorine (*1*). In the seventies, the United States Environmental Protection Agency (USEPA) collected THM data as part of two nationwide surveys: the National Organics Reconnaissance Survey in 1975 (*2*) and the National Organic Monitoring Survey in 1976-77 (*3*).

Toxicological animal studies indicated that chloroform, one of the four THMs (chloroform, bromodichloromethane, chlorodibromomethane and bromoform), was of health concern (*4*). In 1979, the potential health risks from THMs and other DBPs led the USEPA to set an interim maximum contaminant level (MCL) of 0.10 mg/L for total THMs (TTHMs), based on a running annual average calculated from quarterly samples. This rule applies to all community surface and ground water systems, serving at least 10,000 people, that add a disinfectant anywhere in the treatment train (*5*).

Balancing the risks from microbial pathogens and DBPs is a significant challenge facing the drinking water industry. Chlorine disinfection has been successful in virtually irradicating waterborne disease outbreaks such as cholera and typhoid, but other microorganisms that are resistant to traditional disinfection practices still pose challenges to the drinking water industry (*6*). The application of alternative disinfectants, such as ozone and chlorine dioxide, is increasing because of their stronger oxidation potential, as well as lower TTHM formation. Less is known about the DBPs formed by these disinfectants and the health risks associated with them (*7,8*).

The 1996 amendments to the Safe Drinking Water Act required USEPA to develop rules to address the risks posed by microbial pathogens and by DBPs and to balance the risks from the different contaminants (*9*). As a result, the USEPA is developing a group of rules, termed the Microbial/Disinfectants and Disinfection Byproduct (M/DBP) cluster. These include the Enhanced Surface Water Treatment Rule (ESWTR), which is designed to strengthen microbial protection, and the Disinfectants/Disinfection By-Product (D/DBP) Rule, which will minimize the risks from DBPs and residual disinfectants. Some of these rules are being developed in stages to allow for current health effects research and new occurrence and treatment information to be used to support rule development. The objective of this paper is to give some examples of the use of chemical and physical treatment process research, the occurrence of drinking water contaminants and their precursors, methods, and DBP chemistry in the development of DBP related regulations. This paper summarizes some of the main issues that led to changes in the Stage 1 D/DBP regulations between the proposed (*10*) and the final (*11*) rules, but does not address all the contributions to the rule. The overall management of ICR data to support Stage 2 rule development is presented. Although the paper focuses only on the D/DBP Rule, simultaneous compliance with all rules will impact DBP control strategies.

## General Background on DBP Rule Development

In 1992, USEPA initiated a negotiated rulemaking process that was a collaborative effort between the water industry, consumers, environmental, and public health groups, as well as local, state and federal government agencies (*12*). One of the main limitations of the process was the lack of available data to help understand the trade-offs between disinfection for microbial inactivation and controlling DBP formation. Thus, the negotiators agreed to a two-stage rule development process which would provide some level of public health protection in the near-term while allowing for sufficient time to gather and analyze health effects and occurrence information to support development of the long term rules. The Stage 1 D/DBP Rule was to be proposed, promulgated, and implemented concurrently with the Interim ESWTR to minimize microbial risk while utilities altered their treatment to decrease DBP formation.

To support rule development, pertinent information is needed. Specifically, occurrence and health effects studies to indicate what contaminants are prevalent and whether or not they are of public health concern. This includes evaluating differences between chronic or acute health risks and considering end points of concern. Robust analytical methods to measure the contaminants are necessary for compliance monitoring. Best available treatment (BAT) technologies to control for these contaminants and associated costs are also required.

To provide occurrence data for the support of future rule development, the Information Collection Rule (ICR) was promulgated in May 1996 (*13*). Under the ICR, water treatment plants are required to gather occurrence data, treatment plant design and operational data, over an 18-month period. Public water systems (PWSs) that serve ≥ 100,000 persons began monitoring, under the ICR, in July 1997 and ended in December 1998. Under the ICR, some systems are required to conduct DBP precursor removal studies using advanced treatment, namely, granular activated carbon (GAC) adsorption and membrane processes. These studies evaluate the effectiveness of precursor removal technologies, prior to the addition of chlorine disinfection, for the control of the sum of TTHMs and the sum of six haloacetic acids (HAA6), which are mono-, di-, and tri-chloroacetic acid; mono- and dibromoacetic acid; and bromochloroacetic acid.

Data from the ICR was originally meant to support the Interim ESWTR, the ESWTR, and the Stage 2 D/DBP Rule. However, because of the delay in promulgation and implementation of the ICR, data were not available for the Interim ESWTR.

## Stage 1 D/DBP Rule Development

The Stage 1 D/DBP Rule was promulgated in December 1998 (*11*). The purpose of the rule is to improve health protection by decreasing exposure to DBPs. Stage 1 decreases the MCL for TTHM from 0.10 mg/L to 0.080 mg/L and sets a MCL for HAA5 (which is the sum of mono-, di-, and tri-chloroacetic acid and mono- and dibromoacetic acid) at 0.060 mg/L. A summary of Stage 1 MCLs, maximum residual disinfectant levels (MRDLs), BATs and the approved methods used for DBP compliance monitoring are presented in Table I.

**Table I. Summary of Stage 1 of the D/DBP Rule (*11*)**

| DBPs | | | |
|---|---|---|---|
| *Parameter* | *MCL, mg/L* | *BAT* | *Analytical Methods* |
| TTHM | 0.080 | 1 | EPA 502.2, 524.2 & 551.1 (*14*) |
| HAA5 | 0.060 | 1 | EPA 552.1 (*15*) & 552.2 (*14*), SM 6251B (*16*) |
| Chlorite (plants utilizing chlorine dioxide) | 1.0 | 2 | EPA 300.0 (*17*) & 300.1 (*18*), SM 4500-ClO$_2$E (*16*) |
| Bromate (plants utilizing ozone) | 0.010 | 3 | EPA 300.1 (*18*) |
| *Disinfectant Residuals* | | | |
| *Disinfectant* | | | *MRDL, mg/L* |
| Chlorine | | | 4.0 |
| Chloramines | | | 4.0 |
| Chlorine Dioxide | | | 0.8 |

[1]Enhanced coagulation, enhanced softening or GAC10 with chlorine
[2]Optimize treatment processes to reduce disinfectant demand and levels
[3]Optimize ozonation to reduce bromate formation

The Stage 1 D/DBP Rule is designed to increase public health protection without large capital investments or major shifts in disinfection technologies, given the lack of information on the risks associated with the use of alternative disinfectants. It is anticipated that as many as 140 million people will have decreased DBP exposure resulting from a 24 percent decrease in the national average TTHM levels (*19*).

Stage 1 applies to all sizes of community and nontransient noncommunity systems that add a disinfectant to drinking water as part of their treatment. Stage 1 of the Rule also requires surface water systems using conventional treatment to use enhanced coagulation to achieve a required total organic carbon (TOC) removal that is based on raw water quality, (i.e., TOC concentration and alkalinity).

Compliance with Stage 1 is based on running annual averages for all DBPs except for chlorite. Because of its potential acute health risk, it will be monitored daily at the plant, and monthly in the distribution system.

Large surface water systems (population served > 10,000), and large ground water systems under the direct influence of surface water must comply by December 2001. For smaller systems, compliance is delayed until December 2003. To assist states and utilities in implementing the rules, an implementation manual, fact sheets and guidance manuals are available (*20*).

A regulatory impact analysis (RIA) was used during the rule making process to evaluate different technologies for the control of DBPs and to develop costs for different regulatory scenarios. A discussion of the RIA and of technical issues that led to the Stage 1 proposal is presented in Krasner and co-workers (*21*). The final Stage 1 Rule was significantly impacted by all the information that led to the proposal and the information that was consolidated from various research studies and plant performance data, after the proposal. Specifically, research studies on enhanced coagulation (EC) and enhanced precipitative softening (ES), and a reevaluation of allowing disinfection credit for the use of predisinfection prior to precursor removal, were examined. Summaries of some of these studies are presented herein and for more details the reader is referred to the original references.

## Impact of Enhanced Coagulation and Enhanced Precipitative Softening Studies on the 3x3 TOC/Alkalinity Matrix in the Stage 1 D/DBPRule

In the 1994 proposal for Stage 1, USEPA identified the best available technology to comply with the proposed MCLs for TTHM and HAA5 as enhanced coagulation or treatment with GAC with a 10-minute empty bed contact time (GAC10) and a 180-day regeneration frequency (*10*). In addition, USEPA proposed a treatment technique to improve DBP precursor removal. Thus, controlling the formation of these precursors would control for the formation of both known and unknown byproducts. Subpart H systems, (i.e., utilities treating surface water and utilities treating ground water under the direct influence of surface water) that use conventional treatment or precipitative softening, would be required to use EC or ES. The proposed Step 1 TOC percent removals were based on influent TOC and alkalinity concentrations and are presented in Table IIa. The proposed matrix was based on an estimate that 90 percent of affected systems could achieve Step 1 TOC removal through EC or ES. For EC, systems that were unable to meet the percent removals under the Step 1 provision were to use a jar testing protocol (Step 2) to determine how much TOC a PWS could reasonably and practically achieve (*22*). In Step 2, the point of diminishing return (PODR), was defined as the point at which an additional 10 mg/L alum (or equivalent amount of iron) did not decrease residual TOC by at least 0.3 mg/L. If the utility could not meet the requirements, it would request alternative TOC removal criteria from the primacy agency. Details of Step 1 and 2 proposed requirements and exemptions are presented in a paper by Krasner and co-workers (*23*) and in the notice of data availability (NODA) that was published in November 1997 (*24*).

### Table II. Step 1 TOC Removal Matrix for Stage 1 D/DBP Rule

| *Table IIa – Proposed (10)* | | | |
|---|---|---|---|
| *Source Water TOC, mg/L* | *Source Water Alkalinity as mg $CaCO_3$/L* | | |
| | *0—60* | *> 60–120* | *> 120* |
| > 2.0 to 4.0 | 40 % | 30 % | 20 % |
| > 4.0 to 8.0 | 45 % | 35 % | 25 % |
| > 8.0 | 50 % | 40 % | 30 % |
| *Table IIb -- Final (11)* | | | |
| *Source Water TOC, mg/L* | *Source Water Alkalinity as mg $CaCO_3$/L* | | |
| | *0—60* | *> 60–120* | *> 120* |
| > 2.0 to 4.0 | 35 % | 25 % | 15 % |
| > 4.0 to 8.0 | 45 % | 35 % | 25 % |
| > 8.0 | 50 % | 40 % | 30 % |

The following summary presents the most pertinent conclusions from some studies that were used by the USEPA to finalize the Stage 1 treatment technique requirement. For more detail, the reader is referred to the paper by Krasner and co-workers (23), the original references, and the NODA (24).

*University of North Carolina EC Study (25,26)*

This study investigated TOC percent removals for waters that represented each category in the *proposed* 3 x 3 matrix (Table IIa). Jar tests were performed to determine both Step 1 and Step 2 EC TOC removals, coagulant doses and pH values. The waters were also characterized to determine their NOM characteristics by fractionation using XAD-8 resin. The importance of humic/non-humic composition of NOM is because coagulation has been reported to preferentially remove the humic fraction of NOM (27). Furthermore, specific ultraviolet absorbance (SUVA) values were determined. SUVA is defined as the ultraviolet absorbance at 254 nm (m$^{-1}$) divided by the concentration of dissolved organic carbon (DOC) in mg/L. SUVA has been reported to be a good indicator of humic content: SUVA values in the range of 4-5 L/mg-m represent humic material, and SUVA < 3 L/mg-m represents the non-humic fraction (28).

The results of this study showed that 14 out of 31 investigated waters met Step 1 TOC percent removal before the PODR, and these were mostly moderate to high TOC waters with low alkalinity. The study indicated that the low TOC, high alkalinity waters would need to demonstrate alternative performance criteria.

*Metropolitan Water District of Southern California/Colorado University EC Study (29)*

This study investigated EC for two waters: California state project water (SPW) (moderate alkalinity) and Colorado River water (CRW) (high alkalinity). NOM was characterized using XAD-8 resin fractionation (to determine humic/non-humic composition) and ultrafiltration (to determine molecular size). The study showed good removal of high humic content and high molecular weight fractions with increasing alum doses. For these waters, Krasner and co-workers (23) showed that there was a limit beyond which additional coagulant dose would not be practical. For SPW, which had a 52 percent humic fraction and a SUVA value of 2.5 L/mg-m, increasing alum doses reduced the humic fraction with a parallel reduction in SUVA. However, once the SUVA was decreased to 1.7 L/mg-m, an additional 64 mg/L alum dose did not remove more than 0.2 units of SUVA indicating that, for this water, there was a certain SUVA value (# 2.0 L/mg-m) beyond which no increase in alum dose would practically and economically enhance NOM removal (23). Similar behavior was observed for CRW. Even though the two waters had different water quality and NOM characteristics, once the SUVA value in both waters was decreased to below 2.0 L/mg-m, very little reduction in TOC occurred with increasing coagulant dose.

*Malcolm Pirnie Inc. (30) and Colorado University (31)*

In this analysis, EC data from 127 US waters were collected. Cumulative probability distributions of these waters showed a definite trend of increasing SUVA values with increasing TOC or decreasing alkalinity. Using this data set, these researchers developed models to evaluate TOC removal during EC. Nine equations were developed by Chowdhury to predict 90th percentile for the 3 x 3 matrix (*30*).

Furthermore, Edwards (*31*) divided the NOM into two hypothetical DOC fractions, sorbing and non-sorbing DOC, and developed models to predict DOC removal and SUVA removal. Based on the results of Chowdhury (*30*) and Edwards (*31*) and other analyzed data from case-by-case studies and utility surveys, it became apparent to the technical experts and the stakeholders chartered under the Federal Advisory Committee Act (FACA) that the percent removals in the *proposed* 3 x 3 matrix were not going to be achieved by 90 percent of the plants, as previously predicted. Additionally, the group agreed that SUVA, which is a good indicator of humic content of NOM, could be used as a predictor of the amenability of water to enhanced coagulation. Hence, the required percent removals for the 2-4 mg/L TOC row in the 3 x 3 matrix were decreased, as these waters were considered to be more difficult to treat.

It was proposed that waters that had influent raw water SUVA of $\leq$ 2.0 L/mg-m (measured prior to the addition of any oxidant) be exempted from EC requirements, similar to the exemption for raw water TOC concentration $\leq$ 2.0 mg/L. The SUVA value could also be used as an alternative Step 2 requirement. Step 2 PODR would still be used by some waters, as appropriate. Table IIb shows the modified 3 x 3 TOC percent removal matrix as presented in the Stage 1 final rule.

Similarly, for enhanced softening, Clark and Lawler (*32*) presented a paper highlighting the new information for softening plants that became available after the proposal. A survey of ICR softening utilities (49 plants) was conducted and summarized the TOC removal and typical operational challenges faced by these plants (*32*). Only 60 percent of plants in the low TOC, high alkalinity category were able to meet the proposed 20 percent TOC removal. Unlike EC, a Step 2 PODR was not recommended by the technical experts. An evaluation of the softening chemistry and operational challenges determined that the removal of 10 mg/L of magnesium (as $CaCO_3$) can be considered to be enhanced softening. Other limitations include the provision that softening plants do not have to go beyond a minimum alkalinity of 40–60 mg/L as $CaCO_3$, nor have major process changes from lime softening to a lime-soda softening process. Although calcium is removed rapidly with increasing lime dose (until the carbonate is exhausted), magnesium is removed only slightly at a pH below 10.7.

As the pH increases above 10, the major precipitate formed changes from calcium carbonate to magnesium hydroxide presenting operational challenges for plants not designed for magnesium removal. Such challenges include the formation of a more difficult to remove floc (magnesium hydroxide), carryover into the filters, and changes in the sludge. As the lime dose is increased, alkalinity is consumed and soda ash must be added to provide sufficient carbonate, which also results in changes in chemical feed (*32*).

Based on this information, the USEPA decided to change the required TOC percent removal in the low TOC/high alkalinity category from 20 to 15 percent. For Stage 1, ES plants must meet the TOC percent removals shown in the last column in Table IIb. If systems cannot meet these percent removals then they must demonstrate alternative performance criteria. The same exemptions that apply to EC apply to ES

(*11*). Moreover, systems that remove a minimum of 10 mg/L of magnesium hardness as calcium carbonate are exempt from ES requirements. Systems that depress the alkalinity to a minimum level (below 40 mg/L as $CaCO_3$) will be exempt and will not have to change from a lime to a lime-soda ash system.

## Impact of the Evaluation of the Moving the Point of Predisinfection on DBP Control

In the proposed Stage 1 D/DBP Rule, PWSs using conventional treatment could not claim disinfection credit for compliance with the Surface Water Treatment Rule for disinfectants added prior to the removal of precursors, except under some explicit conditions (*10*). This was done to discourage the use of a disinfectant prior to precursor removal in order to minimize DBP formation. An analysis of current predisinfection practices, through a fax survey (*33*), and ICR schematics, indicated that 80 percent of 329 surface water treatment plants were using predisinfection for a variety of reasons. While the majority of plants (222) indicated that predisinfection was used primarily for microbial inactivation, other reasons included algae control, taste and odor control, turbidity control and inorganic oxidation (*24,33*). A breakdown of the results of the survey is shown in Table III.

**Table III. Survey of Predisinfection Practices for Plants Affected by the ICR** (*24,33*)

| Predisinfection Reason | "Yes" Reponses (% of total)* |
|---|---|
| Taste and Odor Control | 114 (35 %) |
| Turbidity Control | 38 (12 %) |
| Algae Growth Control | 177 (54 %) |
| Inorganic Oxidation | 104 (32 %) |
| Microbial Inactivation | 222 (67 %) |
| Other | 27 (8 %) |

* Sum of responses >100% due to utilities using predisinfection for multiple reasons.

This information led the USEPA to reevaluate the predisinfection requirement and ultimately decide to allow PWSs to take credit for the use of disinfectants prior to precursor removal (*11*). Furthermore, the effect of moving the point of disinfectant addition from pre-rapid mix to downstream in the conventional filtration or softening treatment train on DBP formation was evaluated, using a number of surface waters that were representative of the 3 x 3 matrix. Summers and co-workers (*34*) evaluated 16 surface waters and concluded that DBP formation could be decreased with EC, while maintaining predisinfection. Greater benefits however, were observed when the point of chlorination was moved to post-sedimentation, where a median decrease of 17, 25 and 40 percent was observed in total organic halogen (TOX), TTHM and HAA5, respectively (*34*).

### Impact of Method Availability on the Stage 1 D/DBP Rule

When regulations are set, accurate and robust analytical methods are needed to measure regulated contaminants. For Stage 1, TTHM and HAA5 methods are available and can be reliably measured by laboratories nationwide. Therefore, they can be used effectively for compliance monitoring. These methods have minimum detection levels (MDLs) or minimum reporting levels (MRLs) that are significantly lower than the MCLs. For bromate, which is required to be monitored by systems using ozone, two factors contributed to setting the MCL: analytical considerations and the feasibility of using ozone if the MCL was very low. The concern for controlling bromate formation is because it is nephrotoxic in animals (*35*) and in humans (*36*). Based on a multistage model used by the USEPA to extrapolate the results of carcinogenesis experiments with high doses in animals to low doses in humans, the expected level at which there is a potential $10^{-4}$ and $10^{-5}$ risk is 5 and 0.5 µg/L, respectively (*10*). However, Method 300.0, used to measure bromate and other anions, was not sensitive enough to measure bromate lower than 0.010 mg/L. Therefore, a bromate MCL of 0.010 mg/L was proposed and this was not based solely on health risk.

During the time from proposal to the publication of the NODA, Method 300.1 was validated for inorganic compounds, and USEPA believes that more laboratories will be able to effectively measure bromate for compliance monitoring. Method 300.1 uses a higher capacity column, thus allowing a larger volume of sample to be placed on the column, which gives greater sensitivity. Method 300.1 provides performance data documenting that it is feasible for laboratories to reliably quantify at low enough levels to support the 0.010 mg/L MCL.

## The Role of the ICR in the Stage 2 D/DBP Rule Development

A committee of stakeholders was chartered through the FACA in March 1999. This committee will discuss the development of two rules: the Stage 2 D/DBP and the long term 2 ESWTR (LT2ESWTR) Rules. The Stage 2 D/DBP Rule will be proposed and promulgated with the LT2ESWTR to maximize microbial protection if lower DBP levels are set. The following section focuses on the contribution of the ICR data to Stage 2 Rule development.

The 18-month monitoring was designed to gather data for microbial contaminants, DBPs and their precursors, treatment plant design and monthly operating parameters at 300 PWSs (*13*). ICR data validation is based on quality control (QC) criteria defined by the ICR final rule and rule by reference documents. This quality control was done to ensure that the data met the quality necessary to support rule development. After review by the laboratories and utilities submitting data, utility and QC data are uploaded into the ICR Federal Database (ICR FED).

**ICR Data Analysis to Support Stage 2 D/DBP Rule**

During the ICR monitoring, USEPA and the American Water Works Association developed draft ICR data analysis plans that provided an overview of various data analyses approaches and predictive tools that can be used by the technical experts during the negotiations. To characterize plant influent water quality, precursor levels, pathogen and indicator occurrence (monthly monitoring), or DBP distribution occurrence data (quarterly monitoring), vertical analysis, which is an aggregate statistical summary of parameters, will be used. This analysis will also allow for aggregation of the data for comparison among plants. For example, the influent TOC concentrations for all surface water plants, averaged over the 18 months of monitoring, can be statistically evaluated using cumulative probability distributions or box and whisker plots to represent the TOC distribution of all ICR plants. These aggregates can be divided into source water categories, regional, or by disinfectant type. This analysis will provide national estimates of occurrence of DBP precursors and national estimates of DBPs in the finished water and in the distribution system. Treatment process information collected during the ICR will define existing treatment effectiveness for precursor removal and current disinfection practices. A more complex data analysis approach is a horizontal analysis, which looks at plant level analysis of horizontally-linked data (e.g., TOC removal in a conventional plant). Analyses may be based on plant-by-plant analysis followed by vertical aggregations to compare among plants (i.e., TOC removal through the coagulation process among conventional plants). The ICR data analysis will provide a snap-shot in time of the status of the ICR plant occurrence data and existing treatment at the end of the monitoring period. Predictive tools can be used to predict the status of the plants, after compliance with the Stage 1 Rule.

To facilitate data analysis, the data are extracted from the Oracle database, ICR FED, to a more manageable format in a Microsoft Access database. A principal auxiliary database (Aux 1), contains a significant portion of the validated ICR data. However, the quality control data and other data that were not needed immediately to support the rules were not extracted. For horizontal or through-plant analyses, other auxiliary databases (2-6 and 8) will be used. A list of these auxiliary databases is presented in Figure 1. A query tool will be available to facilitate horizontal analyses without in-depth knowledge of database execution.

The first 9 months of validated data were extracted into Aux 1 in July 1999. Other data extractions for the 12 and 18 months are expected as soon as the data are validated and after the utilities and laboratories have had a chance to correct their data. It is expected that the 18 months of validated data will be available in Aux 1 in December 1999. These data will provide the basis for the occurrence information needed to support M/DBP Rule.

**ICR Treatment Study Data Analysis to Support Stage 2 D/DBP Rule**

To determine whether DBP precursor concentrations were high enough to warrant ICR treatment studies (TS), all ground water plants serving $\geq$ 50,000 customers and all surface water plants serving $\geq$ 100,000 customers were required to

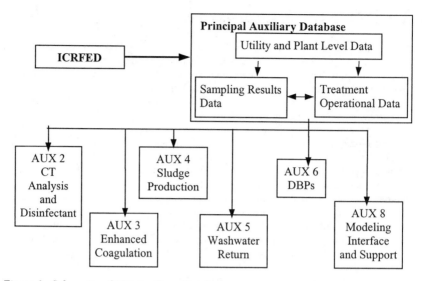

*Figure 1. Schematic of ICR Auxiliary Database*

conduct 12 months of applicability monitoring. The majority of utilities collected monthly TOC samples to fulfill this monitoring requirement. Surface water systems with an average source water influent TOC concentration exceeding 4.0 mg/L and ground water systems (not under the influence of a surface water) with an average finished water TOC concentration exceeding 2.0 mg/L were required to conduct treatment studies. A detailed description of the TS requirements, avoidance criteria, and the applicability monitoring results is presented by Allgeier and co-workers (*37*). Table IV summarizes the results of the TOC applicability monitoring for both surface and ground water plants.

**Table IV. Summary of Applicability Monitoring for ICR Treatment Studies (adapted from *37*)**

| Parameter | Raw Water Source | |
|---|---|---|
| | Surface Water | Ground Water |
| No. of Plants Conducting Monitoring | 308 | 190 |
| Median Annual Average TOC | 2.7 mg/L | 0.6 mg/L |
| Range of Annual Average TOC | < 0.7 to 12.8 mg/L | < 0.5 to 12.8 mg/L |
| TOC Concentration Trigger for Studies | 4.0 mg/L (influent) | 2.0 mg/L (finished) |
| % of Plants Required to Conduct Studies | 22% | 16% |

A total of 498 surface and ground water plants performed the applicability monitoring (Table IV). Only 22 percent of the 308 surface water plants and 16 percent of the 190 ground water plants had average TOC levels that were high enough to warrant conducting treatment studies (*37*).

The GAC and membrane TS serve two purposes: (1) utilities, which may have difficulty meeting future DBP regulations by EC or ES can use the TS results to begin planning for treatment modifications that can achieve higher levels of DBP precursor removal; and (2) the TS data will be used to estimate the feasibility and cost of implementing these processes for various regulatory options considered during the Stage 2 D/DBP Rule development process. While more stringent MCLs will minimize DBP exposure and associated health risks, it is required to determine the costs that these rules will impose on the drinking water industry and consumers prior to rule development.

During the TS, utilities had the option of evaluating either NF membranes or GAC. Sixty-two utilities chose to evaluate GAC and 37 evaluated membranes resulting in a total of 99 studies. The TS final reports were submitted to the USEPA in July 1999. EPA is in the process of reviewing these reports and the data will be uploaded to databases developed to support the treatment studies data analyses.

A technical work group composed of GAC and membrane experts was formed to help USEPA develop strategies for best utilizing and analyzing the data generated during the ICR treatment studies. The technical experts will participate in the data analysis.

## Other Factors That Contribute to Stage 2 D/DBP Rule Development

Information on the composition of DBP precursors, and their reactivity to disinfectants is important in assessing treatment techniques that are feasible for a particular water source. Other important aspects are DBP formation chemistry and the water quality parameters that affect their formation and stability. Studies to characterize NOM have focused on fractionation into the hydrophobic and hydrophilic fractions (using XAD resins) and ultrafiltration to divide the NOM into its molecular size fractions. Although NOM characterization does not directly impact the rules, this information provides the basis for understanding DBP precursors, their reactivity to various disinfectants, and the impact of treatment for the control of DBPs.

A number of chapters in this book discuss recent research in NOM characterization and its reactivity to disinfectants. These researchers investigated NOM fractions (hydrophobic/hydrophilic) and transphilic fractions and their reactivity to chlorination (*38-43*).

## DBP Classes Versus Individual Species

The THM Rule and the Stage 1 D/DBP Rule set MCLs for classes of compounds. One of the main issues that need to be addressed in the development of Stage 2 is whether to regulate individual compounds or just the DBP classes, based on health effects research on THMs and HAAs. Toxicological and reproductive studies indicate that the bromine substituted compounds can present more health risk than the chlorine substituted (44-48). It is hypothesized that the bromine is a better leaving group, which would enhance the mode of action of these compounds.

DBP speciation and the effect of treatment on the distribution of DBP species are important to consider. Separation treatment processes such as coagulation, filtration (including GAC), ultrafiltration and biological filtration remove part of the NOM but do not remove any of the inorganic precursor bromide (49). Nanofiltration and reverse osmosis remove some of the bromide but remove substantially more NOM. The extent of bromide rejection is dependent on the mass transfer characteristics of the membrane (50). Chemical processes, such as ozone oxidation, remove some of the bromide because of its conversion to bromate (51-55). Treatment processes also affect the residual disinfectant demand because of the removal of the compounds that contribute to the demand. Thus, the effect of precursor removal results in a change in the ratio of bromide to NOM and bromide to chlorine, which in turn shifts DBP speciation (55–58).

Many studies have evaluated the shift in DBP speciation with a change in the bromide to DOC or bromide to chlorine ratio (55-62). In general, increases in the bromide to DOC or bromide to chlorine ratio result in a shift to the bromine-substituted DBPs. Shifts in DBP speciation are also a function of NOM characteristics, and some organic compounds, such as acetyl-containing compounds favor bromine substitution (61,63). Ozonation, under conditions of constant bromide to DOC ratios, was found to increase the bromine incorporation factor, upon subsequent chlorination, when compared to the non-ozonated water (61).

If the Stage 2 Rule sets a MCL for an individual DBP, particularly the bromine-substituted DBPs, utilities treating waters with medium to high bromide levels will need to evaluate the shift in DBP speciation based on their selected treatment technology. The ICR will provide occurrence information on the individual species as well as the DBP classes.

## DBP Identification, Stability, and Methods

DBPs selected for monitoring under the ICR were based on occurrence information from limited data sets, analytical method and performance standards availability. THMs, HAAs, haloacetonitriles (HANs), chloral hydrate (CH) and the halopropanones are some of the DBPs monitored under the ICR, although some of them are not candidates for regulation. However, it was believed that monitoring for these compounds under the ICR would provide occurrence information for these parameters. Many analytes, such as HANs and CH, were measured by the same laboratories performing other DBP analyses and had to go through the laboratory approval process. For low molecular weight aldehydes, cyanogen chloride (CNCl) and low level bromate, some of the more problematic analytes, the samples were sent to a central laboratory to minimize analytical error. This was because of the following: (1) stability issues (e.g., CNCl and aldehydes); (2) the lack of performance evaluation standards which limited USEPA's ability to check laboratory performance; and (3) low-level bromate was not considered to be feasible by routine analyses.

The low-level bromate that is measured for the ICR data is done using the selective anion concentration method with an MDL of 0.2 µg/L (64). Research studies have attempted to improve the detection limit for bromate to allow for its measurement closer to the risk levels. Method 300.1 (18) (an ion chromatography (IC) method) has an MDL of 1.4 µg/L and has been validated by performance data. Other developed methods which may be considered in Stage 2 because of their sensitivity are (1) a method that uses IC but improves the detection limit using a post-column reagent (o-dianisidine), which when coupled to Method 300.1, lowers the MDL for bromate to 0.1 µg/L (65), and (2) an inductively coupled plasma mass spectrometry method that can measure bromate at an MDL of 0.3 µg/L (66). Similarly, other low-level bromate measurement techniques have been reported (67,68).

Cyanogen chloride is an important by-product resulting from the chloramination of drinking waters. While it was believed to be a good surrogate for by-products of chloramination, its instability does not make it a candidate for regulation. Its instability has also been a problem in health effects studies as it could not be determined whether its toxicity was from the parent compound or from its decomposition products. CNCl and cyanogen bromide are very prone to hydrolysis. CNCl is believed to provide a host for the hydroxide ion and the hydrolysis is catalyzed in the presence of free chlorine (69). Xie and Reckhow (70) reported that a 1 mg/L free chlorine residual resulted in about a 90 percent destruction of CNCl. Sodium sulfite, a chlorine quenching agent, also enhanced the decomposition of CNCl. A shift to cyanogen bromide with a corresponding decrease in CNCl at higher bromide concentrations was reported (70). The mechanism for the decomposition of CNCl is believed to involve the transference of the hydroxo group to the carbon atom of CNCl with the liberation of a chlorine molecule (69).

The importance of DBP method availability was significant in setting the MCL for haloacetic acids. For Stage 1, only HAA5 is regulated. At the time the rule was finalized, methods and analytical standards were only available for five of the nine HAAs. Currently, methods exist for measuring all nine haloacetic acid species. Therefore, the ICR data will provide occurrence information on bromochloroacetic acid, which is the sixth species in HAA6, and voluntary monitoring for the other more bromine-substituted HAAs, namely, bromodichloracetic acid (BDCAA), dibromochloroacetic acid (DBCAA) and tribromoacetic acid (TBAA). The delay in method availability stemmed from the lack of pure standards to allow for accurate quantification. The stability of BDCAA, DBCAA, and TBAA was examined in the neutral pH region at 25 °C, with and without a biocide (sodium azide) (71). Although the biocide improved the stability, it was still an issue for these three species. TBAA is very susceptible to hydrolysis and is believed to yield bromoform. In TBAA, a bromine substituted tetrahedral carbon atom is attached to a planar carbonyl functional group. It can be hypothesized that the driving force for the instability of the compound is to relieve steric hindrance. Bromine is a bigger atom than chlorine and this provides a good leaving group to yield bromoform. Trichloroacetic acid (TCAA) is probably more stable because chlorine is a smaller atom, although it can be hydrolyzed at high pH to yield chloroform (72).

The sum of the halogenated byproducts is generally measured using the TOX parameter. The known DBPs make up a fraction of the TOX (20 to 50 percent). Mass spectral techniques, coupled with gas or liquid chromatography, and Fourier transfer infrared spectrometry have been used in an attempt to identify other DBPs, particularly the more polar and higher molecular weight. An overview of drinking water disinfection byproducts is presented in Richardson (73). Such research is pertinent to strengthening our understanding of DBP formation and the risk associated with DBPs.

## Summary and Conclusions

The USEPA is addressing concerns to balance risks between microbial pathogens and DBPs. The D/DBP Rule is developed in stages which enables the agency to use the latest science (health effects research, occurrence, treatment technologies, costs and other information) to support rule development. To ensure that the concerns of all interested parties are addressed, the Office of Ground Water and Drinking Water (USEPA) is using a collaborative effort between the USEPA and stakeholders (i.e., FACA) for the rule-making process.

In the Stage 1 D/DBP Rule, the enhanced coagulation requirements in the proposed 3 X 3 TOC removal matrix for surface water systems were modified to allow for 90 percent of the systems to comply with the step 1 TOC removal requirement. Waters with low TOC and moderate-to-high alkalinity were expected to be more difficult to treat. Based on results from surveys and individual research studies, which included NOM fractionation of source waters to characterize their humic content, the TOC removal requirement for these difficult to treat waters was decreased by 5 percentage points to achieve the goal of selecting percent removals that can be "reasonably" met by 90 percent of PWSs practicing enhanced coagulation. An important modification that took place between the proposal and the final Stage 1 Rule was the application of the correlation between SUVA and the amenability of a water to TOC removal by coagulation. Enhancing coagulation while maintaining disinfection for operational and inactivation reasons, can result in decreased DBP formation especially when the disinfectant is applied post-sedimentation.

The ICR data that has been collected by 300 PWSs will be analyzed to support the development of Stage 2. The ICR treatment studies data will generate a large database of precursor removal process performance and cost information over a wide range of water qualities

Ongoing research to better understand DBP precursors and their control and to optimize existing analytical methods and develop new ones are important to further our understanding of DBPs, their formation and control.

# Acknowledgments

The authors acknowledge Stuart W. Krasner for his helpful comments and the contributions of Steven C. Allgeier, Eric Bissonette and Patricia Snyder Fair for their review of this paper. Alison Gusses' contribution to this work was supported in part by an appointment to the Internship Program at the Office of Ground Water and Drinking Water administered by the Oak Ridge Institute for Science and Education through an interagency agreement between the U.S. Department of Energy and the USEPA.

This paper does not necessarily reflect the views and policies of the USEPA, nor does the mention of trade names or products constitute endorsement or recommendation of their use.

# REFERENCES

1.  Rook, J. J. *Water Treat. Exam.* **1974**, *23* (2), 234-243.
2.  Symons, J. M.; Bellar, T.A.; Carswell, J.K.; DeMarco, J.; Kropp, K.L.; Robeck, G.G.; Seeger, D.R.; Slocum, C.J.; Smith, B.L.; Stevens, A.A. *J. AWWA* **1975**, *67* (11), 634-647.
3.  Brass H. J.; Feige, M.A.; Halloran, T.; Mello, J.W.; Munch, D.; Thomas, R.F. In *Drinking Water Quality Enhancement Through Source Protection;* Pojasek, R.B., Ed.; Ann Arbor Sci. Publ.: Ann Arbor, MI, 1977; pp 393-416.
4.  National Research Council, *Drinking Water and Health (Vol 1)*, National Academy of Sciences: Washington, D.C., 1977.
5.  USEPA. *Fed. Reg.* **1979**, *44* (231), 68624-68707.
6.  Craun, G. F.; Hubbs, S.A.; Frost, F.; Calderon, R.L; Via, S. *J. AWWA* **1998**, *90* (9), 81-91.
7.  Glaze, W. H.; Weinberg, H. S. *Identification and Occurrence of Ozonation By-Products in Drinking Water*; AWWARF and AWWA: Denver, CO, 1993.
8.  Richardson, S. D.; Thurston Jr., A. D.; Collette, T. W.; Schenk Patterson, K.; Lykins Jr., B. W.; Majetich, G.; Zang, Y. *Environ. Sci. Technol.* **1994**, *28* (4), 592-599.
9.  United States 104th Congress. *Safe Drinking Water Act Amendments of 1996*, Public Law 104-182, Washington, D.C., 1996.
10. USEPA. *Fed. Reg.* **1994**, *59* (145), 38668-38829.
11. USEPA. *Fed. Reg.* **1998**, *63* (241), 69390-69476.
12. Means III, E. G.; Krasner, S. W. *J. AWWA* **1993**, *85* (2), 68-73.
13. USEPA. *Fed. Reg.* **1996**, *61* (94), 24354-24388.
14. USEPA. *Methods for the Determination of Organic Compounds in Drinking Water- Supplement III*, EPA/600/R-95/131, 1995.
15. USEPA. *Methods for the Determination of Organic Compounds in Drinking Water- Supplement II*, EPA/600/R-92/129, 1992.
16. *Standard Methods for the Examination of Water and Wastewater*; APHA, AWWA, and WEF: Washington, D.C., 1995; 19th Edition, pp 6-67 - 6-76, 4-57 - 4-58.

44

17. USEPA. *Methods for the Determination of Inorganic Substances in Environmental Samples*, EPA/600/R-93/100, 1993.
18. USEPA. *USEPA Method 300.1, Determination of Inorganic Anions in Drinking Water by Ion Chromatography*, Revision 1.0, EPA/600/R-98/118, 1997.
19. USEPA. *Stage 1 Disinfectants and Disinfection Byproducts Rule*, Fact Sheet, EPA 815-F-98-010, 1998.
20. *United States Environmental Protection Agency, Office of Ground Water and Drinking Water*, URL http://www.epa.gov/safewater/mdbp/mdbp.html.
21. Krasner, S.W.; Owen, D.M.; Cromwell III, J.E. In *Water Disinfection and Natural Organic Matter: Characterization and Control;* Minear, R.A.; Amy, G.L., Eds.; American Chemical Society: Washington, D.C., 1996, pp 10-23.
22. Krasner, S. W.; Amy, G. *J. AWWA* **1995**, *87* (10), 93-107.
23. Krasner, S. W.; Chowdhury, Z. K.; Edwards, M. A.; Bell, K. A. In *American Water Works Association Water Quality Technology Conference Proceedings*, Denver, CO, 1997.
24. USEPA. *Fed. Reg.* **1997**, *62* (212), 59388-59484.
25. Singer, P.C. et al. *Enhanced Coagulation and Enhanced Softening for the Removal of Disinfection By-Product Precursors: An Evaluation;* prepared for AWWA Government Affairs Office, Washington, D.C., 1995.
26. White, M.C.; Thompson, J.D.; Harrington, G.W.; Singer, P.C. *J. AWWA* **1997**, *89* (5), 64-77.
27. Owen, D.M.; Amy, G.L.; Chowdhury, Z.K. *Characterization of Natural Organic Matter and Its Relationship to Treatability;* AWWARF and AWWA: Denver, CO, 1993.
28. Edzwald, J.K.; van Benschoten, J.E. In *4th International Gothenburg Symposium on Chemical Treatment Proceedings*, Madrid, Spain, 1990.
29. Krasner, S. W.; Amy, G.; Zhu, H. W. In *American Water Works Association Enhanced Coagulation Research Workshop Proceedings*, Denver, CO, 1994.
30. Chowdhury, Z. In *American Water Works Association Water Quality Technology Conference Proceedings*, Denver, CO, 1997.
31. Edwards, M. *J. AWWA* **1997**, *89* (5), 78-89.
32. Clark, S. C.; Lawler, D. F. In *American Water Works Association Water Quality Technology Conference Proceedings*, Denver, CO, 1997.
33. McGuire, M.J. *Analysis of Fax Survey Results;* prepared for the AWWA Government Affairs Office by McGuire Environmental Consultants, Inc., Washington, D.C., 1997.
34. Summers, R.S.; Solarik, G.; Hatcher, V.A.; Isabel, R.S.; Stile, J.F. In *American Water Works Association Water Quality Technology Conference Proceedings*, Denver, CO, 1997.
35. Kurokawa, Y.; Maekawa, A.; Takahashi, M.; Hayashi, Y. *Environ. Health Perspect.* **1990**, 87, 309-335.
36. Gradus, B-E.D., et al. *Am. J. Nephrol.* **1984**, *4*, 188.
37. Allgeier, S.C.; Shukairy, H.M.; Westrick, J.J. *J. AWWA* **1998**, *90* (11), 70-82.
38. Frimmel, F.H; Hesse, S.; Kleiser, G. In *American Chemical Society Division of Environmental Chemistry Preprints of Extended Abstracts*, Anaheim, CA, 1999, pp 206-208.

39. Wu, W.W.; Chadik, P.A.; Davis, W.M.; Delfino, J.J.; Powell, D.H. In *American Chemical Society Division of Environmental Chemistry Preprints of Extended Abstracts,* Anaheim, CA, 1999, pp 213-216.

40. Dickenson, E.R.V.; Amy, G.L. In *American Chemical Society Division of Environmental Chemistry Preprints of Extended Abstracts,* Anaheim, CA, 1999, pp 216-217.

41. Leenheer, J.A.; Croué, J-P.; Benjamin, M.; Korshin, G.V.; Hwang, C.J.; Bruchet, A.; Aiken, G.R. In *American Chemical Society Division of Environmental Chemistry Preprints of Extended Abstracts,* Anaheim, CA, 1999, pp 220-221.

42. Rostad, C.E.; Katz, B.G.; Leenheer, J.A.; Martin, B.S.; Noyes, T.I. In *American Chemical Society Division of Environmental Chemistry Preprints of Extended Abstracts,* Anaheim, CA, 1999, pp 222-223.

43. Karanfil, T.; Kilduff, J.C.; Kitis, M.; Wigton, A. In *American Chemical Society Division of Environmental Chemistry Preprints of Extended Abstracts,* Anaheim, CA, 1999, pp 226-229.

44. Pegram, R. A. In *Disinfection By-products in Drinking Water: Critical Issues in Health Effects Research Workshop Report,* Chapel Hill, NC, 1995, pp 68-69.

45. Bull, R. J. In *Disinfection By-products in Drinking Water: Critical Issues in Health Effects Research Workshop Report,* Chapel Hill, NC, 1995, pp 29-30.

46. Linder, R.E.; Klinefelter, G.R.; Strader, L.F.; Narotsky, M.G.; Suarez, J.D.; Roberts, N.L; Perreault, S.D. *Fundam. & Appl. Toxicol.* **1995,** *28,* 9-17.

47. Linder, R.E.; Klinefelter, G.R.; Strader, L.F.; Veeramachaneni, D.N.R.; Roberts, N.L; Suarez, J.D. *Repr. Toxicol.* **1997,** *11* (1), 47-56.

48. Linder, R.E.; Klinefelter, G.R.; Strader, L.F.; Suarez, J.D.; Roberts, N.L. *Repr. Toxicol.* **1997,** *11* (5), 681-688.

49. Summers, R.S.; Benz, M.A., Shukairy, H.M; Cummings, L. *J. AWWA* **1993,** *85* (1), 88-95.

50. Allgeier, S.C. Master's Thesis, University of Cincinnati, Cincinnati, OH, 1995.

51. Krasner, S.W.; Glaze, W.H.; Weinberg, H.S.; Daniel, P.A.; Najm, I.N. *J. AWWA* **1993,** *85* (1), 73-81.

52. Kruithoff, J.C.; Schippers, J.C. *Water Supply* **1993,** *11* (121).

53. Von Gunten, U.; Hoigné, J. In *Disinfection By-Products in Water Treatment: The Chemistry of their Formation and Control;* Minear, R.A.; Amy, G.L., Eds.; CRC Press: New York, 1996; pp 187-206.

54. Siddiqui, M.S.; Amy, G.L.; Rice, R.G. *J. AWWA* **1995,** *87* (10), 58-70.

55. Shukairy, H.M.; Miltner, R.J.; Summers, R.S. *J. AWWA* **1994,** *86* (6), 72-87.

56. Amy, G.L.; Tan, L.; Davis, M.K. *Wat. Res.* **1991,** *25* (2), 191-202.

57. Symons, J.M.; Krasner, S.W.; Simms, L.A.; Sclimenti, M. *J. AWWA* **1993,** *85* (1), 51-62.

58. Symons, J.M.; Speitel Jr., G.E.; Diehl, A.C.; Sorensen Jr., H.W. *J. AWWA* **1994,** *86* (6), 48-60.

59. Shukairy, H.M.; Miltner, R.J.; Summers, R.S. *J. AWWA* **1995,** *87* (10), 71-82.

60. Symons, J.M.; Krasner, S.W.; Sclimenti, M.J.; Simms, L.A.; Sorensen Jr., H.W.; Speitel Jr., G.E.; Diehl, A.C. In *Disinfection By-Products in Water Treatment: The Chemistry of their Formation and Control;* Minear, R.A.; Amy, G.L., Eds.; CRC Press: New York, 1996; pp 91-130.

61. Shukairy, H.M; Summers, R.S. In *Disinfection By-Products in Water Treatment: The Chemistry of their Formation and Control;* Minear, R.A.; Amy, G.L., Eds.; CRC Press: New York, 1996; pp 311-335.
62. Krasner, S.W.; Sclimenti, M.J.; Chinn, R.; Chowdhury, Z.K.; Owen, D.M. In *Disinfection By-Products in Water Treatment: The Chemistry of their Formation and Control;* Minear, R.A.; Amy, G.L., Eds.; CRC Press: New York, 1996; pp 59-90.
63. Merlet, N. Ph.D. Thesis, Université de Poitiers, France, 1986.
64. Hautman, D.P. In *American Water Works Association Proceedings: 1992 Water Quality Technology Conference,* Toronto, Ontario, Canada, 1992, pp 993-1007.
65. Wagner, H.P.; Pepich, B.V.; Hautman, D.P.; Munch, D.J. *J. Chromatography A* **1999,** *850,* 119-129.
66. Creed, J.T.; Magnuson, M.L.; Pfaff, J.D.; Brockhoff, C. *J. of Chromatography A* **1996,** *753* (2), 261-267.
67. Weinberg, H.S.; Yamada, H. *Anal. Chem.* **1998,** *70,* 1-6.
68. Eschigo, S.; Yamada, H.; Minear, R.A. In *American Chemical Society Division of Environmental Chemistry Preprints of Extended Abstracts,* Anaheim, CA, 1999, pp 249-251.
69. Bailey, P.L.; Bishop, E. *J. Chem. Soc.* **1973,** *9,* 912-916.
70. Xie, Y.; Reckhow, D.A. In *American Water Works Association Proceedings: 1992 Water Quality Technology Conference,* Toronto, Ontario, Canada, 1992, pp 1761-1777.
71. Brophy, K.S.; Weinberg, H.S.; Singer, P.C. In *American Chemical Society Division of Environmental Chemistry Preprints of Extended Abstracts,* Anaheim, CA, 1999, pp 256-259.
72. Reckhow, D.A.; Singer, P.C. In *Water Chlorination Chemistry, Environmental Impact and Health Effects;* Jolley, R.L.; Bull, R.J.; Davis, W.P.; Katz, S.; Roberts Jr., M.H.; Jacobs, V.A., Eds.; Lewis Publishers, Inc.: Chelsea, MI, 1985; Vol. 5, pp 1229-1257.
73. Richardson, S.G. In *Encyclopedia of Environmental Analysis and Remediation;* John Wiley and Sons: New York, 1998; Vol. 3, pp 1398.

Chapter 4

# Extensions and Verification of the Water Treatment Plant Model for Disinfection By-Product Formation

Gabriele Solarik[1], R. Scott Summers[1], Jinsik Sohn[1],
Warren J. Swanson[2], Zaid K. Chowdhury[2], and Gary L. Amy[1]

[1]Center for Drinking Water Optimization, Department of Civil,
Environmental and Architectural Engineering, University of Colorado
at Boulder, Campus Box 421, Boulder, CO 80309–0421
[2]Malcolm Pirnie, Ind., 432 North 44th Street, Suite 400,
Phoenix, AZ 85008

## Introduction

The U.S. Environmental Protection Agency (USEPA) Water Treatment Plant (WTP) model was developed in 1992 and used to support the Disinfectant/Disinfection By-product (D/DBP) Reg/Neg process in 1993-94 *(1)*. The model predicts (1) the behavior of water quality parameters that impact the formation of disinfection by-products (DBPs) and (2) the formation of DBPs. This version of the model was limited in several ways: (a) many existing process, inactivation, DBP formation, and disinfectant decay algorithms within the WTP model were limited and/or outdated; (b) new process, inactivation, DBP formation, and disinfectant decay algorithms needed to be added; and (c) multiple points of chlorination, parallel treatment trains, and the distribution system were not modeled.

## Objectives

The overall objectives of this project are to modify existing model algorithms and to extend the model with new algorithms. The specific objective of this paper is to report the results from the modification, extension, and verification of the model for conventional and softening treatment with chlorination. New algorithms were developed and verified for advanced treatment processes and alternative disinfectants. However, these results will be presented in more detail in future publications.

## Approach

The WTP model uses empirical correlations to predict central tendencies for natural organic matter (NOM) removal, disinfection, and DBP formation in a treatment plant. NOM characteristics such as total organic carbon (TOC), ultraviolet absorbance at 254 nm (UVA), and specific UVA (SUVA) are used to predict trihalomethane (THM) and haloacetic acid (HAA) formation. The original model and its verification are discussed by Harrington et al. *(2)*.

The algorithms were generally developed using multiple linear regression, and the following regression parameters are given for the equations: the multiple correlation coefficient ($R^2$), which measures the strength of the correlation by indicating the proportion of the variability in the dependent variable that is explained by all predictor variables combined; the adjusted correlation coefficient ($R^2_{adj}$), which is the multiple correlation coefficient adjusted by the number of predictor variables; the standard estimate of error (SEE), which measures the amount of scatter in the vertical direction of the data (i.e., around the dependent variable) about the regression plane; the F-statistic, which can be used to assess the goodness of fit using the F-test of the variance accounted for by regression; and the number of data points (n) used in equation development *(3)*.

Algorithm equations, together with data ranges for the input parameters, are given in this paper. These data ranges represent the boundary conditions within which the equations were developed and should be used. However, the WTP model does not restrict the use of the equations outside these boundary conditions. It is the user's responsibility to apply the model in an appropriate manner.

All verification results presented herein were developed using independent data, i.e., data that were not used in the development of the predictive equations. The predicted results used only measured input variables, unless otherwise noted. The errors in predictions of the verification data set are reported either as the standard error (SE), which is an estimate of the variance or as the average error (AE), which expresses the absolute difference between the measured and predicted value as a percentage of the measured value. The equations for the SE and the AE are as follows:

$$SE = \sqrt{\left( \frac{\Sigma(\text{Measured - Predicted})^2}{n} \right)}$$

$$AE = 100 \left( \frac{|\text{Measured - Predicted}|}{\text{Measured}} \right)$$

# Results and Discussion

## Conventional Treatment by Coagulation and Softening

*Coagulation and Softening pH*

The WTP model equations to predict pH changes due to coagulation and softening were not revised and are calculated based on raw water alkalinity, coagulant dose, and carbonate chemistry *(4)*. The model uses only equilibrium considerations and does not take into account the kinetics of processes such as calcium carbonate precipitation or carbon dioxide dissolution. The model assumes that the water treatment plant is, in effect, a closed system.

Figure 1 shows the verification of settled water pH by coagulation (both alum and iron), using independent bench-scale jar test batch data *(5,6)*. The line on this graph, and all subsequent verification graphs (unless otherwise noted), indicates the line of equality, i.e., any data points on this line indicate a perfect prediction. For alum coagulation, comparison between the measured and predicted data shows that pH is, in general, underpredicted, i.e., the model overpredicts the depression of pH by coagulant addition. The SE for pH verification for the waters coagulated with alum is 0.5 units, and the model underpredicted settled water pH by an AE of 5 percent. Similar results were found in the 1992 version of the model *(2)*. For iron coagulation, the pH predictions are within the ranges shown for alum coagulation; however, too few data are available to conclude any consistent trends.

Coagulation pH is used as an input parameter into the models calculating settled water TOC and UVA. Thus the underprediction of pH can lead to a propagation of error in settled water quality. The consistent underprediction of coagulation pH can be remedied by recalibration of the model or by the possible inclusion of an open carbonate chemistry system, as suggested by Tseng and Edwards *(7)*. Prior to use or recalibration, the model results will be compared to results from pilot- and full-scale plants, where a flow-through system is used. Following the verification with pilot- and full-scale data, the pH models will be recalibrated if necessary.

*TOC*

In the 1992 version of the WTP model, TOC removal by alum and iron coagulation was predicted by an empirical equation based on the raw water TOC, coagulant dose, and coagulant pH. In the current version of the WTP model, TOC removal by coagulation is predicted using the semi-empirical sorption model proposed by Edwards *(8)*. The model proposed by Edwards was based on dissolved organic carbon (DOC); however, the author showed it to predict TOC removal nearly as well. The model does not consider particulate organic matter (POC), which, together with DOC comprises TOC.

The model divides TOC into fractions that are sorbable and nonsorbable by the coagulant. The nonsorbable fraction cannot be removed by coagulation, and TOC removal is attributed solely to the sorbable fraction. The model is designed to handle

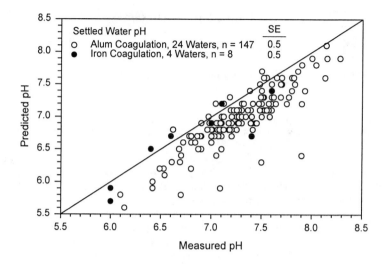

*Figure 1. Verification of Settled Water pH Prediction*

both alum and iron coagulation. The model uses similar input parameters as were used previously (raw water TOC, coagulant dose, and coagulation pH), but also uses calculated model coefficients and the raw water SUVA. The raw water SUVA is a more refined indicator of NOM characteristics and indicates the humic content in the water. Coagulation typically preferentially removes humic NOM. Thus, a water with a low humic content, indicated by a low SUVA, is expected to have a large nonsorbable fraction, whereas a water with a higher humic content and a higher SUVA will have a lower nonsorbable fraction *(9)*.

Additional information and detail about the equations can be found in *(8)*.

Coagulation:
$$TOC_{settled} = TOC_{sorb,eq} + TOC_{nonsorb}$$

where
$TOC_{settled}$ = settled water TOC (mg/L)
$TOC_{sorb,\ eq}$ = sorbable TOC remaining in solution at equilibrium (mg/L)
    = $f(pH_{coag}$, coag. type, $Dose_{coag}$, $SUVA_{raw}$, $TOC_{raw})$ (based on Langmuir adsorption model)

$TOC_{nonsorb}$ = non-sorbable TOC concentration (mg/L)
    = $TOC_{raw} * F_{nonsorb}$
  where
    $F_{nonsorb} = K_1 * SUVA_{raw} + K_2$    ($K_1$ and $K_2$ are coagulant dependent; $K_1$ has a negative value)

Alum: $(R^2_{adj} = 0.982, SEE = 0.40\ mg/L, n = 608)$
$TOC_{settled}$ = settled TOC (mg/L): $1.0 \le TOC_{settled} \le 26$
$TOC_{raw}$ = raw water TOC (mg/L): $1.8 \le TOC_{raw} \le 26.5$
$SUVA_{raw}$ = raw water SUVA (L/mg·m): $1.32 \le SUVA_{raw} \le 6.11$
$Dose_{coag}$ = coagulant dose (mmol Al/L): $0 \le Dose_{coag} \le 1.51$
$pH_{coag}$ = coagulation pH: $5.5 \le pH_{coag} \le 8.0$

Iron: $(R^2_{adj} = 0.988, SEE = 0.47\ mg/L, n = 250)$
$TOC_{settled}$ = settled TOC (mg/L): $0.9 \le TOC_{removed} \le 26$
$TOC_{raw}$ = raw water TOC (mg/L): $2.3 \le TOC_{raw} \le 26.5$
$SUVA_{raw}$ = raw water SUVA (L/mg·m): $1.26 \le SUVA_{raw} \le 6.11$
$Dose_{coag}$ = coagulant dose (mmol Fe/L): $0 \le Dose_{coag} \le 1.22$
$pH_{coag}$ = coagulation pH: $3.0 \le pH_{coag} \le 8.0$

TOC removal by softening is predicted using an empirical equation, developed by this project, based on raw water TOC, softening pH, and lime and coagulant doses from the American Water Works Association (AWWA) Water Industry Technical Action Fund (WITAF) database *(10)*. It is able to predict TOC removal for softening plants with or without coagulant addition. For plants that do not use a coagulant together with lime, the coagulant dose term in the equation goes to a value of one.

Softening:
$TOC_{removed} = 4.657 \times 10^{-4} (TOC_{raw})^{1.3843} (pH_{sft})^{2.2387} (Dose_{lime})^{0.1707} (1 + Dose_{coag})^{2.4402}$
$(R^2 = 0.957, R^2_{adj} = 0.955, SEE = 0.352\ mg/L, F = 486, n = 92)$
$TOC_{removed}$ = TOC removed by softening (mg/L): $0.1 \le TOC_{removed} \le 6.8$
$TOC_{raw}$ = raw water TOC (mg/L): $0.9 \le TOC_{raw} \le 14.1$
$pH_{sft}$ = pH of softening: $8.9 \le pH_{sft} \le 12.5$
$Dose_{lime}$ = applied lime dose (mg/L): $33 \le Dose_{lime} \le 410$
$Dose_{coag}$ = applied coagulant dose (meq/L): $0 \le Dose_{coag} \le 0.138$

The use of these models was verified using independent bench-scale jar test data from 24 waters (n=157) for alum coagulation, four waters (n=8) for iron coagulation, and four waters (n=28) for softening*(5, 6)*. The results are shown in Figure 2. The verification results used measured pH values. For coagulation at high TOC, the model has the tendency to underpredict settled water TOC and for softening, the model tends to underpredict settled water TOC. The SE of the settled water TOC is 0.54 mg/L for alum coagulation, 0.37 mg/L for iron coagulation, and 0.90 mg/L for softening. The AE for alum and iron coagulation was 16 percent, and 80 percent of the errors were within 21 percent of the measured value. The AE for softening was 24 percent, and 80 percent of the errors were within 30 percent of the measured values.

*UVA*

In the 1992 version of the WTP model, UVA removal equations were limited by small data sets. The new equations predicting UVA removal by coagulation and

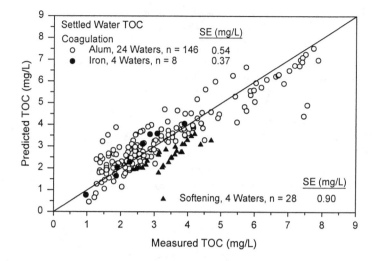

*Figure 2. Verification of Settled Water TOC*

softening are based on data analysis performed on the AWWA/WITAF database *(10)*, thereby significantly extending the data sets used for equation development. For softening, the SUVA data range includes two abnormally high values (above 11.2 L/mg-m). 90 percent of the data used to develop the equation for UVA removal had a SUVA between 1.8 and 5.2 L/mg-m.

Coagulation:
$$\mathrm{UVA_{removed}} = 5.716(\mathrm{UVA_{raw}})^{1.0894}(\mathrm{Dose_{coag}})^{0.305}(\mathrm{pH_{coag}})^{-0.9513}$$

*($R^2 = 0.900$, $R^2_{adj} = 0.900$, SEE = 0.040 1/cm, F = 3372, n = 1127)*
$\mathrm{UVA_{removed}}$ = UVA removed by coagulation (1/cm): $0.000 \le \mathrm{UVA_{removed}} \le 0.691$
$\mathrm{UVA_{raw}}$ = raw water UVA(1/cm): $0.015 \le \mathrm{UVA_{raw}} \le 0.751$
$\mathrm{Dose_{coag}}$ = applied coagulant dose (meq/L): $0.008 \le \mathrm{Dose_{coag}} \le 0.151$
$\mathrm{pH_{coag}}$ = pH of coagulation: $3.0 \le \mathrm{pH_{coag}} \le 8.3$

Softening:
$$\mathrm{UVA_{removed}} = 0.01685(\mathrm{TOC_{removed}})^{0.8367}(\mathrm{SUVA_{raw}})^{1.2501}$$

*($R^2 = 0.984$, $R^2_{adj} = 0.983$, SEE = 0.032 1/cm, F = 1019, n = 36)*
$\mathrm{UVA_{removed}}$ = UVA removed by softening (1/cm): $0.014 \le \mathrm{UVA_{removed}} \le 0.874$
$\mathrm{TOC_{removed}}$ = TOC removed by softening (mg/L): $0.1 \le \mathrm{TOC_{removed}} \le 6.8$
$\mathrm{SUVA_{raw}}$ = raw water SUVA (L/mg-m): $1.8 \le \mathrm{SUVA_{raw}} \le 12.5$

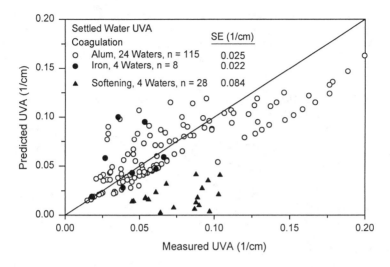

*Figure 3. Verification of Settled Water UVA*

Settled water UVA verification is shown in Figure 3. UVA removal equations were verified using the same databases that were used to verify TOC removal equations *(5,6)*. As is the case with TOC, the majority of settled water UVA is underpredicted for coagulation at high UVA values, i.e. the equations tend to overpredict UVA removal. Furthermore, the errors in settled water UVA predictions are greater for softening than for coagulation. The SE of UVA is 0.025 1/cm and 0.022 1/cm for alum and iron coagulation, respectively, and 0.084 1/cm for softening. The data set used for verification of UVA removal by softening is very limited. Additional verification will be performed prior to any recalibration of this algorithm, either with bench-scale data, if available, to check the adequacy of the equation, or with full-scale data to check the model.

*Impact of Coagulation pH on Settled Water TOC and UVA*

The underprediction of pH, and its use as an input parameter to settled water TOC and UVA predictions, may lead to a propagation of error in settled water TOC and UVA predictions. Figure 4 shows the impact of coagulation pH on settled water TOC and UVA predictions for a specific water quality condition, indicating some sensitivity of the TOC and UVA prediction to coagulation pH. For an underprediction of pH by 0.5 units, TOC and UVA predictions can vary by up to 10 and 5 percent, respectively, for this specific condition. Figure 5 shows the comparisons of measured and predicted settled water TOC, using predicted pH values for 24 waters for alum coagulation and four waters for iron coagulation using the same independent data sets as were used in Figure 2 *(5, 6)*. The same general trends can be seen as in Figure 2: settled water TOC is predicted relatively well with an SE of approximately 0.5 mg/L, except at the higher TOC values, where settled water

**54**

TOC is generally underpredicted, i.e., removal is overpredicted. Figure 6 compares the TOC predictions, using measured pH values and using predicted pH values as inputs into the TOC predictions. For both iron and alum coagulation, settled water TOC predictions for coagulated waters are insensitive to errors in settled water pH

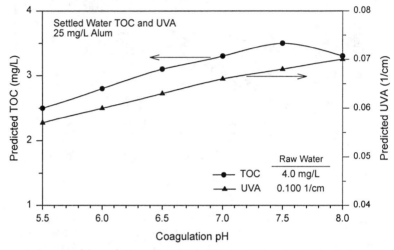

*Figure 4. Impact of Coagulation pH on Settled Water TOC and UVA Predictions*

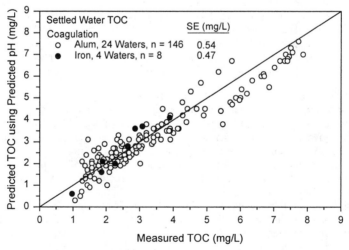

*Figure 5. Verification of Settled Water TOC using Predicted Coagulation pH*

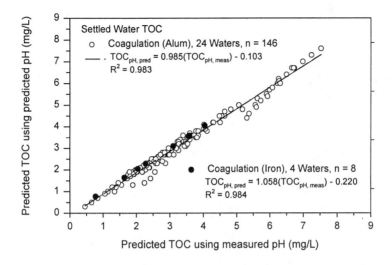

*Figure 6. Impact of Measured Versus Predicted Coagulation pH on Settled Water TOC*

predictions (which were, on average, 0.5 pH units), as indicated by the slope of approximately 1 and the high $R^2$ values. Similar results were found for the same evaluation of UVA predictions (slope of 1.02 and $R^2 > 0.99$).

**Chlorination**

The 1992 version of the WTP model used two equations to predict chlorine decay: (a) second-order reaction with respect to chlorine concentration was assumed for times less than 5 hours, and (b) a first-order reaction was assumed for times between 5 and 120 hours. The reaction constants were correlated with raw water quality parameters. The WTP model chlorine decay reactions have been updated with recent work by Koechling et al. *(11)* and Isabel et al. *(12)*, who have shown that a single Monod-type kinetic reaction well predicts chlorine decay. Separate chlorine decay equations were developed for raw and treated waters. The equations are based on bench-scale data and have the following form:

$$\frac{dC_t}{dt} = -\frac{k_2 * SUVA_0 * C_t}{\alpha_1 + C_t}$$

where $SUVA_0$ is the initial SUVA (UVA/DOC), and $k_2$ and $\alpha_1$ are kinetic rate parameters. The model takes the following form when integrated:

$$C_t = \alpha_1 * \ln \frac{C_0}{C_t} - k_2 * SUVA_0 * t + C_0$$

where $C_t$ is the chlorine residual concentration at any reaction time t, and $C_0$ is the initial chlorine dose. The kinetic rate parameter $\alpha_1$ is related to the initial DOC and UVA and for a chlorine-to-TOC ratio of 2, $\alpha_{1@2}$, takes the following form:

$$\alpha_{1@2} = 4.98 * UVA_0 - 1.91 * DOC$$

Isabel et al. *(12)* found $\alpha_1$ to be independent of treatment and developed the following correction factor for $\alpha_1$, making it applicable to other chlorine-dose-to-TOC ratios:

$$\alpha_1 / \alpha_{1@2} = 0.503(Cl_2/TOC)$$

The relationship for $k_2$ is a power function with respect to initial UVA and where a and b are fitted parameters dependent on treatment and chlorine dose. Thus the model takes into account the changes in chlorine decay rates for differently treated waters.

$$-k_2 = a * UVA_0{}^b$$

**DBP Formation**

The WTP Model simulates the formation of THMs and HAAs under conditions of full-scale treatment plants. Individual THMs (four) and HAAs (six), as well as sums, total THMs (TTHMs), and HAA5 or HAA6, are modeled. One of the biggest challenges to DBP modeling is an assessment of the impact of different treatment processes on the formation of DBPs. For many advanced treatment processes, like GAC, membranes, and biofiltration, there is little DBP formation data at the plant level. Multiple points of chlorination also present a challenge, as the DBP formation algorithms are based on single doses of chlorine and not chlorine residuals.

The 1992 version of the WTP model used THM equations that were developed from raw water chlorination studies using very high chlorine doses in some cases. These equations were used to predict TTHM formation in both raw and treated waters. In addition, these equations predicted concentrations on a molar basis and had to be converted to a mass basis using a second empirical correlation. At the time of development of the 1992 version of the WTP model, only limited data were available to develop predictive equations for HAAs. HAAs were predicted either by correlations with predicted THM formation or by empirically developed equations from laboratory studies using one set of reaction conditions and one chlorine dose.

*Raw and Treated Waters*

New empirical equations predicting THMs (total and four species), HAA5, HAA6, and six species based on low to moderate chlorine doses applied to raw/untreated water were developed by Amy et al. *(13)*. Water quality parameters such as TOC, UVA, Br, pH, and temperature, as well as applied chlorine dose and reaction time, are used to model DBP formation.

Raw water models:

$$TTHM_{raw} = 0.0412(TOC_{raw})^{1.098}(Cl_2)^{0.152}(Br_{raw})^{0.068}(T)^{0.609}(pH_{raw})^{1.601}(t)^{0.263}$$

*($R^2_{adj}$ = 0.90, F =1198, n = 786)*

$$HAA6_{raw} = 9.98(TOC_{raw})^{0.935}(Cl_2)^{0.443}(Br_{raw})^{-0.031}(T)^{0.387}(pH_{raw})^{-0.655}(t)^{0.178}$$

*($R^2_{adj}$ = 0.87, F = 831, n = 738)*

$TTHM_{raw}$ = raw water TTHM ($\mu$g/L)
$HAA6_{raw}$ = raw water HAA6 ($\mu$g/L)
$TOC_{raw}$ = raw water TOC (mg/L): $1.2 \leq TOC_{raw} \leq 10.6$
$Cl_2$ = applied chlorine dose (mg/L): $1.51 \leq Cl_2 \leq 33.55$
$Br_{raw}$ = raw water bromide concentration ($\mu$g/L): $7 \leq Br_{raw} \leq 600$
T = temperature (°C): $15 \leq T \leq 25$
$pH_{raw}$ = raw water pH: $6.5 \leq pH_{raw} \leq 8.5$
t= reaction time (hour): $2 \leq t \leq 168$

Both DBP equations are positively correlated with TOC, chlorine dose, temperature, and time. The TTHM formation equation is positively correlated with the bromide concentration and pH, as would be expected. Research has shown that THM formation shifts to the more brominated (higher mass) species at higher bromide concentrations *(14)* and THM formation is base-catalyzed *(15)*. For HAA6, the formation equation is negatively correlated to bromide as increases in bromide concentration will shift HAA speciation towards species that are not enumerated in HAA6 *(16)*. The predictive equation for HAA6 is also negatively correlated to reaction pH, as increasing the pH has been shown to decrease the formation of trichloroacetic acid, which is generally the most abundant species in HAA6 *(17)*.

Boundary conditions for reaction time, temperature, and pH need to be noted. Due to a lack of data, the equations were only developed for reaction times longer than two hours, temperatures between 15 and 25°C, and pH values between 6.5 and 8.5. For cases where the WTP model is used to simulate colder or warmer treatment conditions, the equations will be used outside the boundary conditions. For treatment scenarios with very short reaction times, less than two hours, use of the model will also be used outside the boundary conditions.

New DBP formation equations for treated water were based on work performed by Amy et al. *(13)* using both iron and alum coagulated waters. In the WTP model, these equations are applied to coagulated water, as well as water that is softened, treated by granular activated carbon (GAC), treated by membranes, and ozonated and biotreated. The DBP prediction equations for treated waters use the combined TOC and UVA (TOC*UVA) input parameter to model DBP formation. The TOC*UVA input parameter accounts for the impact of treatment on NOM removal as well as NOM characteristics, i.e., NOM reactivity. The boundary conditions for these equations are similar to those for raw water. Insufficient pH and temperature-dependent data were available to develop DBP formation equations for treated waters. Instead, temperature and pH factors were developed from raw water data and applied to the treated water equations. However, these factors are only valid in the 15 to 25°C temperature range and the 6.5-8.5 pH range. For enhanced coagulation and softening, the pH boundary conditions may be exceeded.

Treated water models:

$$TTHM = 23.9(TOC*UVA)^{0.403}(Cl_2)^{0.225}(Br)^{0.141}(1.027)^{(T-20)}(1.156)^{(pH-7.5)}(t)^{0.264}$$

*(R² = 0.919, R²_{adj} = 0.917, SEE = 0.218 μg/L, F = 798, n = 288)*

$$HAA6 = 41.6(TOC*UVA)^{0.328}(Cl_2)^{0.585}(Br^-)^{-0.12}(0.932)^{(pH-7.5)}(1.021)^{(T-20)}(t)^{0.150}$$

*(R² = 0.936, R²_{adj} = 0.937, SEE = 0.191 μg/L, F = 1040, n = 288)*

TTHM = treated water TTHM (μg/L): $13 \leq TTHM \leq 690$
HAA6 = treated water HAA6 (μg/L): $12 \leq HAA6 \leq 643$
TOC = treated water TOC (mg/L): $1.00 \leq TOC \leq 7.77$
UVA = treated water UVA (1/cm): $0.016 \leq UVA \leq 0.215$
$Cl_2$ = applied chlorine dose (mg/L): $1.11 \leq Cl_2 \leq 24.75$
Br = treated water bromide concentration (μg/L): $23 \leq Br \leq 308$
T = temperature (°C): 20
pH = treated water pH: 7.5
t = reaction time (hour): $2 \leq t \leq 168$

Figures 7 and 8 show the verification of DBP predictions for coagulated and softened waters for TTHM and HAA6, respectively. DBP formation data collected from various studies yielded a total of 47 coagulated waters and four softened waters and were used for verification of DBP formation equations *(5, 18-20)*. For coagulation, TTHM formation prediction after coagulation is generally underpredicted, while HAA6 formation is generally overpredicted. However, 90 percent of the errors are within 24 μg/L of measured values for TTHM and within 18 μg/L for measured values for HAA6. The general underprediction of TTHM and overprediction of HAA6 may be linked to the reaction pH. For many of these waters, the reaction pH was less than 7.0. At lower pH, TTHM formation is decreased, while HAA6 is increased *(15, 17)*.

For softening, only a very limited data set was available for verification. The errors in TTHM and HAA6 predictions are larger than for coagulation and both are

generally overpredicted.  However, due to the limited data, these results are difficult
to interpret.

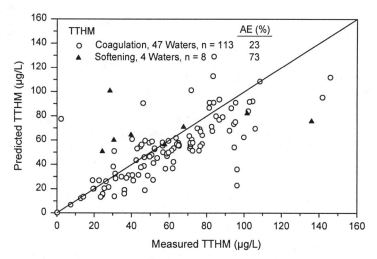

*Figure 7.  Verification of TTHM Formation (Coagulation/Softening)*

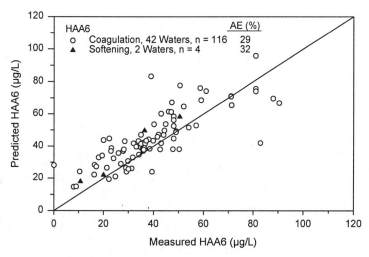

*Figure 8.  Verification of HAA6 Formation (Coagulation/Softening)*

*Prechlorination*

The 1992 version of the model simplified prechlorination, i.e., chlorine added prior to coagulation. It assumed that DBPs formed during the coagulation process from chlorine addition could be modeled using raw water DBP formation models. In a recent prechlorination study performed by Summers et al. *(5)*, the authors concluded that coagulation was effective in decreasing the DBPs formed with prechlorination relative to the chlorination of raw water, i.e., DBP precursors were removed by coagulation in the presence of chlorine.

To better predict DBP formation for prechlorination plants, an empirical prechlorination factor was developed to account for the decrease in DBP formation that occurs when chlorine is added either pre- or post-rapid mixing (RM), as compared to raw water DBP formation using data from 20 waters *(5)*. This relationship is used to modify the DBP formation that would be predicted by the raw water DBP formation model. The percent decrease in DBP formation (compared to raw water DBP formation) that can be attributed to coagulation was related to TOC removal by coagulation, to account for the higher precursor removals at higher coagulant doses.

Decrease in TTHM Formation (%) = 0.853(Percent TOC Removal)

Decrease in HAA6 Formation (%) = 0.794(Percent TOC Removal)

For the 20 waters used in the equation development, the average decreases for TTHM and HAA6 were predicted by the above equations to be 24 and 22 percent, respectively. This corresponds well with the average TTHM and HAA6 decreases reported by Summers et al. *(5)* of 23 and 18 percent, respectively. The prechlorination equations were not verified due to a lack of availability of prechlorination data.

## DBP Modeling Under Different Chlorination Scenarios

In the new WTP model, DBP formation under three chlorination scenarios is modeled.

1. Prechlorination only (Figure 9): A single point of chlorination prior to rapid mixing. For this approach, DBP formation is modeled in two separate, additive stages. First, the raw water DBP formation model is proportionally adjusted with the prechlorination factor for DBP formation through sedimentation. Formation after sedimentation is modeled using the treated water model with settled water quality (TOC, UVA, pH) and chlorine residual. Since the water has already been in contact with chlorine, and the fraction of the NOM that reacts very rapidly with the chlorine has most likely been consumed, only the relative formation during the reaction time between the plant effluent and sedimentation is added to the formation predicted in the first step by the raw water model.

When chlorine is added before or during coagulation, UVA values will be lower than by coagulation only, due to UVA oxidation by the chlorine. This is taken into account by the following equation developed from the database of Summers et al. *(5)* for 20 waters:

UVA reaction due to chlorine addition
$$UVA_{Pre-Cl2} = 0.7437 (UVA_{no\,Cl2}) + 0.0042$$

*($R^2 = 0.930$, $R^2_{adj} = 0.930$, SEE = 0.006 1/cm, F = 991, n = 76)*
$UVA_{Pre-Cl2}$ = settled UVA after prechlorination (1/cm):
$$0.015 \leq UVA_{Pre-Cl2} \leq 0.120$$
$UVA_{no\,Cl2}$ = settled UVA without prechlorination (1/cm):
$$0.017 \leq UVA_{no\,Cl2} \leq 0.150$$

2.  Post-chlorination only (Figure 10A): A single point of chlorination after sedimentation. The treated water model is applied using settled water quality and chlorine dose.

3.  Pre- and post-chlorination (Figure 10B): Two points of chlorination--prior to rapid mixing and after sedimentation. For this approach, the raw water

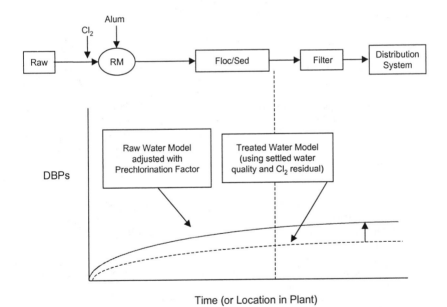

*Figure 9. DBP Modeling: Pre-Chlorination Only*

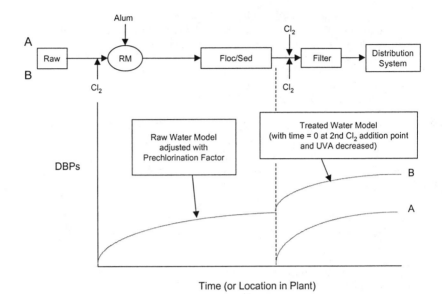

*Figure 10. DBP Modeling: (A) Post-Chlorination and (B) Pre- and Post-Chlorination*

model proportionally adjusted with the prechlorination factor is applied for DBP formation before the filter. After rechlorination, the treated water model is applied using settled water quality, with time = 0 and the UVA decreased by prechlorination. The rechlorination dose is added to the chlorine residual at that point to yield the effective dose for input to the DBP formation model. DBP formation at the second chlorination point is added to that resulting from the first chlorination point to model the cumulative formation.

DBP formation is modeled as cumulative formation through processes and with multiple points of chlorination. Currently, separate equations for DBP formation in the distribution system do not exist. The distribution system is considered to be merely an extension of the plant, and DBP formation is assumed to follow the same formation kinetics and rates.

DBP formation equations were developed for disinfection with free chlorine only. For chloramines, the WTP model assumes that 20 percent of that formed with free chlorine is formed when chloramine disinfection is used. This assumption is based on work by Amy et al. *(21)* and was used in the 1992 version of the WTP model.

*Giardia* and *Cryptosporidium*

In this revised version of the WTP model, physical removal of *Giardia* is achievable through sedimentation and filtration, direct filtration, and membranes. For *Cryptosporidium*, removal is achievable through sedimentation/filtration and membranes. Log removals affiliated with these processes are currently user-defined. For inactivation, the model uses the C*t values required by the Surface Water Treatment Rule (SWTR) Guidance Manual *(22)*. For chlorine disinfection, *Giardia* inactivation is predicted by regressions of the SWTR Guidance Manual tables. For alternate disinfectants, *Giardia* inactivation is determined by linear interpolations for temperature ranges in the SWTR Guidance Manual tables. As a temporary measure, until sufficient data are available, the C*t values for *Cryptosporidium* inactivation are calculated by multiplying the *Giardia* C*t values by a factor of 10 for ozone and 7.5 for chlorine dioxide.

**Advanced DBP Precursor Removal Processes**

Advanced DBP precursor removal processes have also been modified and/or created for the WTP model and will be discussed in future publications. Equations for TOC removal by GAC were developed using the logistics function to predict a breakthrough curve *(23)*. The parameters of the logistics functions were developed to take into account the impact of influent TOC and pH, empty bed contact time (EBCT) *(24)*, and the impact of single versus multiple blended contactors *(23)*. UVA removal by GAC is predicted by using linear correlations between GAC effluent TOC and UVA.

New empirical equations were developed for TOC, total hardness, alkalinity, and bromide rejection by membranes to include recovery and molecular weight cut-off as well as feed water quality parameters. The new rejection equations predict permeate water quality parameters and use similar formats to those used in the original algorithm.

A new algorithm was developed for ozonation. Empirical equations were developed for ozone decay, the oxidation of UVA absorbing compounds by ozonation, and bromate formation. Additional equations were developed to include staging effects and multiple chambers. Bromate formation models were developed based on semi-batch and true batch experiments *(25)*. From simulation results derived from three full-scale ozonation plants, the true-batch model showed better agreement as compared to the semi-batch model.

TOC removal by biological filtration in the WTP model occurs only after ozonation. Attempts were made to develop empirical correlations; however, only relationships with poor predictive abilities could be developed. Instead, the WTP model assumes 15 percent TOC removal for anthracite/sand filters and 20 percent TOC removal for GAC filters. These values represent average removals as reported in over 20 literature studies. The model assumes no UVA removal through the biofilter.

A more detailed discussion of algorithms for advanced treatment processes will be presented in future publications.

## Future Work

Additional algorithms need to be updated or added in the WTP model. These include:

- *Giardia* and *Cryptosporidium* inactivation
- Chlorine and chloramine decay (in plant and distribution system)
- DBP formation after chloramines
- DBP formation and degradation in the distribution system
- Chlorine dioxide decay and inactivation
- Chlorite formation

Prior to the public release and use of the WTP Model, it will be verified and recalibrated (if needed) using Information Collection Rule (ICR) data and treatment studies data *(26)*. Based on these results, the individual model algorithms may be corrected by a constant correction factor for cases where a consistent under- or over-prediction occurs, or a model algorithm may be recalibrated.

## Summary

The WTP model has been updated to include many new algorithms. This paper reports on those that represent conventional and softening treatment plants and includes: (a) NOM removal processes by coagulation and softening; (b) chlorination and multiple points of chlorination; and (c) DBP formation.

The model consistently underpredicts coagulation pH by an average of 5 percent. The average errors in coagulated TOC and UVA were about 15 percent, while errors for softened TOC and UVA predictions were higher. Errors in TOC and UVA predictions are mainly associated with the algorithm itself, as the predictions were shown to be relatively insensitive to errors in settled water pH predictions. For DBP formation predictions, TTHM formation is generally underpredicted for treated waters, while HAA6 formation is generally overpredicted. The treated water DBP formation equations, which were developed using coagulated water data, predict coagulated DBP formation well, with average errors of 23 and 29 percent for TTHM and HAA6, respectively.

## Acknowledgments

This research was funded through cooperative agreements between the USEPA Technical Support Center, Office of Ground Water and Drinking Water, and the University of Cincinnati (CR-825790) and the University of Colorado at Boulder (CR-826722). The views expressed here are those of the authors and do not necessarily reflect the views of the USEPA.

The authors would like to acknowledge the help of all those involved in developing the new version of the WTP model, especially Dr. Michelle M. Frey.

# Literature Cited

1. Roberson, J.A.; Cromwell III, J.E.; Krasner, S.W.; McGuire, M.J.; Owen, D.M.; Regli, S.; Summers, R.S. *J.—Am. Water Works. Assoc.* **1995**, *87* (10), 46-57.

2. Harrington, G.W.; Chowdhury, Z.K.; Owen, D.M. *J.—Am. Water Works. Assoc.* **1992**, *84* (11), 78-86.

3. Crow, E.L.; Davis, F.A.; Maxfield, M.W. *Statistics Manual*; Dover Publications Inc.: Mineola, NY, 1960; pp 147-190.

4. Stumm, W.; Morgan, J.J. *Aquatic Chemistry: An Introduction Emphasizing Chemical Equilibria in Natural Waters*; John Wiley and Sons: New York, NY, 1981.

5. Summers, R.S.; Solarik, G.; Hatcher, V.A.; Isabel, R.S.; Stile, J.F. Impact of Point of Chlorine Addition and Coagulation. Final Project Report. USEPA Office of Groundwater and Drinking Water: Cincinnati, OH, 1998.

6. Dryfuse, M. M.S. Thesis, University of Cincinnati, Cincinnati, OH, 1995.

7. Tseng, T.; Edwards. M. *J.—Am. Water Works. Assoc.* **1999**, *91* (4), 159-170.

8. Edwards, M. *J.—Am. Water Works. Assoc.* **1997**, *89* (5), 78-89.

9. White, M.C.; Thompson, J.D.; Harrington, G.W.; Singer, P.S. *J.—Am. Water Works. Assoc.* **1997**, *89* (5), 64-77.

10. Tseng, T.; Edwards, M.; Chowdhury, Z.K. American Water Works Association National Enhanced Coagulation and Softening Database, 1996.

11. Koechling M.T.; Rajbhandari A.N.; Summers R.S. *Proceedings, American Water Works Association Annual Conference*, Dallas, TX, June 1998, pp 363-373.

12. Isabel, R.S.; Summers, R.S.; Koechling, M.T.; Anzek, M. Unpublished Work, 1999.

13. Amy, G.; Siddiqui, M.; Ozekin, K.; Zhu, H.W.; Wang, C. Empirically Based Models for Predicting Chlorination and Ozonation By-Product: Haloacetic Acids, Chloral Hydrate, and Bromate. EPA Report CX 819579. USEPA Office of Groundwater and Drinking Water: Cincinnati, OH, 1998.

14. Symons, J.M.; Krasner, S.W.; Simms, L.A.; Sclimenti, M. *J.—Am. Water Works. Assoc.* **1993**, *85* (1), 51-62.

15. Johnson, J.D.; Jensen, J.N. *J.—Am. Water Works. Assoc.* **1986**, *78* (4), 156-162.

16. Cowman, G.A.; Singer, P.C. *Environ. Sci. Technol.* **1996**, *30* (1), 16-24.

17. Stevens, A.A.; Moore, L.A.; Miltner, R.J. *J.—Am. Water Works. Assoc.* **1989**, *81* (8), 54-60.

18. Summers, R.S.; Hooper, S.M.; Shukairy, H.M.; Solarik, G.; Owen, D.M. *J.—Am. Water Works. Assoc.* **1996**, *88* (6), 80-93.

19. J. M. Montgomery Consulting Engineers. Effect of Coagulation and Ozonation on Disinfection Byproduct Formation. Final Project Report. American Water Works Association: Washington, D.C., 1992.

20. Summers, R.S.; Owen, D.M.; Chowdhury, Z.K.; Hooper, S.M.; Solarik, G.; Gray, K. Removal of DBP Precursors by GAC Adsorption. American Water Works Association (AWWA) Research Foundation and AWWA: Denver, CO, 1998.

21. Amy, G.L.; Greenfield, J.H.; Cooper, W.J. In *Water Chlorination: Chemistry, Environmental Impact and Health Effects*; Jolley, R.L.; Condie, L.W.; Johnson, J.D.; Katz, S.; Minear, R.A.; Mattice, J.S.; Jacobs, V.A., Eds.; Lewis Publishers: Chelsea, MI, 1990; Vol. 6.

22. Guidance Manual for Compliance with the Surface Water Treatment Requirements for Public Water Systems; Criteria and Standards Division, USEPA Office of Drinking Water: Washington, D.C., 1989.

23. Chowdhury, Z.K.; Solarik, G.; Owen, D.M.; Hooper, S.M.; Summers, R.S. *Proceedings,* American Water Works Association Annual Conference, Toronto, Canada, June 1996, Vol. D, pp 629-650.

24. Hooper, S.M.; Summers, R.S.; Solarik, G.; Hong, S. *Proceedings,* American Water Works Association Annual Conference, Toronto, Canada, June 1996.

25. Ozekin, K. Ph.D. Dissertation, University of Colorado, Boulder, CO, 1994.

26. USEPA. National Primary Drinking Water Regulations: Monitoring Requirements for Public Drinking Water Supplies: Cryptosporidium, Giardia, Viruses, Disinfection Byproducts, Water Treatment Plant Data and Other Information Requirements; Final Rule. *Fed. Reg.* **1996**, *61* (94), 24354.

# Natural Organic Matter Characterization and Reactivity

Chapter 5

# Comprehensive Isolation of Natural Organic Matter from Water for Spectral Characterizations and Reactivity Testing

Jerry A. Leenheer[1], Jean-Philippe Croué[2], Mark Benjamin[3], Gregory V. Korshin[3], Cordelia J. Hwang[4], Auguste Bruchet[5], and George R. Aiken[6]

[1]U.S. Geological Survey, Building 95, MS 408, Federal Center, Denver, CO 80225
[2]Laboratoire de Chimie de l Environment, UPRESA CNRS 6008, Ecole Supérieure d'Ingenieurs de Poitiers, Université de Poitiers, 86022 Poitiers Cedex, France
[3]Department of Civil Engineering, MS FX 10, University of Washington, Seattle, WA 98195–2700
[4]Metropolitan Water District of Southern California, 700 Moreno Avenue, La Verne, CA 91750–3399
[5]Suez Lyonnaise des Eaux, CIRSEE, France
[6]U.S. Geological Survey, Boulder, CO 80303

A variety of approaches were tested to comprehensively isolate natural organic matter (NOM) from water. For waters with high NOM concentrations such as the Suwannee River, Georgia, approaches that used combinations of membrane concentrations, evaporative concentrations, and adsorption on nonionic XAD resins, ion exchange resins and iron oxide coated sand isolated over 90% of the NOM. However, for waters with low NOM concentrations, losses of half of the NOM were common and desalting of NOM isolates was a problem. A new comprehensive approach was devised and tested on the Seine River, France in which 100 L of filtered water was sodium softened by ion exchange and vacuum evaporated to 100 mL. Colloids (32% of the NOM) were isolated using a 3,500 Dalton membrane by dialysis against 0.1 $M$ HCl and 0.2 $M$ HF to remove salts and silica. On the membrane permeate, hydrophobic NOM (42%) was isolated using XAD-8 resin and hydrophilic NOM (26%) was isolated using a variety of selective desalting precipitations. The colloid fraction was characterized by IR and NMR spectroscopy as N-acetylamino sugars.

**68**

# Introduction

The objective of comprehensive isolation and characterization of natural organic matter (NOM) from water has recently gained added importance because of the need to determine NOM reactivity with various disinfecting agents in water that produce disinfection by-products, and because of the need to understand NOM chemistry to design better water treatment processes for NOM removal. Comprehensive isolation and characterization is not necessarily defined as 100% recovery of the NOM, but rather it is defined as recovery of representative portions of all compound-class fractions of NOM in a form free from inorganic salts that is suitable for various spectral characterizations and reactivity studies.

This study began in 1995 with a research project funded by the American Water Works Association Research Foundation (AWWARF) entitled "Isolation, Fractionation, and Characterization of Natural Organic Matter in Drinking Water" The objective of this research was to evaluate and optimize existing methods for isolating and characterizing NOM in various water supplies, and to evaluate certain new methods, such as NOM adsorption on iron oxide coated sands. The results of this study (1) showed that good recoveries of NOM could be obtained using resin adsorbents from high NOM concentration, low salinity waters such as the Suwannee River in Georgia. However in waters with low NOM concentrations and greater inorganic salt concentrations such as the South Platte River in Colorado, two separate attempts to fractionate and isolate NOM using various resin adsorbents with complete desalting only resulted recovery of 67% and 48% of the dissolved organic carbon (DOC). Other techniques such as vacuum evaporation and freeze-drying gave high DOC recoveries, but all the inorganic salts were coisolated. Membrane isolation techniques such as reverse osmosis, nanofiltration, and ultrafiltration gave partial recoveries of both NOM and inorganic salts. NOM acids were quantitatively desorbed and base eluted from iron oxide-coated sands, but sulfate was coisolated in this procedure. In all of these procedures, isolation and purification of the polar NOM fractions was especially problematic.

To address the problem of polar NOM isolation, characterization, and reactivity, a second AWWARF-funded research project entitled "Characterization of the Polar Fraction of NOM with Respect to DBP Formation" was begun in 1998. This project sampled both the untreated water inputs and treated waters at various points in water treatment plants to determine changes in NOM as a result of water treatment. As the project title indicates, the emphasis was on polar NOM fractions; therefore, new isolation approaches had to be devised to better address recovery and purification (desalting) of polar NOM. The objective of this study is present a new comprehensive NOM isolation approach that overcomes problems with past approaches (2-4) that have resulted in low and biased NOM recoveries for some waters. The two major problems to be addressed by this study are better recovery and purification of the colloidal NOM fraction, and better desalting of the hydrophilic NOM fractions.

70

# Missing NOM

The low DOC recovery from the South Platte River study (*1*) led to a reassessment of the comprehensive isolation approach that was primarily based on resin sorbents. The following evidence indicated that the "missing NOM" has hydrophilic base/neutral characteristics, and it might be in the small colloidal size range. Solute size is a serious limitation for sorbent resins as previous research by Aiken et al. (*5*) indicated a significant decrease in column distribution coefficients, $K_D$ for XAD-4 resin sorption of polyacrylic acid between 2,000 and 5,000 Daltons. A sequential ultrafiltration study (*6*) of NOM in seven waters indicated 20-40% of the NOM had molecular mass greater than 5,000 Daltons, suggesting that certain resin sorbents might size-exclude a significant percentage of NOM. In studies of the NOM in the Blavet River in France (*1*), the [13]C-NMR spectra of the RO residue (Figure 1) suggested the presence of much more carbohydrate carbon (60-90 ppm) than was in the fraction isolates from the sorbent resins. A pyrolysis, gas chromatography/mass spectroscopy study of the same RO residue indicated that about one-third of the NOM was comprised of amino sugars that were poorly recovered as found by analysis of the fractions from resin sorbents(*1*) . A final form of evidence of the nature of the "missing NOM" was found in the silica gel that precipitated during vacuum evaporation of sodium-softened Colorado River water. This silica gel was dissolved in dilute hydrofluoric acid, and the resulting fluosilicic acid and excess hydrofluoric acid were removed by dialysis using a 3,500 Dalton membrane. A significant amount of N-acetyl amino sugars were isolated and identified by FT-IR spectrometry.

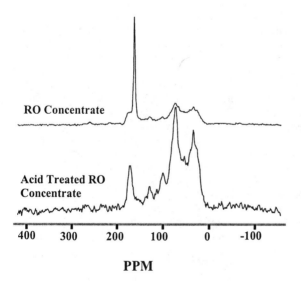

Figure 1. [13]C-NMR spectra of RO concentrates from the Blavet River, France

The new comprehensive isolation approach combined RO concentration (for large samples), vacuum evaporation, and membrane dialysis (to recover colloidal NOM) before resin sorbents were used. Previous studies on various waters had not identified any alternative to vacuum evaporation for the recovery of the hydrophilic neutral fraction. Borate-form anion-exchange resins (7) were tested to recover carbohydrates in the hydrophilic neutral fraction, but recoveries were low and massive amounts of silica also were recovered on these resins. Sodium softening (for RO concentration) also greatly enhanced the efficiency of salt removal in the zeotrophic distillation procedure because calcium and magnesium salts are poorly removed by that procedure.

For membrane dialysis, 2,000 and 3,500 Dalton membranes were tested on salt-saturated concentrates. Osmotic pressure caused the 2,000 Dalton membrane to burst; therefore, the 3,500 Dalton dialysis was chosen for the colloid size cutoff. Sulfate was partially retained by the 3,500 Dalton membrane at neutral pH. At pH 1 where sulfate was partially converted to bisulfate that has a smaller hydrated radius, sulfate was quantitatively removed. Therefore, membrane dialysis is a three-step procedure whereby: (1) salts and low MW NOM are removed by dialysis against 0.1 $M$ HCl, (2) silica gel is dissolved by dialysis against 0.2 $M$ HF, and (3) excess HF is removed by dialysis against distilled water.

Lastly, the procedures for recovery of low molecular weight hydrophilic NOM were optimized. Removal of sodium salts by the zeotrophic distillation procedure gave much less coprecipitation of hydrophilic NOM than when di- and trivalent cations were present. Addition of the 1.0 $M$ HCl wash of the barium sulfate precipitate resulted in a 10-30% increase in recovery of the low molecular weight NOM fraction. Addition of a nitrate removal step by reduction of nitric acid to gaseous nitrogen dioxide by formic acid resulted in NOM isolates that were less hygroscopic and less likely to be altered by reaction with nitric acid.

## Analytical Methods

### Field Processing Procedures

A variety of filters, RO membranes, NF membranes, and water processing units were evaluated in the first study (1); only the equipment used for the Sacramento/ San Joaquin Delta waters will be presented. Water sampled at various points in the water treatment plant was filtered through 25 μm and 1 μm porosity glass fiber cartridge filters in series (3) and passed through a sodium softener unit. The large 1,000 L samples were collected in a holding tank, and concentrated 17-18-fold through a spiral-wound RO membrane (Filmtech TW30-4040). The 17-18-fold concentration was limited by the dissolved silica concentration because greater concentration factors have been found (8) to cause silica precipitation and fouling of the RO membrane with possible NOM loss by coprecipitation. The water samples were then placed in 20 L high-density polyethylene cubitainers, chilled in ice, and shipped in coolers to the laboratory for NOM fractionation and isolation.

**Laboratory Equipment**

Glass chromatography columns with Teflon end caps were obtained from Spectra/Chrom. Column dimensions for 1 L resin bed-volume were 5 cm ID x 60 cm; 500 mL resin-bed volume were 5 cm ID x 30 cm; 80 mL resin-bed volumes were 1.5 cm ID x 30 cm; and 20 mL resin-bed volumes were 1 cm ID x 28.5 cm. Excess volume in each column was left vacant to allow for resin bed expansion during different chemical treatments. A reciprocating, ceramic piston pump (FMI Lab Pump Model QD) was used to pump water and reagents through the 1 L and 500 mL resin-bed columns connected with 3 mm ID, 6 mm OD FEP Teflon tubing at a flow rate of 250 mL/m. A smaller pump (FMI Lab Pump Model RP-SY) was used to pump water and reagents through the 80 mL resin-bed column at a flow rate of 20 mL/m and through the 20 mL resin-bed column at a flow rate of 10 mL/m. These smaller columns were connected to the pump with 1.5 mm OD, 3 mm ID Teflon tubing. Selection, preparation, regeneration, capacities, and packing of resin adsorbents has been discussed previously (2,3).

Spectra/Por 3 regenerated cellulose dialysis membrane was used. It had a molecular weight cutoff of 3,500 Daltons. Prior to use, an appropriate length of membrane was cut to accommodate the volume to be dialyzed, and the membrane was washed by soaking it is 4 L of deionized water overnight. For dialysis, the bottom of the dialysis tubing was closed with a plastic snap closure, the solution was poured into the resulting bag using a funnel, and the top of the bag was closed with another closure. Dialysis was then conducted in 4 L Teflon beakers and the solution was stirred with a magnetic stirrer.

For evaporations, one freeze-dryer and two vacuum rotary evaporators were used. The large Buchi Rotavapor 150 had 10 L evaporation and condensation flasks. Its evaporation rate for water was about 2 L per hour for a water-bath temperature of 60 °C at 750 mm mercury vacuum. A small Buchi Rotavapor R had 1 L evaporation and condensation flasks, and its evaporation rate for water was about 0.5 L per hour for a water-bath temperature of 60 °C at 750 mm mercury vacuum. The freeze dryer was a Labconco Lyph-Lock 6 Model.

For centrifuge separations, a Sorvall Superspeed RC-2B Refrigerated Centrifuge was used with an angle head. Centrifuge bottles were 250 mL high density polyethylene that were spun at 7,000 rpm for 30 m.

**Fourier Transform-Infrared (FT-IR) Spectrometry**

Infrared spectra were collected using 2 to 5 mg of NOM fraction isolates in KBr pellets. The Perkin Elmer System 2000 FT-IR used a pulsed laser source and a deuterated triglycine sulfate detector. The instrument was setup to scan from 4,000 to 400 $cm^{-1}$ averaging 10 scans at 1.0 $cm^{-1}$ intervals with a resolution of 4.0 $cm^{-1}$. All spectra were normalized after acquisition to a maximum absorbance of 1.0 for comparative purposes.

## Solid-State Cross Polarization Magic Angle Spinning (CPMAS) [13]C-Nuclear Magnetic Resonance (NMR) Spectrometry

CPMAS [13]C-NMR spectra were obtained on 20 to 200 mg of NOM samples. The acid and neutral NOM fractions were in the hydrogen form, and the base fractions were in the ammonium-salt form. Freeze-dried samples were packed in sapphire rotors. CPMAS [13]C-NMR spectra were obtained on a 200-megahertz (MHz) Chemagnetics CMX spectrometer with a 7.5-mm-diameter probe. The spinning rate was 5000 Hz. The acquisition parameters included a contact time of 5 ms, pulse delay of 1 s, and a pulse width of 4.5 μs for the 90° pulse. Variable contact time studies by Alemany et al. (9) indicate these are the optimum parameters for quantitatively determining different carbon structural group contributions to the NOM NMR spectra.

## Comprehensive NOM Isolation Scheme

This procedure was used with and without reverse osmosis preconcentration. The only change was that the sodium-saturated MSC-1 resin did not have to be used in Step 1 because a presoftening step is used before RO concentration. If additional acid, base, and neutral fractions that correspond to the preparative DOC fractionation procedure (2,3) are desired, this DOC fractionation can be easily adapted to the dialysis permeate sample with the desalting procedure of Step 5 applied to the hydrophilic acid fraction separated by the Duolite A-7 anion exchange resin. The stepwise comprehensive NOM isolation procedure follows:

*Step 1*

Pass a 100 L sample through the 1 L bed-volume column of MSC-1 ion exchange resin in the sodium form followed by 1 L of deionized water rinse. Vacuum evaporate the sample to 100 mL. Rinse the rotary evaporation flask with 10 mL of 1 *M* HF to recover colloids coprecipitated with the silica that adheres to the flask. Combine this rinse with the 100 mL sample. Acidify the sample with HCl to pH 1. Place the salt, silica, and NOM slurry in a 3,500 Dalton dialysis bag, and dialyze for 24 hours against three 4-L portions of 0.1 *M* HCl . Place the dialysis bag in 4 L of 0.2 *M* HF and dialyze until the silica gel precipitate is dissolved. Lastly, dialyze for 24 hours against two 4-L portions of deionized water to remove residual HF and fluosilicic acid. Freeze-dry the contents of the dialysis bag to isolate the colloid fraction.

*Step 2*

Pump the 12 L of the combined permeate sample (from the 0.1 *M* HCl dialysis) through a 1-L bed-volume of XAD-8 resin and a 0.5-L bed-volume column of XAD-4 resin in series. Follow the sample with a pH-2 formic acid rinse until the conductivity of the effluent is the same as the conductivity of the column influent.

*Step 3*

Desorb the XAD-8 resin column with 750 mL of 75% acetonitrile/25% water followed by 1.0 L of deionized water rinse. Evaporate and freeze-dry to isolate the hydrophobic NOM fraction.

*Step 4*

Desorb the XAD-4 resin column with 500 mL of 75% acetonitrile/25% water followed by 500 mL of deionized water rinse. Evaporate and freeze-dry to isolate the transphilic NOM fraction. "Transphilic" is a new operational adjective meaning intermediate or transitional polarity between hydrophobic and hydrophilic properties of NOM.

*Step 5*

Adjust the column effluent from step 2 to pH 2 with sodium hydroxide. Vacuum-evaporate this effluent to a salt slurry and add glacial acetic acid in a volume approximately equivalent to the slurry volume. Vacuum-filter the salt with a 47-mm disk glass fiber filter (1-μm porosity) and rinse the salt cake with 2, 25 mL portions of glacial acetic acid. Repeat the evaporation and filtration steps two more times (or until no more salt can be removed without coprecipitating NOM as evidenced by color in the salt cake). As the quantity of salt decreases during each successive step of the zeotrophic distillation procedure, the volume of acetic acid rinse of the salt cake should be decreased proportionately. Dilute the final filtrate (30 to 50 mL) with an equal volume of deionized water and add 30 mL of 1 $M$ barium formate to remove sulfate as barium sulfate. Centrifuge out the precipitate and wash the precipitate with 20 mL of 1.0 $M$ HCl. Vacuum-evaporate the HCl supernatant rinse to dryness using multiple additions of anhydrous acetonitrile to remove HCl from the concentrate sample. Combine the dried residue from the HCl wash step with the supernatant from first centrifuge separation and pass this solution through a 100-mL column of MSC-1H resin followed by a 200-mL rinse of deionized water. This procedure removes barium and other metal cations. Vacuum-evaporate this solution to dryness using multiple additions of anhydrous acetonitrile to remove HCl, HBr, and part of the $HNO_3$ from the concentrated sample. Add 50 mL of anhydrous formic acid to the residue in the vacuum evaporation flask, gently warm, and immediately evaporate when the solution turns reddish-brown and nitrogen dioxide begins to outgas. As a safety note, this procedure should not be attempted with large quantities of nitric acid it may explosively decompose. Redry the sample using multiple additions of acetonitrile to remove the formic acid. Add 50 mL of methanol, warm, and evaporate to remove boric acid. Redissolve the residue in 50 mL of 50:50 acetic acid/water and add 5 mL of saturated silver acetate in 50:50 acetic acid/water to precipitate the last traces of bromide and chloride. Remove this precipitate by centrifugation and pass the supernatant through a 20-mL bed-volume column of MSC-1H resin followed by a 20-mL rinse. Vacuum-evaporate to dryness. Take up the residue in 50 mL of deionized water and freeze-dry to isolate the hydrophilic acid plus neutral fraction.

*Step 6*
    Rinse the MSC-1H column with 500 mL of deionized water and desorb with 100 mL of 3 *M* ammonium hydroxide followed by a 200-mL rinse of deionized water. Evaporate and freeze-dry to isolate the hydrophilic base fraction.

## Results and Discussion

    Spectral characterizations and interpretations using FT-IR and $^{13}$C-NMR spectrometry are extensively explained in reference *1*.

### Seine River Water

    The Seine River was sampled June 8, 1998, at the Vigneux treatment plant that is south and upstream of Paris; 100 L of water were sampled after sodium softening and before reverse osmosis concentration to eliminate any possible NOM loss by RO concentration. Because of NOM mass limitations with this small sample, isolation of a transphilic NOM fraction was omitted by deleting step 4 of the isolation procedure. Following is a histogram (Figure 2) of the isolated NOM fractions.

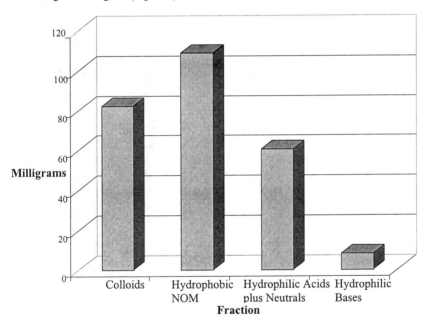

Figure 2. Masses of NOM Fractions from the Seine River.

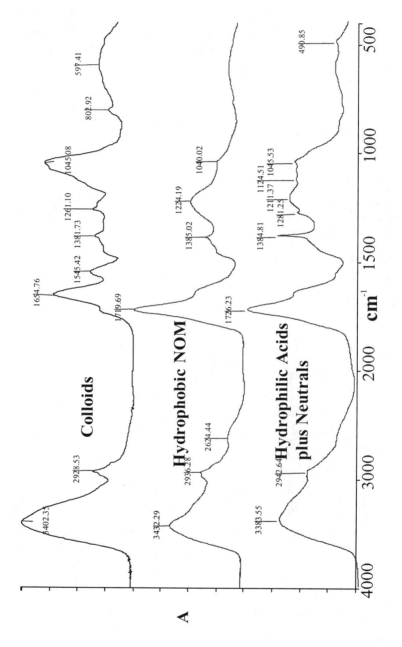

Figure 3. FT-IR spectra of NOM fractions from the Seine River

The colloid fraction accounts for 32% of the NOM mass for the Seine River and thus accounts for much of the missing NOM. The FT-IR spectra are shown in Figure 3. The amide 1 (1655 cm⁻¹) and amide 2 (1545 cm⁻¹) peaks, the methyl peak at 1382 cm⁻¹, the broad OH peak at 3402 cm⁻¹ and the broad C-O peak at 1045 cm⁻¹ are all indicative of n-acetyl amino sugars for the colloid fractions. The hydrophobic and hydrophilic fractions are both typical of aquatic fulvic acids as shown by the dominant carboxyl peak near 1725 cm⁻¹, with the hydrophilic acid plus neutral fraction having more hydrophilic alcohol groups near 1045 cm⁻¹ than the hydrophobic NOM fraction. The absence of inorganic salt peaks with the exception of a trace of nitric acid in the hydrophilic acid plus neutral fraction (peak at 1385 cm⁻¹) is an indication of how well the NOM desalting procedures work. This result demonstrates that FT-IR is an excellent characterization tool for aquatic NOM.

The $^{13}$C-NMR spectra for the Seine River NOM fractions are shown in Figure 4.

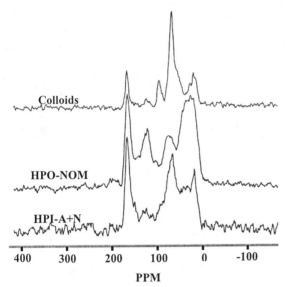

Figure 4. $^{13}$C-NMR of NOM isolated from the Seine River, France

As with the FT-IR spectra, the colloid fraction is very different from the other NOM fractions. The carbon distribution of this fraction is dominated by the carbohydrate peaks near 75 and 100 ppm, and the N-acetyl group of the amino sugars is indicated by the amide carbonyl peak at 175 ppm and the methyl group near 20 ppm. The C-N linkage is the shoulder near 55 ppm. The hydrophobic and hydrophilic NOM fractions are dominated by the carboxyl-group peak near 175 ppm. Without the FT-IR spectral data, it would not have been possible to determine from

[13]C-NMR spectra whether the 175 ppm peak is amide (in the case of colloids) or carboxyl (in the case of the hydrophobic and hydrophilic NOM fractions).

## Sacramento/San Joaquin Delta Water after Alum and Ozone Treatment

The effect of various water-treatment processes on NOM concentration and composition was studied using Sacramento/San Joaquin Delta waters sampled November 11, 1998, from the city of Martinez, California water-treatment plant. This water was first treated with 0.2-0.3 mg/L ozone followed by addition of 26 mg/L alum. Following coagulation and sedimentation, the clarified water was contacted with 0.7–0.9 mg/L ozone. At the time of sampling, 882 L of this water after the second ozonation step was concentrated by reverse osmosis to 54 L at the time of sampling. The RO concentrate was processed through the DOC fractionation scheme described above. Comparison of NOM compositions of the untreated water with the treated water is shown in the histograms of Figure 5.

This water treatment decreased the DOC concentration from 2.5 mg/L to 1.6 mg/L, and the NOM mass distribution was significantly altered. The colloidal NOM was significantly decreased (most likely by losses on the RO membrane used for NOM preconcentration) and the hydrophilic NOM fraction significantly increased by the water treatment process (most likely by NOM oxidation by ozone).

Carbon, hydrogen, and nitrogen analyses are shown in Table I for the major NOM fractions in the treated and untreated waters. As the hydrophilic character of the fraction increases, carbon percentage decreases and nitrogen percentage increases. The C, H, N percentages of colloids isolated from the untreated water sample are lower than in colloids from the treated water sample because of the presence of inorganic clay colloids indicated by the FT-IR spectra (not shown) in colloids from the untreated water sample.

Table I. Elemental analyses (C, H, N) for NOM fractions isolated from
Sacramento/San Jaoquin Delta waters before and after treatment with alum and
ozone (ND=Not Determined).

| Sample   NOM Fraction | Carbon (%) | Hydrogen (%) | Nitrogen (%) |
|---|---|---|---|
| Untreated Water | | | |
| Colloids | 37.25 | 4.98 | 3.73 |
| Hydrophobic NOM | 51.80 | 5.20 | 1.85 |
| Transphilic NOM | 43.64 | ND | 3.04 |
| Hydrophilic Acids plus Neutrals | 36.19 | ND | 4.25 |
| Alum and Ozone-Treated Water | | | |
| Colloids | 44.00 | 6.45 | 9.12 |
| Hydrophobic NOM | 52.83 | 5.61 | 1.84 |
| Transphilic NOM | 45.11 | 5.47 | 4.10 |
| Hydrophilic Acids plus Neutrals | 36.83 | 4.40 | 3.88 |

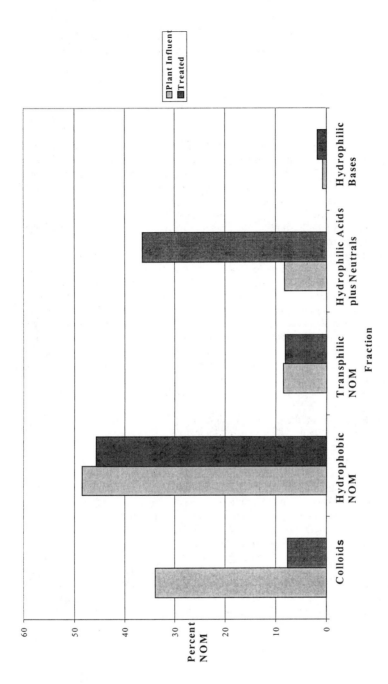

Figure 5. NOM composition changes in Sacramento/San Joaquin Delta water after alum and ozone treatment.

Organic carbon recoveries were 87% for the untreated water and 68% for the treated water without including the hydrophilic base fraction that had insufficient mass for elemental analysis. Organic carbon losses for these waters have known and unknown components. The sodium softening step removed 14% of the organic carbon for both the untreated and treated waters based upon TOC difference analysis. This 14% loss completes the carbon mass balance for the untreated water. Elution of the sodium resin with 3 $M$ NH$_4$OH did not recover this retained carbon for the untreated water sample which indicates this NOM has hydrophobic neutral or strong base characteristics that are causing these NOM fractions to be retained by the resin matrix. A better design for the preparative NOM fractionation procedure would have been to isolate the hydrophobic neutral fraction first before ion-exchange resins are used (3). This loss points out a serious limitation for RO preconcentration approaches where sodium softening is frequently required.

For the water treated with alum and ozone, 18% of the organic carbon loss must still be accounted after adjusting for the organic carbon loss on the sodium exchange resin. The major volatile acids resulting from ozone treatment are acetic and formic acids. Direct ion chromatographic analyses of these two acids accounted for 2.6% of the organic carbon that would be lost in the evaporation steps in the NOM isolation procedure. This leaves 15.4% organic carbon loss unaccounted. In the untreated water, 24.8% of the DOC is in the colloid fraction whereas in the treated water only 5.2% of the DOC is in the colloid fraction. A recent study (10) found that colloidal NOM comprised of amino sugars and polysaccharides derived from bacteria irreversibly fouled NF membranes used for drinking water treatment of Ohio River water. It is likely that colloidal NOM also fouls the RO membranes used to preconcentrate the NOM for the treated water water of this study, and this may account for most the remaining 15.4% of the DOC loss.

The FT-IR spectra of the treated NOM fractions of Figure 5 are shown in Figure 6.

The IR spectrum of the colloid fraction in Figure 6 is similar to that for the colloid fraction in Figure 4, which is indicative of N-acetylamino sugars. A continuous increase in hydrophilic character going from hydrophobic to hydrophilic NOM fractions is indicated by the increase in the broad C-O peak (1,000 to 1,100 cm$^{-1}$) of alcohols. The hydrophilic base fraction contains ammonium carboxylates as indicated by the ammonium peaks at 3201 and 1404 cm$^{-1}$, and by the carboxylate peak at 1594 cm$^{-1}$. There are no inorganic solute peaks in any of the spectra.

The $^{13}$C-NMR spectra for the alum- and ozone-treated Sacramento-San Joaquin Delta Water NOM fractions are shown in Figure 7. The low aromatic carbon contents (100-160 ppm) of all NOM fractions are consistent with selective oxidation of aromatic carbon by ozone. The presence of oxalic acid is clearly shown in the hydrophilic NOM fraction by the peak at 163 ppm. The polarity transition from hydrophobic to transphilic to hydrophilic NOM is shown by the decrease in the aliphatic (hydrophobic) hydrocarbon peak (0-60 ppm) and the increase in the alcohol and carbohydrate peaks (60-110 ppm) that are hydrophilic.

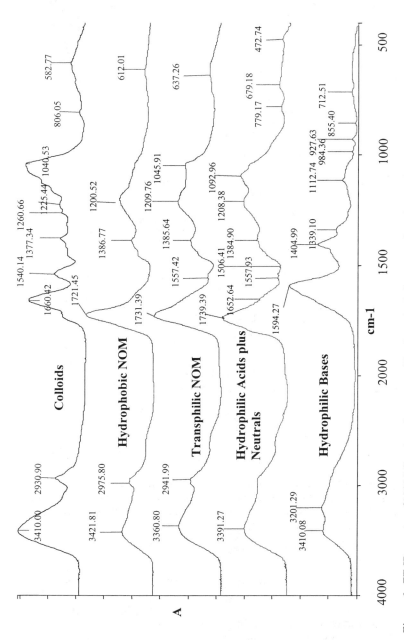

Figure 6. FT-IR spectra of NOM fractions Sacramento/San Joaquin Delta water after alum and ozone treatment.

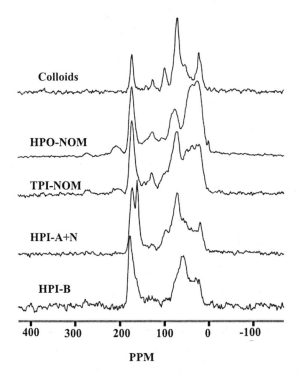

Figure 7. $^{13}$C-NMR spectra of NOM fractions isolated from the Sacramento/San Joaquin Delta Water after alum and ozone treatment

## Conclusions

The comprehensive approach to NOM isolation presented in this report identifies most of the "missing NOM" from previous studies as being in the colloid fraction isolated by membrane dialysis. This colloid fraction was spectrally characterized as N-acetyl amino sugars. The desalting procedures resulted in NOM isolates that can be characterized by FT-IR spectrometry without salt interference.

## Acknowledgments

This research originated from two projects funded by the American Water Works Association Research Foundation. Robert L. Wershaw and Ted I. Noyes of the U.S. Geological Survey assisted in the acquisition and interpretation of the

$^{13}$C-NMR spectra. Trade names used in this study are for identification purposes only and do not constituted endorsement by the U.S. Geological Survey.

# References

1. AWWARF. *Isolation, Fractionation, and Characterization of Natural Organic Matter in Drinking Water;* American Water Works Association: Denver, CO (in press).
2. Leenheer, J. A. *Environmental Science and Technology* **1981**, *15*, 578.
3. Leenheer, J. A.; Noyes, T. I. *U. S. Geological Survey Water-Supply Paper 2230*, Washington, DC, 1984.
4. Aiken, G. R.; Leenheer, J. A. *Chemistry and Ecology* **1993**, *8*, 135.
5. Aiken, G. R.; Thurman, E. M.; Malcolm, R. L.; Walton, H. F. *Analytical Chemistry* **1979**, *51*, 1799.
6. AWWARF. *Characterization of Natural Organic Matter and Its Relationship to Treatability*; American Water Works Association: Denver, CO, 1993, p. 57.
7. Leenheer, J.A. Proceedings of the Natural Organic Matter Workshop, Poitiers, France, 1997, pp. 4-1 to 4-5.
8. Kaakinen, J. W.; Eisenhauer, R. J.; Van Hoek, C. *Desalination* **1977**, *23*, 357.
9. Alemany, L. B; Grant, D. M.; Pubmire, R. J.; Alger, T. D.; Zilm, K. W. *J. Am. Chem. Soc.* **1983**, *105*, 2142.
10. Speth, T. F.; Summers, R. S.; Gusses, A. M. *Environmental Science and Technology* **1998**, *32*, 3612.

Chapter 6

# Technology-Related Characterization of Hydrophilic Disinfection By-Products in Aqueous Samples

Fritz H. Frimmel, Sebastian Hesse, and Georg Kleiser

Engler-Bunte-Institut, Universität Karlsruhe, Engler-Bunte-Ring 1, D-76131 Karlsruhe, Germany

Although much benefit has been gained using ozone or chlorine as oxidants and disinfectants in water treatment, both of these substances react with natural organic matter (NOM) to form disinfection by-products (DBPs). Only little is known about their reactions with water constituents, e.g., with high molecular organic compounds such as humics. The combination of size exclusion chromatography and DOC-detection gives more information on the reactions of ozone, OH-radicals, and chlorine with NOM. The objective of this paper is to show results of LC/DOC-analysis of different raw water NOM before and after oxidation and disinfection, respectively. The results were compared with water quality parameters, such as bacterial regrowth potential and DBP formation potential. It will be shown that in all cases of oxidation and disinfection low molecular acids were produced and high molecular compounds were removed. The low molecular organic substances were responsible for higher bacterial regrowth potential. The DBP-formation potential was reduced after preoxidation with ozone. In case of more OH-radical induced pre-oxidation DBP-formation was higher than after ozonation.

## Introduction

Final disinfection with chlorine is an important barrier for microorganisms in drinking water treatment. Nevertheless chlorine disinfection is in great discussion because in some cases toxic by-products are formed during chlorination. Therefore the potential of organic water constituents (NOM) to form disinfection by-products (DBPs) is a key quality parameter in drinking water supply. Numerous small molecules have been identified as final DBPs (1). The first identified DBP in tap water was chloroform (2). Trihalomethanes are the most frequently identified and usually the most concentrated DBPs found. In addition to THM more than 80 halogenated carboxylic acids have been identified, of which haloacetic acids (HAAs) are known to induce liver tumors in animals. 3-Chloro-4-(dichloromethyl)-4-oxobutenoic acid (MX) is the most mutagenic by-product in drinking water ever identified and has been shown to be an animal carcinogen (3). To be able to quantify also high molecular chlorinated organic compounds the sum parameter adsorbable organic halogen (AOX) was introduced by Sontheimer and Kühn (4). Using this parameter a wide range of DBPs, which are different in molar mass and ratio of halogenation (Figure 1), can be quantified. Important precursors of halogenated compounds in drinking water treatment are humic substances (5-7). Therefore it is necessary to get more information on their behavior during water treatment and disinfection.

Legislation has set maximum concentration values for some of the toxic DBPs. In German drinking water the THM concentration is limited to 10 µg/L (8). The guideline of the EU for important parameters regarding disinfection and oxidation of drinking water is shown in Table I (9).

### Table I. EU-Drinking Water Directive (9)

| Parameter | Parametric value |
| --- | --- |
| THM | 0.1 mg/L |
| oxidizability | 5.0 mg/L (as $O_2$) |
| E. coli | 0/100 mL |
| Enterococci | 0/100 mL |

Relatively little is known about the higher molecular DBPs. There is also practically no information on the molecular properties of larger DBPs, and on how the formation is influenced by the different ways of water treatment. A reason for this has to be seen in the lack of suitable analytical methods.

A powerful detection system for dissolved organic carbon (DOC) was developed giving reliable results in the low µg/L range (10). In combination with gel chromatography the DOC detector gives insight into the molecular size distribution of NOM and the changes due to different treatment steps including disinfection. Aqueous samples can be analyzed directly without further pre-treatment. Furthermore other nondegradative detection systems (e.g., UV/VIS absorbance, fluorescence) can be applied simultaneously and led to carbon normalized data, which allow quantification of bleaching effects and structural changes.

Detailed information on the changes of the molecular character of NOM can be obtained by the examination of oxidation and disinfection procedures. Oxidation before disinfection is done to remove DBP-precursors. Usually ozone is used as oxidant. Ozone is able to split double bonds according to the mechanism found by Criegee, which is shown in Figure 2 (11). The products of this process are carboxylic acids and aldehydes, which can also be regarded as disinfection and oxidation by-products, respectively.

The paper presents results on changes of NOM before and after oxidation and disinfection processes. As oxidation methods ozonation and the oxidation with PEROXONE (ozone/hydrogen peroxide-process) were examined in detail. In the

86

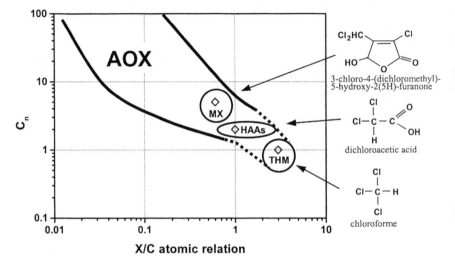

Fig. 1. Halogenation ratio and molar mass of different DBPs.

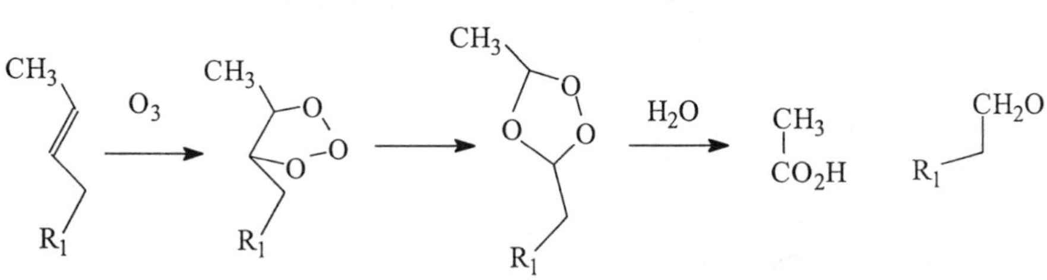

Fig. 2. Mechanism of the reaction of ozone with double bonds according to Criegee.

PEROXONE-process the oxidation is carried out by OH-radicals, which have a higher oxidation potential than ozone. Furthermore the oxidation and DBP-formation reactions of chlorine were examined.

## Experimental Details

Samples of different origin were taken in the years 1994 to 1998. Brown water: Lake Hohloh (HO9)/Black Forrest, Germany (DOC: 28 mg/L, $\kappa$: 26 $\mu$S/cm, pH 4.3); lake water: Lake Constance/Germany (DOC: 1.2 mg/L, $\kappa$: 324 $\mu$S/cm, pH 7.8); river water: Ruhr (DOC: 2.3 mg/L, $\kappa$: 473$\mu$S/cm, pH 7.6); reservoir water: Rapp-Bode-reservoir/Harz, Germany (DOC: 2.0 mg/L, $\kappa$: 158 $\mu$S/cm, pH 6.8).

*DOC*: Dissolved organic carbon (DOC) was determined after membrane filtration (polycarbonate 0.45 $\mu$m) using an UV-DOC analyzer (Dohrmann, DC 80). The relative standard deviation of this method is $\pm$ 3.8%.

The *molecular size distribution* was determined according to Huber and Frimmel (*10*, *12*) (column 1: 15 cm length, 1.5 cm radius; column 2: 25 cm length, 2 cm radius; eluent: 0.028 mol/L phosphate buffer at pH 6.6; flow rate: 1 mL/min; injection volume: 2 mL) using TSK-HW-40S gel (Toyopearl, TosoHass) and DOC-detection (DOC Analyzer, Gräntzel). The sample was injected directly into the analytical system without pretreatment. To avoid problems with the exact determination of the molecular size, in the interpretation of the data it is only distinguished between "high molecular" and "low molecular".

*Chlorination*: Chlorination was done in 0.1 mol/L phosphate buffer at pH 7 using sodium hypochlorite. The dosage of chlorine was 10 mg/L $Cl_2$. The water of Lake Hohloh was diluted to a DOC concentration of 3 mg/L to achieve ratios of $Cl_2$/DOC comparable to the other raw waters used. After 48 h the excess of sodium hypochlorite was destroyed (sodium sulfite). If residual hydrogen peroxide was present, additional hypochlorite was added to oxidize the hydrogen peroxide. Aliquots of the solution were used for THM and AOX analysis.

*THM-formation potential (THM-FP)*: THM were analyzed using a GC/ECD-system with purge and trap injection (Chrompack, CP9000). The procedure was carried out as follows: injection of 10 mL solution; 10 min purging, column: CPSiL 13CP, 25 m x 0.32 mm (1.2 $\mu$m diameter), temperature program: 40 °C (3 min), 1 °C/min to 52 °C, 10 °C/min to 162 °C. The standard deviation is below $\pm$ 5% (*13*).

*AOX-formation potential (AOX-FP)*: The determination of AOX was done with a pyrolysis unit at 950 °C in combination with a micro-coulometer (Euroglas) according to DIN 38409. The standard deviation of the method is $\pm$ 8%.

*Assimilable organic carbon (AOC)*: The sample was sterile filtrated (0.2 $\mu$m), and 25 mL of a nutrient solution were added to a volume of 275 mL of the sample. After addition of the inoculum (microorganisms from the filtrate of the sample) the increase in turbidity was determined measuring light scattering (12° forward) in the sample. After 60 h the assimilability is given as the difference of the DOC value. The regrowth rate was calculated as the differential of the change in turbidity at the point of the highest lead. The replication factor is defined as the quotient of the turbidity at the beginning and the end of the measurement (*14*).

*Ozonation and PEROXONE-process*: Ozonation was carried out in a 1.4 L stirred vessel reactor at a temperature of 10° C. For the PEROXONE process hydrogen peroxide was added before starting the ozonation. The ozone gas was dispersed with a glass frit ($O_3$: 1 g/g DOC). The ozone in-gas-concentration was 7.0 mg/L and the gas flow rate was 20 L/h. The absorbed ozone mass was calculated by measuring the difference between in- and out-gas concentration.

*General principle*: The general principle of the experimental set-up is shown in Figure 3. To examine the effects of chlorination the water was chlorinated without further pre-treatment. The influence of oxidation processes was examined using as first step ozone and PEROXONE, respectively. After the oxidation chlorination was performed as second step. For comparison the values of oxidized samples were compared with the values of directly chlorinated raw water as reference value.

# Results

## Effect of chlorination on NOM

The disinfection with chlorine affects the NOM by two different means. On the one hand chlorine reacts with NOM to DBPs. On the other hand chlorine acts as oxidant. The parameter Dissolved Organic Carbon (DOC) is often used to describe the effect of treatment processes. There was no significant change of the DOC after chlorination. Consequently, the NOM is not oxidized to carbon dioxide. However, the chlorination causes drastical changes in structure and functionality of NOM. This is shown in the LC-DOC chromatogram of a brown water before and after chlorination (Figure 4). Here, the high molecular fraction (e.g., humic substances), which elutes at shorter retention times $t_R$ (16 min to 25 min), is removed, whereas new low molecular fractions eluting at $t_R > 25$ min are formed. Further, the intensity of the UV detection at the wavelength of 254 nm is reduced drastically after chlorination. In more detail, the $UV_{254nm}$ chromatograms in Figure 4 show a partly complete elimination of $UV_{254nm}$ absorbance of the high molecular fraction, which is characterized by a high quantity of aromatic structures (e.g., humic substances). On the other hand the formed low molecular fractions after chlorination can be assumed to have less aromatic structures and double bonds, resulting in a decrease of $UV_{254nm}$ absorbance.

The chlorination causes a decrease in the concentration of humics and an increase of low molecular compounds. These formed low molecular compounds (low molecular organic acids, and low molecular amphiphilics) can be responsible for a promotion of bacterial growth in supply nets, when the disinfectant is consumed by NOM. A higher bacterial regrowth potential was already found in chlorinated samples, if free chlorine was removed (*15*).

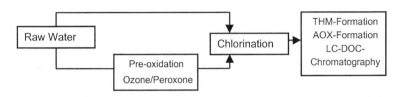

Fig. 3. General experimental procedure.

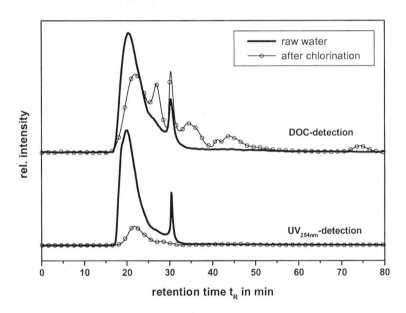

Fig. 4. DOC and $UV_{254nm}$ chromatograms of brown water (DOC: 3 mg/L) before and after chlorination (addition of chlorine: 10 mg/L).

### Effect of ozone and PEROXONE on NOM

The ozonation of river water causes no decrease of DOC as a result of no significant mineralization of NOM to carbon dioxide. Comparable with the chlorination experiment of brown water given in Figure 4 the oxidation of the river water using ozone shows a drastical decrease of humics and an increase of low molecular organic substances (Figure 5). The difference between the chromatograms of Figure 4 and Figure 5 are due to the use of 2 different columns. In Figure 5 the high molecular fraction elutes at retention times $t_R$ from 25 min to 35 min, and the low molecular fraction at $t_R > 35$ min.

The ozonation also results in a reduction of the $UV_{254nm}$ absorbance in all fractions to nearly the same extent. The decrease of the DOC concentration of the high molecular substances and the increase of the low molecular substances as well as the decrease of UV absorbance can be explained by the ozonation mechanism found by Criegee (11). Ozone breaks double bonds and aromatic rings of high molecular substances and forms low molecular substances like carboxylic acids or aldehydes.

The bacterial growth in river water before and after oxidation with ozone and PEROXONE is shown in Figure 6. In this method the turbidity is used to determine the bacteria concentration. For regrowth experiments the ozonation and the PEROXONE process were carried out with an absorbed ozone mass of 1 mg/mg DOC. For the PEROXONE process 0.5 mol hydrogen peroxide per mol ozone were added before the absorption of ozone. This is equivalent to the stoichiometric relation of the formation of OH-radicals from ozone and hydrogen peroxide.

The treatment with ozone and PEROXONE results in an increase in the bacteria concentration compared to raw water. In the PEROXONE sample a shift to longer times of the lag-phase occurred because of residual hydrogen peroxide. Obviously, the bacteria need more time to adapt their metabolism to the new environmental conditions. The ratios of the replication factors after ozonation and the PEROXONE process in relation to those of raw water are shown in Table II. It can be assumed, that the replication factor defined as the ratio of the turbidity at the end to that of the beginning of the experiment represents the quantity of bioavailable organic carbon and consequently the regrowth potential of the water. Therefore after oxidation the regrowth potential was always higher than in raw water. In case of river water (River Ruhr) and reservoir water the regrowth potential after the PEROXONE process is significantly lower than after ozonation. In case of surface water (Lake Constance) there was no significant change of the regrowth potential after ozone or PEROXONE treatment.

#### Table II. Ratio of replication factors after oxidation to raw water.

| Treatment | Lake Constance | River Ruhr | Reservoir Water |
|-----------|----------------|------------|-----------------|
| Ozone     | 1.6            | 2.2        | 2.1             |
| Peroxone  | 1.7            | 1.5        | 1.3             |

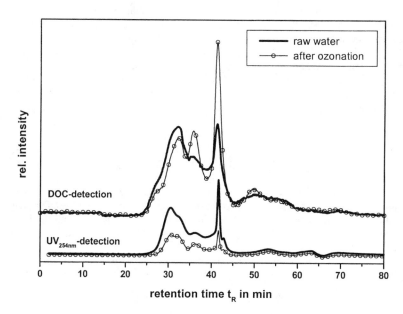

Fig. 5. DOC and UV$_{254nm}$ chromatograms of river water (DOC: 2.0 mg/L, spectral absorbance$_{254\ nm}$: 5.6/m) before and after ozonation (absorbed ozone mass: 1 mg O$_3$ per mg DOC).

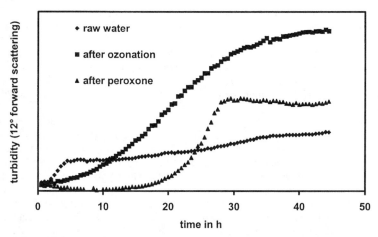

Fig. 6. Bacterial regrowth curves of river water (River Ruhr) and after treatment with ozone and PEROXONE.

All in all, more bioavailable substances are formed during ozonation than in the PEROXONE process. It can be assumed that NOM is cleaved directly by ozone to low molecular compounds (equation [1]), which are available for microorganisms. However, in the PEROXONE process there is a competition between direct reactions of ozone with NOM and the reaction of ozone with hydrogen peroxide (equation [2]). A higher addition of hydrogen peroxide effects a higher yield of OH-radicals. On the other hand hydrogen peroxide scavenges OH-radicals as shown in equation (3) and the efficiency of the NOM-oxidation is reduced (*17*).

$$NOM + O_3 \rightarrow NOM_{OX} \qquad (1)$$
$$2\,O_3 + H_2O_2 \rightarrow 2\,OH\bullet + 3\,O_2 \qquad (2)$$
$$H_2O_2 + OH\bullet \rightarrow H_2O + HO_2\bullet \qquad (3)$$

OH-radicals can react in a similar way with NOM, but can further cause an increase of OH-/O-groups in the molecule without reducing the molecular mass. Besides this, OH-radicals also react with inorganic scavengers, e.g., hydrogen carbonate, and reduce the efficiency of the oxidation, and therefore the formation of bioavailable compounds.

The assimilable organic carbon is consists of low molecular substances. For minimizing bacterial regrowth the quantity of low molecular substances has to be observed.

## Influence of ozone and PEROXONE on DBP formation

Pre-oxidation is applied to reduce formation of chlorinated DBP in the following disinfection step. Usually pre-oxidation is done with ozone. OH-radicals as oxidants are known to be more reactive but less selective than ozone (*16*). The influence of the oxidation with ozone and PEROXONE on THM- and AOX-formation is shown in Figure 7 and Figure 8, respectively.

After ozonation the AOX-formation potential of river water (River Ruhr), reservoir water (Rapp-Bode) and surface water (Lake Constance) was reduced to 86%, 71%, and 56% of the initial AOX formation potential, respectively. For the same samples THM-formation potential was reduced to 80%, 67%, and 48%, respectively.

The changes in the AOX-formation potential values are very similar to those of the THM-formation potential after oxidation processes. An increasing addition of hydrogen peroxide caused higher AOX- and THM-formation potential compared to ozonation treatment of the raw waters. A ratio of hydrogen peroxide to ozone below 2 shows no significant change of AOX- and THM-formation potential compared to the values before the treatment.

Ozone seems to be more efficient than OH-radicals for the removal of (chlorinated) DBP precursors. As a result higher DBP concentrations were achieved using PEROXONE as advanced oxidation process: Firstly, the OH-radicals can be scavenged by inorganic compounds, e.g., bicarbonate, hydrogen peroxide or ozone, and therefore they can not react with NOM. Secondly, ozone reacts more selectively

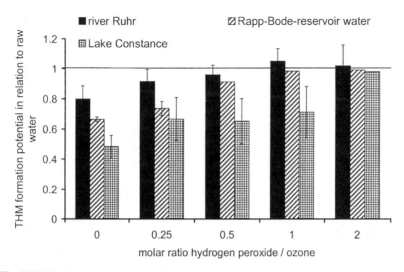

Fig. 7. THM-formation potential after treatment with ozone and PEROXONE with different molar ratios of hydrogen peroxide to ozone in relation to raw water (absorbed ozone mass: 1 mg $O_3$ per mg DOC, addition of hydrogen peroxide before ozonation).

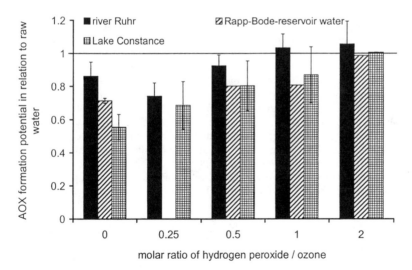

Fig. 8. AOX-formation potential after treatment with ozone and PEROXONE with different molar ratios of hydrogen peroxide to ozone in relation to raw water. (absorbed ozone mass: 1 mg $O_3$ per mg DOC, addition of hydrogen peroxide before ozonation).

than OH-radicals with those centers in NOM, which are responsible for a higher DBP-formation potential.

To refer to the results of LC-DOC analysis the formation of small molecules as effect of oxidation does not necessarily result more precursors for THM. The direct formation from high molecular substances as a result of a reaction with chlorine with these substances is as well a mechanism to form low molecular DBPs.

## Conclusions

In this investigation pre-oxidation with ozone is shown to be a powerful method to decrease the formation of halogenated DBPs. The pre-oxidation with OH-radicals in the PEROXONE process was less efficient than ozonation. The direct reactions of ozone seem to remove selectively those centers of high-molecular organic compounds (e.g., humic substances) which readily react with chlorine. The formed low molecular compounds seem to be not account for the formation of halogenated compounds. The highest concentration of these substances was achieved after ozonation, but ozonation also resulted in the lowest THM- and AOX-formation potential. However, low molecular substances are responsible for high bacterial regrowth potential. The reactions with OH-radicals result in a reduced formation of low molecular products.

OH-radicals can also react with aliphatic side chains of high-molecular compounds yielding an increasing number of alcoholic or ketonic groups in NOM. These centers can react with chlorine and form DBPs, but they do not necessarily increase the bacterial regrowth potential. Only little is known about the health effect of the by-products of ozonation or OH-radical induced oxidation. Further investigations in this field are necessary.

To decrease the DBP formation a high removal of NOM during the treatment processes should be achieved. A more detailed look at bacteria and assimilable organic carbon can help to reduce the necessary concentration of chlorine for disinfection. Lower concentrations of chlorine reduce the DBP formation. The chlorination of raw water with a high content of NOM should be avoided. Besides chlorinated DBPs low molecular compounds are also formed during chlorination, which result in a higher bacterial regrowth after the consumption of chlorine.

The simulation of DBP formation could be an interesting method to improve the knowledge about the parameters that affect DBP formation (18). More detailed informations on formation of inorganic DBPs such as bromate and iodate are also needed.

## Acknowledgment

We thank the Deutsche Verein des Gas- und Wasserfachs (DVGW) and the Bundesministerium für Bildung, Wissenschaft, Forschung und Technologie (BMBF) for supporting this work. Further thanks to Margit Müller for the useful comments to the manuscript.

# References

1. Richardson, S.D. In *The Encyclopedia of Environmental Analysis & Remediation;* Meyers R.A.; John Wiley & Sons: New York, 1998, 3, pp 1398-1412.
2. Rook, J.J. *Water Treat. Exam.* **1974**, *23 (2)*, 234-243
3. Hemming, J.; Holmbom, B.; Reunanen, M.; Kronberg, L. *Chemosphere* **1986**, 15, 549-556.
4. Kühn, W. AWBR-Bericht **1975**, 93-103.
5. Thurmann, E. M. *Organic geochemistry of natural waters;* Martinus Nijhoff/ Dr. W. Junk Publishers: Boston, MA, 1985.
6. Suffet, I. H. and MacCarthy, P. *Aquatic Humic Substances;* American Chemical Society: Washington D. C. 1989.
7. Frimmel, F. H.; Christman, R. F. *Humic Substances and their Role in the Environment;* John Wiley & Sons, Chichester, 1988.
8. Bundesminister für Jugend, Familie, Frauen und Gesundheit: Bekanntmachung der Neufassung der Verordnung über Trinkwasser und Wasser für Lebensmittelbetriebe (Trinkwasserverordnung - TrinkwV); *Bundesgesetzblatt* **1990**, 1, 2612 – 2629.
9. Council Directive 98/83/EC on the Quality of Water intended for Human Consumption; *Official Journal of the European Comunities*, no. L330, Dec. 05, 1998, p. 32.
10. Huber, S. A. Frimmel, F. H.; *Fres. J. Anal. Chem.* **1992**, *342*, 198-200.
11. Criegee, R. *Angew. Chemie* **1975**, *87*, 765-771.
12. Huber, S. A. Frimmel, F. H.; *Vom Wasser* **1996**, *86*, 277-290.
13. Frimmel, F.H. Schmiedel, U.; *Fres. J. Anal. Chem.* **1993**, *346*, 707-710.
14. Hambsch, B. Werner, P.; Frimmel, F.H.; *Acta Hydrochim. Hydrobiol.* **1992**, *20*, 9-14.
15. Hambsch, B. Schmiedel, U.; Werner, P.; Frimmel, F. H.; *Acta Hydrochim. Hydrobiol.* **1993**, *21*, 167-173.
16. Hoigné, J. In *The Handbook of Environmental Chemistry;* Part C; J. Hrubec; Springer, Berlin, Germany, 1998; 5, 84-143.
17. Glaze, W.H.; Kang, J.-W.;Chapin, D. H. *Ozone Sci. Eng.,* **1987**, 9, pp 335-343.
18. Summers, R. S.; Chowhury, Z. K.; Swanson, W.; Solarik, G.; Sohn, J.; Amy, G.L. Simulating DPB formation under practical conditions: The water treatment plant model. In *Preprints of Extended Abstracts of the 217th ACS National Meeting,* Anaheim, American Chemical Society, 1999, pp 203-205.

Chapter 7

# Occurrence of Amino Acids in Two Drinking Water Sources

Russell Chinn and Sylvia E. Barrett

Metropolitan Water District of Southern California,
700 Moreno Avenue, La Verne, CA 91750-3399

Measurement of 16 proteinaceous amino acids (free and combined) in two different surface-water sources has been achieved by the application of a carbamate derivatizing agent, 6-aminoquinolyl N-hydroxysuccinimidyl carbamate (AQC). This highly active derivatizing agent has several properties that are useful in quantifying amino acid concentrations as low as 1 µg/L, which is a typical level in drinking water sources. The AQC analytical method has been used to measure amino acid levels in water from the Colorado River and the California State Water Project. The analytical results show that four of the same free amino acids appear among the top five (in terms of high levels present) in both water sources. These similarly high-concentration free amino acids are serine (average 5.1 µg/L), alanine (average 4.5 µg/L), proline (average 6.4 µg/L), and glycine (average 2.3 µg/L). Also, three of the same *combined* amino acids rank among the four highest concentrations found in each of the two water sources: proline (average 26.9 µg/L), glycine (average 9.4 µg/L), and glutamic acid (average 10.6 µg/L).

## Background

Individual components of the natural organic matter (NOM) in water have typically been very difficult to elucidate. Amino acids, which constitute a small portion of NOM, are a specific group of compounds present in water that can be quantified at environmental levels by forming intensely fluorescent amino acid derivatives (*1*). A fluorescence derivatizing agent, 6-aminoquinolyl-N-hydroxy-succinimidyl carbamate (AQC), has provided a powerful technique for use in the biotechnology field (*2–4*). AQC is a fluorophoric reagent that makes it possible to detect low levels of amino acids after they have been separated by reverse-phase high-performance liquid chromatography (HPLC).

Amino acids can readily react with chlorine to form an assortment of disinfection by-products (DBPs) during drinking water treatment (5). Chlorination of amino acids has been shown to produce mutagens (6) and can result in the formation of odorous aldehydes (7). The U.S. Environmental Protection Agency has promulgated Stage 1 of the Disinfectants/DBP Rule, which will require more stringent regulation of DBPs in drinking water (8). Because of new DBP regulations, information on the presence of DBP precursors such as amino acids has become increasingly important.

## Relevance of Studying Amino Acids

The study of NOM has become more important as researchers try to understand the formation of various DBPs; however, because of the complexity of NOM, it has been a very difficult material to analyze. Some studies rely on very general aggregate types of measurements, such as dissolved organic carbon (DOC), ultraviolet absorbance at 254 nm, biodegradable organic carbon (BDOC), etc. These measurements are useful in studying and improving water treatment technology. However, specific information about the reactive substances in NOM is important (9). The amino acids are one of the few specific groups of compounds in NOM that can be readily measured by using commercially available standards. Amino acids are an important part of the nitrogen content of NOM, but a minor part of the carbon pool of NOM (1). However, amino acids can represent a large portion of the DOC and BDOC in a nanofiltration permeate (10).

## Organic Nitrogen Content of Water

Some broad classes of organic nitrogen compounds that are present in natural waters are peptides, polypeptides, proteins, and amino sugars (11–13). These larger amino compounds can be broken down by acid hydrolysis into their constituent amino acids, which then can be measured by a method for free amino acids. Acid hydrolysis of the NOM is one way to determine indirectly the peptide and protein content of a particular water. Another related class of compounds found in pond and river water are the heterocyclic DNA and RNA bases (adenine, thymine, uracil, and guanine), which can sometimes be analyzed by methods similar to those used for amino acids (11).

Environmental amino acid levels have been related to algal blooms, which may also affect the level of other NOM components (11, 14). They are also significant in the microbial nutrient web and play an active role in the various cycling processes occurring in biologically active aquatic systems (15). The marine environment is where much of the amino acid cycling behavior has been studied (16). There is little published information on the naturally occurring levels of amino acids in various drinking water sources (*17*), and this could be partially because of the relatively low levels present in the environment and difficulties with analytical methods.

## Amino Acids and Drinking Water Treatment

One important characteristic of amino acids is the active amine functionality, which can readily react with chlorine to form an assortment of different by-products. Each amino acid has different chemical moieties that will dictate the production of various reaction products. The reactions are dependent on time, oxidant concentration, pH, temperature, ratio of active nitrogen to oxidant dose, bromide concentration, and the levels of each type of reactive compound in a particular sample matrix. All of these considerations are important when studying DBP formation in drinking water treatment plants. The reaction kinetics and chlorine behavior between ammonia and organic amines have also been studied under various conditions, especially with amino acid compounds (*6, 18–20*).

Knowing the free and combined amino acid levels can help in determining the potential types and levels of DBPs that can form during drinking water treatment. Chlorination of alanine can produce acetaldehyde and acetonitrile (*20, 21*). Chlorination of glutamic acid can yield aldehydes and nitriles (*22*). Proline is an unusual heterocylic amino acid that contains a secondary amine as part of a pyrrolidine ring. Two important proline chlorination by-products found are dichloro- and trichloroacetic acid (*23*). Tyrosine has been found to be a precursor to 3-chloro-4-(dimethylchloro)-5-hydroxy-2(5H)-furanone, or MX (*6*). In addition, other researchers have studied the formation of chlorination by-products from amino acids (*24-27*).

Furthermore, a common problem that water utilities have in measuring chlorine residuals is caused by a positive interference exerted by chlorinated organic amines. This area has been the subject of study because many chlorinated organic amines have little or no bactericidal activity, yet they interfere with the residual chlorine measurement. This fact could present a problem in water disinfection processes (*28*). Chlorinated organic amines can exhibit a breakpoint curve phenomenon similar to that of ammonia, so their chlorine demand is not stoichiometric relative to the active amine. This is a possible explanation for studies that have found much higher chlorine demands than the molar ratio would indicate (*5*).

## Amino Acid Analysis

HPLC is the preferred method for amino acid analysis. One of the more common HPLC techniques used in low level amino acid analysis employs fluorescence detection of derivatized amino acids. Various fluorophoric reagents are used to react with the amino group to allow for highly sensitive detection.

One common reagent used is *o*-phthalaldehyde (OPA) with 2-mercaptoethanol. Drawbacks of the OPA methods are (a) that the derivatives are unstable and (b) that without a special conversion procedure, OPA does not react with secondary amines such as proline. Other fluorescence derivatizing agents commonly used in amino acid analysis—e.g., 4-dimethyl-aminoazobenzene sulfonyl chloride (DABS-Cl), phenylisothiocyanate (PITC), 9-fluorenylmethyl chloroformate, 5-[dimethyl-amino] naphthalene-1-sulfonyl chloride (dansyl chloride)—also have some weaknesses. Undesirable attributes include the formation of multiple products from a single amino acid, large interfering peaks formed by excess reagent, and poor quantitative reproducibility.

The method employed in this study uses the derivatizing agent AQC. A convenient synthetic method to prepare these types of carbamate compounds has been successfully developed (29). AQC does not have several of the drawbacks associated with the reagent chemistries discussed above and has been found to be a highly effective method for environmental applications. One of the analytical characteristics that makes AQC effective is its ability to be used on samples containing high salt content, which usually interferes with the derivatizing process and requires desalting of the sample.

# Experimental Approach

## Sampling Locations

The Metropolitan Water District of Southern California (Metropolitan) uses Colorado River water (CRW) and state project water (SPW) as its drinking water sources. During the winter, samples for analysis were taken from the F. E. Weymouth Filtration Plant influent. The samples were taken when the plant influent was either 100% SPW or 100% CRW.

In addition, Metropolitan operates an oxidation demonstration plant (ODP). The ODP has a maximum treatment capacity of 5.5 million gallons per day. Samples were taken in March at the plant influent and at one of the filter effluents representing a biologically active filter sample. The filters contained 2 in. of granular activated carbon, 8 in. of sand, and a layer of gravel. The plant had been operating for just over a year, and the filters had reached a steady-state condition when sampled. During the time of sampling, the ODP was operating with 100% CRW.

## Analytical Method

Each water sample was prepared by rotary evaporation to concentrate 500 mL of sample, which took approximately 3 h at 67°C under 15 mm Hg vacuum. The reduced pressure allowed faster concentration and minimized the heating of the sample, which could have enhanced the thermal degradation of NOM components. The concentrate (~5-15 mL) was transferred to a graduated cylinder, and the volume was brought up to 20 mL with 20 mM HCl; therefore, a concentration factor of 25 was achieved with this procedure.    An aliquot of the concentrate was evaporated to dryness with a gentle stream of a helium gas, hydrolyzed under vacuum at 107°C for 24 h, and allowed to cool.    Hydrolysis was performed in 8 mL hydrolysis tubes (Pierce, Rockford, IL) with 6 N constant boiling hydrochloric acid (HCl).    A test of the hydrolysis procedure with leucine enkephalin (formula weight 555.6) indicated that the method could achieve an average recovery of 71% for the constituent amino acids in that compound.    A reagent kit (Waters Corp., Milford, MA) supplied the derivatizing reagent, buffer (borate), and a mixed eluant buffer system.

A 16-component amino acid standard (Pierce Standard-H) plus ammonia was diluted in reagent water to yield a 10 pmol/mL stock solution. The standard contains common protein hydrolysate amino acids. Tryptophan was not included in this study because it requires special analytical considerations.    A derivatized calibration stock standard was prepared by taking 10 µL of amino acid stock solution, 70 µL of sample buffer, and 20 µL of 10 mM AQC solution and allowed to react for 10 min at 55°C.    Dried samples were reconstituted with 20 µL of 20 mM HCl.    The 20 µL reconstituted sample was derivatized with 60 µL of sample buffer and 20 µL of AQC solution.    The buffer should maintain a pH between 8 and 10 when added to the sample.

The three eluants used to achieve chromatographic separation of the amino acid derivatives were 100% acetonitrile, distilled water, and a mixed acetate-phosphate (a-p) buffer.    The initial concentration of the a-p buffer was 100%; at 0.5 min, it was 99% a-p buffer and 1% acetonitrile; at 0.8 min, it was 95% a-p buffer and 5% acetonitrile; at 19 min, it was 91% a-p buffer and 9% acetonitrile; at 30.5 min, it was 83% a-p buffer and 17% acetonitrile; and at 34 min there was no a-p buffer, and the eluant was 40% water and 60% acetonitrile.    For the remaining 6 min of the run, 100% a-p buffer was used.

All three eluants were filtered through a 0.2 micron filter before use.    The HPLC system (HP1090; Hewlett-Packard Co., Avondale, PA) is capable of low-pressure tertiary gradient mixing and high-pressure pumping through a heated 37°C column oven.    A 3.9 mm diameter, 150 mm length C18 column containing 4-micron spherical silica packing (Novapak Water Corp.) was connected with 0.009 in. internal diameter stainless steel tubing.    A 5-micron prefilter was used at the column inlet and a manual injection valve (7125; Rheodyne, Cotati, CA) fitted with a 20 µL sample loop was used for sample introduction.    The linear gradient flow was set at 1 mL/min, and the run time was 45 min per analysis, including reconditioning time. The wavelengths used for the dual-beam fluorescence detector (LC-240; Perkin-Elmer Corp., Norwalk, CT) were 240 nm for excitation and 395 nm for emission. The detector flow cell volume was 7 µL.

## Results and Discussion

The results for 16 different amino acids are shown in Table I for SPW and in Table II for CRW. Duplicate analyses of separate aliquots of sample water were conducted except for the CRW hydrolyzed sample, for which no duplicate was analyzed because of an instrument problem. Because of the long sample preparation time, no further replicate samples were processed. Procedural blank determinations showed no appreciable background interference. Free and total amino acid levels were determined from the same sample aliquots.

The relative rank among the 16 different amino acids is shown, where a ranking of 1 represents the highest concentration. The data are presented on a weight basis rather than on a molar basis. The rankings may be slightly different when calculated on a molar basis. Individual amino acid percentages with respect to the corresponding sum of amino acids were calculated on a weight basis. The free amino acids were determined before acid hydrolysis, and after acid hydrolysis the sample was analyzed for total amino acids. The combined amino acids were calculated by subtracting the free amino acids from the corresponding total amino acids.

In other research (1), the major amino acids found in hydrolyzed NOM isolates included glycine, glutamic acid, aspartic acid, and serine. In the hydrolyzed SPW, these four amino acids were ranked 2, 4, 6, and 3, respectively, whereas in hydrolyzed CRW these four amino acids were ranked 4, 3, 8, and 7, respectively. In hydrolyzed CRW, arginine, proline, glutamic acid, and glycine were ranked 1, 2, 3, and 4 on a weight basis. However, on a molar basis, proline, glycine, arginine, and glutamic acid were ranked 1, 2, 3, and 5. The same amino acids are ranked high on either a weight or molar basis; only the relative order is somewhat different.

It is important to note when interpreting this data set that the amino acid levels represent concentrations for those particular samples, and these concentrations will change seasonally because of varying hydrology and biological activity such as algal blooms. Table III shows the five amino acids at the highest levels in each type of water. The last column in Table III shows the ratio of free to combined amino acids. The last column in Table III shows that glutamic acid had the lowest free to combined ratio in both waters, which means that very little free glutamic acid is in the water; but it occurs in much larger amounts as a combined amino acid.

### Most Abundant Amino Acids

It can be seen by the sums of the five combined amino acids at the highest concentrations in Table III that they represent the bulk of the combined amino acids (average of 82%) for both waters. The three free amino acids at the highest concentrations for SPW and CRW in Table III make up approximately 50% of the free amino acid content in each water. For the five free amino acids at the highest concentrations, four individual amino acids—serine, alanine, proline, and glycine—

#### Table I. Amounts of Amino Acids (μg/L) Found in SPW

| No. | L-amino acid | Free Avg | Free Rank | Free aa% | Combined Avg | Combined Rank | Combined aa% | Total Avg | Total Rank | Total aa% |
|---|---|---|---|---|---|---|---|---|---|---|
| 1 | Aspartic acid | 1.6 | 7 | 5.3 | 3.1 | 5 | 4.5 | 4.7 | 6 | 4.8 |
| 2 | Serine | 4.9 | 1 | 16.3 | 3.3 | 4 | 4.8 | 8.2 | 3 | 8.4 |
| 3 | Glutamic acid | 0.5 | 12 | 1.7 | 6.5 | 3 | 9.4 | 7 | 4 | 7.2 |
| 4 | Glycine | 2.3 | 4 | 7.7 | 8.6 | 2 | 12.4 | 10.9 | 2 | 11.2 |
| 5 | Histidine | 1.2 | 8 | 4.0 | 0.1 | 13 | 0.1 | 1.3 | 14 | 1.3 |
| 6 | Arginine | 1.7 | 6 | 5.7 | 1.8 | 9 | 2.6 | 3.5 | 8 | 3.6 |
| 7 | Threonine | 1.2 | 8 | 4.0 | 2.6 | 6 | 3.7 | 3.8 | 7 | 3.9 |
| 8 | Alanine | 4.6 | 2 | 15.3 | 1.1 | 11 | 1.6 | 5.7 | 5 | 5.9 |
| 9 | Proline | 4.4 | 3 | 14.7 | 35.4 | 1 | 51.0 | 39.8 | 1 | 40.9 |
| 10 | Tyrosine | 2.2 | 5 | 7.3 | nd | 14 | nd | 1.9 | 12 | 2.0 |
| 11 | Valine | 0.9 | 10 | 3.0 | 1.5 | 10 | 2.2 | 2.4 | 11 | 2.5 |
| 12 | Methionine | 1.7 | 6 | 5.7 | nd | 14 | nd | nd | 16 | nd |
| 13 | Lysine | 1.1 | 9 | 3.7 | 1.9 | 8 | 2.7 | 3 | 9 | 3.1 |
| 14 | Isoleucine | 0.5 | 12 | 1.7 | 1.1 | 11 | 1.6 | 1.6 | 13 | 1.6 |
| 15 | Leucine | 0.6 | 11 | 2.0 | 2 | 7 | 2.9 | 2.6 | 10 | 2.7 |
| 16 | Phenylalanine | 0.6 | 11 | 2.0 | 0.4 | 12 | 0.6 | 1 | 15 | 1.0 |
| Sums | | 30.0 | | | 69.4 | | | 97.4 | | |

Avg = Average amounts based on duplicate analysis
nd = Component not detected (estimated detection limit = 1 μg/L)
Free = unhydrolyzed sample; Total = hydrolyzed sample
Combined = Total minus Free
aa% = Percentage of individual amino acid to sum of amino acids

**Table II. Amounts of Amino Acids (μg/L) Found in CRW**

| No. | L-amino acid | Free | | | Combined | | | Total | | |
|---|---|---|---|---|---|---|---|---|---|---|
| | | Avg | Rank | aa% | Avg | Rank | aa% | Value* | Rank | aa% |
| 1 | Aspartic acid | 1.8 | 7 | 4.4 | 2.7 | 8 | 3.1 | 4.5 | 8 | 3.6 |
| 2 | Serine | 5.2 | 3 | 12.8 | 3 | 7 | 3.4 | 8.2 | 7 | 6.5 |
| 3 | Glutamic acid | nd | 12 | nd | 14.7 | 3 | 16.8 | 14.7 | 3 | 11.7 |
| 4 | Glycine | 2.3 | 5 | 5.7 | 10.3 | 4 | 11.8 | 12.6 | 4 | 10.0 |
| 5 | Histidine | 0.8 | 8 | 2.0 | nd | 13 | nd | nd | 14 | nd |
| 6 | Arginine | 8.3 | 2 | 20.5 | 20.6 | 1 | 23.5 | 28.9 | 1 | 22.9 |
| 7 | Threonine | 2.3 | 5 | 5.7 | 6.7 | 5 | 7.7 | 9 | 6 | 7.1 |
| 8 | Alanine | 4.4 | 4 | 10.9 | 5.5 | 6 | 6.3 | 9.9 | 5 | 7.9 |
| 9 | Proline | 8.5 | 1 | 21.0 | 18.4 | 2 | 21.0 | 26.9 | 2 | 21.3 |
| 10 | Tyrosine | 1.9 | 6 | 4.7 | nd | 13 | nd | 1.1 | 13 | 0.9 |
| 11 | Valine | 0.8 | 8 | 2.0 | 0.9 | 12 | 1.0 | 1.7 | 12 | 1.3 |
| 12 | Methionine | 1.8 | 7 | 4.4 | nd | 13 | nd | 1.8 | 11 | 1.4 |
| 13 | Lysine | 0.3 | 11 | 0.7 | nd | 13 | nd | nd | 14 | nd |
| 14 | Isoleucine | 0.6 | 10 | 1.5 | 1.2 | 11 | 1.4 | 1.8 | 11 | 1.4 |
| 15 | Leucine | 0.8 | 8 | 2.0 | 2 | 9 | 2.3 | 2.8 | 9 | 2.2 |
| 16 | Phenylalanine | 0.7 | 9 | 1.7 | 1.5 | 10 | 1.7 | 2.2 | 10 | 1.7 |
| Sums | | 40.5 | | | 87.5 | | | 126.1 | | |

Avg = Average amounts based on duplicate analysis

nd = Component not detected (estimated detection limit = 1 μg/L)

Free = unhydrolyzed sample; Total = hydrolyzed sample

Combined = Total minus Free

aa% = Percentage of individual amino acid to sum of amino acids

*No duplicate analysis for this sample

are common to both waters. These four amino acids are the lowest-molecular-weight amino acids (glycine, 75 g/mol; alanine, 89 g/mol; serine, 105 g/mol; and proline, 115 g/mol). Previous findings (*10, 23, 30*) have shown that the lower-molecular-weight free amino acids tend to be more prominent.

Very little has been reported on environmental levels of proline, and it is one of the highest occurring amino acids found in these samples. A possible reason that proline is sometimes not reported in the literature is that when direct OPA derivatization is used, secondary amines such as proline are not determined. Looking at the combined amino acids for SPW, proline contributes over one-half (51.2%) of the sum of amino acid concentrations, and in CRW it is the second highest concentration of the combined amino acids (21%). Both waters contain proline, glycine and glutamic acid as part of the five highest concentrations for the combined amino acids, making up 60.5% for CRW and 73% in SPW. Proline is the only amino acid that appears in both waters as one of the three highest concentrations of amino acids in the free or combined forms.

## Comparing CRW Versus SPW

Comparing the sums for SPW in Table I and CRW in Table II, CRW had approximately 30% higher levels of amino acids than SPW. For both SPW and CRW, the sum of the free amino acid concentrations is about 30% of the total amino acid concentration. This is a relatively high percentage compared to the percentages observed in other studies (*31*).

The biggest individual difference in CRW and SPW was the amount of arginine present in CRW. CRW (28.9 µg/L) contains more than eight times the arginine than in SPW (3.5 µg/L). Arginine is an aliphatic monocarboxylic acid with four amine groups so it is considered to be a basic amino acid.

Both waters had an average of 2 µg/L of tyrosine, which, as discussed above, has been found to be a precursor to MX, the potent mutagen (*6*).

## Interferences in Determining Chlorine Residuals

Even though individual amino acid concentrations are low, their total amount could cause a measurable difference in the total chlorine residual measurement because of the formation of compounds that can be measured as part of the chlorine residual. The average sum of the free amino acids in CRW and SPW was 0.035 mg/L. Assuming an average chlorine demand of 3:1 mg/mg gives a theoretical chlorine demand of 0.1 mg/L. The calculated average chlorine demand of 3:1 mg/mg is based on work by Hureiki and colleagues (*5*), in which they determined chlorine demand for each of the free amino acids. This demand must be satisfied before a free-chlorine residual can be obtained. Free amino acids can contribute to what has been called "unknown" organic material in the water, which may react with chlorine to cause a higher-than-expected total chlorine residual to be measured (*28*).

**Biologically Active Filtration**

The ODP data in Table IV show five of the more prominent free amino acids that exhibited sizable increases in concentration after ozonation and biofiltration. The average increase for those five amino acids was 158%, which indicates that a significant rise in amino acid concentration can occur after ozonation and biofiltration.

The increase in free amino acids can result from (a) a breakdown of the peptide linkages of proteins during ozonation and (b) the release of amino acids from the biologically active filters (*20*). Alternatively, ozonation can oxidize amino acids to aldehydes and organic acids (*32*), and these compounds can be degraded during biologically active filtration (*33*). Based on the results of this test, the mechanisms for the release of free amino acids were more dominant than the oxidation/degradation pathway. An increase in amino acid levels could increase the likelihood of regrowth problems in the distribution system and possibly reduce post-filtration disinfection.

## Conclusions

Amino acids are important components of NOM in water. They can be present in both the free and combined forms. Because chlorination of amino acids can lead to DBP formation, taste-and-odor problems, and interferences in measuring chlorine residuals, it is important to determine the amount of amino acids in the source water. Use of AQC as a derivatizing agent allowed for the measurement of secondary amines, such as proline, which was a major amino acid in the source waters studied.

## Literature Cited

1. Krasner, S. W.; Croué, J.-P.; Buffle, J.; Perdue, E. M. *J.—Am. Water Works Assoc.* **1996,** *88* (6), 66–79.
2. Reverter, M.; Lundh, T.; Lindberg, J. E. *J. Chromatogr., B: Biomed. Sci. Appl.* **1997,** *696* (1), 1–8.
3. Weiss, T.; Bernhardt, G.; Buschauer, A.; Jauch, K.-W.; Zirngibl, H. *Anal. Biochem.* **1997,** *247* (2), 294–304.
4. Merali, S.; Clarkson, A. B., Jr. *J. Chromatogr., B: Biomed. Sci. Appl.* **1996,** *675* (2), 321–326.
5. Hureiki, L.; Croué, J. P.; Legube, B. *Wat. Res.* **1994,** *28* (12), 2521–2531.

**Table III: The Top Five Most Abundant Amino Acids in Metropolitan's Waters**

*SPW*

| Rank | Free | | Combined | | Total | | Free/Combined % | |
|------|------|------|----------|------|-------|------|-----------------|------|
| | *AA* | *aa%* | *AA* | *aa%* | *AA* | *aa%* | *AA* | *aa%* |
| 1 | Ser | 16.3 | Pro | 51.2 | Pro | 40.9 | Iso | 45.5 |
| 2 | Ala | 15.3 | Gly | 12.4 | Gly | 11.2 | Leu | 30 |
| 3 | Pro | 14.7 | Glu | 9.4 | Ser | 8.4 | Gly | 26.7 |
| 4 | Gly | 7.7 | Ser | 4.8 | Glu | 7.2 | Pro | 12.4 |
| 5 | Tyr | 7.3 | Asp | 4.5 | Ala | 5.9 | Glu | 7.7 |
| Sums | | 61.3 | | 82.3 | | 73.6 | | |

*CRW*

| Rank | Free | | Combined | | Total | | Free/Combined % | |
|------|------|------|----------|------|-------|------|-----------------|------|
| | *AA* | *aa%* | *AA* | *aa%* | *AA* | *aa%* | *AA* | *aa%* |
| 1 | Pro | 21 | Arg | 23.5 | Arg | 22.9 | Arg | 40.3 |
| 2 | Arg | 20.5 | Pro | 21 | Pro | 21.3 | Leu | 40 |
| 3 | Ser | 12.8 | Glu | 16.8 | Glu | 11.7 | Thr | 34.3 |
| 4 | Ala | 10.9 | Gly | 11.8 | Gly | 10 | Gly | 22.3 |
| 5 | Thr/Gly | 5.7 | Thr | 7.7 | Ala | 7.9 | Glu | 0 |
| Sums | | 70.9 | | 80.8 | | 73.8 | | |

AA = Amino acid; aa% = percent of individual amino acid to sum of amino acids
Free = unhydrolyzed sample; Total = hydrolyzed sample
Combined = Total minus Free
Free/Combined = ratio of Free to Combined amino acids

**Table IV. The Effect of Ozonation and Biofiltration on the Concentration of Five Free Amino Acids ($\mu$g/L) in CRW**

| Location | His | Arg | Val | Leu | Phe |
|----------|-----|-----|-----|-----|-----|
| Plant influent | 5.8 | 2.2 | 8.2 | 2.7 | 2.0 |
| Plant effluent | 15.9 | 4.7 | 13.6 | 9.6 | 5.6 |
| % Increase | 174 | 114 | 66 | 256 | 180 |

6. Horth, H.; Fielding, M.; James, H. A.; Thomas, M. J.; Gibson, T.; Wilcox, P. In *Water Chlorination: Chemistry, Environmental* Impact *and Health Effects*; Jolley, R. L., Condie, L. W., Johnson, J. D., Katz, S., Minear, R. A., Mattice, J. S., Jacobs, V. A., Eds.; Lewis Publishers: Chelsea, MI, 1990; Vol. 6, pp 763–781.

7. Bruchet, A.; Costentin, E.; Legrand, M. F.; Mallevialle, J. *Wat. Sci. Technol.* **1992,** *25* (2), 323–333.

8. U.S. Environmental Protection Agency. *Fed. Reg.* **1998,** *63* (241), 69390–69476.

9. Owen, D. M.; Amy, G. L.; Chowdhury, Z. K. *Characterization of Natural Organic Matter and Its Relationship to Treatability*; AWWA Research Foundation and AWWA: Denver, CO, 1993.

10. Agbekodo, K. M.; Legube, B.; Coté, P. *J.—Am. Water Works Assoc.* **1996,** *88* (5), 67–74.

11. Ram, N. M.; Morris, J. C. *Environment Int.* **1980,** *4,* 397–405.

12. Aiken, G. R.; McKnight, D. M.; Wershaw, R. L.; MacCarthy, P. In *Humic Substances in Soil, Sediment, and Water: Geochemistry, Isolation, and Characterization*; Wiley-Interscience: New York, 1985; pp 303–326.

13. Bruchet, A.; Rousseau, C.; Mallevialle, J. *J.—Am. Water Works Assoc.* **1990,** *82* (9), 66–74.

14. Hoehn, R. C.; Barnes, D. B.; Thompson, B. C.; Randall, C. W.; Grizzard, T. J.; Shaffer, P. T. B. *J.—Am. Water Works Assoc.* **1980,** *72* (6), 344–350.

15. Fuhrman, J. *Marine Ecol. Prog. Ser.* **1990,** *66,* 197–203.

16. Jorgensen, N. O. G.; Kroer, N.; Coffin, R. B.; Yang, X.; Lee, C. *Marine Ecol. Prog. Ser.* **1993,** *98,* 135–148.

17. Labouyrie-Rouillier, L.; Croué, J.-P.; Legube, B. In *Extended Abstracts*, Fourth International BOM Conference, International Working Group on Biodegradable Organic Matter in Drinking Water, University of Waterloo, ON, Canada, 1996.

18. Snyder, M. P.; Margerum, D. W. *Inorg. Chem.* **1982,** *21* (7), 2545–2550.

19. Yoon, J.; Jensen, J. N. *Environ. Sci. Technol.* **1993,** *27* (2), 403–409.

20. Le Cloirec, C.; Martin, G. In *Water Chlorination: Chemistry, Environmental Impact and Health Effects*; Jolley, R. L., Bull, R. J., Davis, W. P., Katz, S., Roberts, M. H., Jacobs, V. A., Eds.; Lewis Publishers: Chelsea, MI, 1985; Vol. 5, pp 821–834.

21. Stanbro, W. D.; Smith, W. D. *Environ. Sci. Technol.* **1979,** *13* (4), 446–451.

22. Pereira, W. E.; Hoyano, Y.; Summons, R. E.; Bacon, V. A.; Duffield, A. M. *Biochim. Biophys. Acta* **1973,** *313,* 170–180.

23. de Leer, E. W. B.; Erkelens, C.; de Galan, L. In *Water Chlorination: Chemistry, Environmental Impact and Health Effects*; Jolley, R. L., Condie, L. W., Johnson, J. D., Katz, S., Minear, R. A., Mattice, J. S., Jacobs, V. A., Eds.; Lewis Publishers: Chelsea, MI, 1990; Vol. 6, pp 763–781.

24. Trehy, M. L.; Yost, R. A.; Miles, C. J. *Environ. Sci. Technol.* **1986,** *20* (11), 1117–1122.

25. Nweke, A.; Scully, F. E. *Environ. Sci. Technol.* **1989,** *23* (8), 989–994.
26. de Leer, E.W.B.; Baggerman, T.; Schaik, P.; Zuydeweg, C.W.S.; de Galan, L. *Environ. Sci. Technol.* **1986,** *20* (12), 1218–1223.
27. McCormick, E. F.; Conyers, B.; Scully, F. E. *Environ. Sci. Technol.* **1989,** *27* (2), 255–261.
28. Wolfe, R. L.; Ward, R.; Olsen, B. H. *Environ. Sci. Technol.* **1985,** *19* (12), 1192–1195.
29. Noriyuki, N.; Iwaki, K.; Kinoshita, T.; Takeda, K.; Ogura, H. *Anal. Chem.* **1986,** *58* (12), 2372–2375.
30. Lytle, C. R.; Perdue, E. M. *Environ. Sci. Technol.* **1981,** *15* (2), 224–228.
31. Thurman, E. M. *Developments in Biochemistry: Organic Geochemistry of Natural Waters*; Nijhoff-Junk Publishers: Dordrecht, Netherlands, 1985; Ch. 6, pp 151–180.
32. Lacheur, R. M.; Glaze, W. H. *Environ. Sci. Technol.* **1996,** *30,* 1072–1080.
33. Coffey, B. M.; Krasner, S. W.; Sclimenti, M. J.; Hacker, P. A.; Gramith, J. T. In *Proceedings,* AWWA Water Quality Technology Conference, New Orleans, LA, Nov 12-16, 1995; American Water Works Association: Denver, CO, 1996; Part I, pp 357–389.

Chapter 8

# The Effect of Structural Characteristics of Humic Substances on Disinfection By-Product Formation in Chlorination

Wells W. Wu[1], Paul A. Chadik[1], William M. Davis[2],
Joseph J. Delfino[1], and David H. Powell[3]

Departments of [1]Environmental Engineering Sciences and [3]Chemistry,
University of Florida, Gainesville, FL 32611
[2]Environmental Laboratory, ES–P, Waterways Experiment Station,
Vicksburg, MS 39180

The influence of structural characteristics of humic substances on disinfection by-product (DBP) formation was investigated for seven humic substances isolated from aquatic and terrestial sources. The structural characterizations included elemental analysis, total acidity titration, ultraviolet/visible (UV/VIS) spectroscopy, and $^{13}C$ nuclear magnetic resonance (NMR) spectroscopy. The aqueous humic substances were chlorinated at pH 7.0 and 8.5, with and without the presence of the bromide ion to produce trihalomethanes and haloacetic acids.

C/H and C/N ratios, correlated well with DBP formation for the humic substances investigated. Based on the results of computation with respect to total acidity, the contribution of carboxylic groups to the C/H ratio of the humic substances studied was not significant. The decrease in UV absorbance at 254 nm ($\Delta UV254$) and UV absorbance at 254 nm varied linearly with the formation of both trihalomethanes and haloacetic acids, with the $\Delta UV254$ consistently showing better correlation. Both aromatic content and phenolic character determined by $^{13}C$-NMR correlated well with DBP formation.

The relationship between aromatic content determined by $^{13}C$-NMR and other techniques were evaluated. Aromaticity data by $^{13}C$-NMR correlated well with those obtained by other methods that are independent of NMR spectroscopy. This paper presents techniques other than NMR that would be attractive alternatives for predicting DBP formation.

**109**

# Introduction

Chlorine, the most widely used disinfectant in the United States, reacts with humic substances to form disinfection by-products (DBPs) (*1*). DBP formation during chlorination has been correlated with several non-specific parameters characteristic of humic substances such as dissolved organic carbon and chlorine demand (*2*); however other structural characteristics, less frequently used in characterizing DBP formation, may provide additional insight regarding the type and amount of specific DBPs formed in a given chlorination condition. The nature and extent of interaction between humic substances and disinfectant species depend on numerous structural factors. For example, disinfectants such as chlorine are electrophiles and tend to react with electron-rich sites such as aromatic structures in humic substances. Accordingly, methods to characterize these electron-rich sites in terms of aromaticity may be useful in predicting and understanding DBP formation (*3*).

The objective of this research was to investigate the relationship between selected structural characteristics of seven humic substances and the DBPs formed under a variety of chlorination conditions. The structural characterizations used in this research included elemental analysis, total acidity titration, ultraviolet/visible (UV/VIS) spectroscopy, and $^{13}C$ nuclear magnetic resonance (NMR) spectroscopy. The target DBPs were four trihalomethanes (THMs) and nine haloacetic acids (HAAs).

# Materials and Methods

## Humic Substances

Seven humic substances were extracted and purified from water, soil, sediment, and commercial sources (*4*): four aquatic humic substances, Santa Fe River (SFR), St. Marys River (St MR), Newnans Lake (NL DOC) and a surficial groundwater source from Orange Heights (OH DOC); a soil humic substance from Orange Heights soil (OHS); a sediment humic substance from Newnans Lake bottom sediment (NLS); and a commercial source (Aldrich). The extraction technique involved acidification and concentration on a macroporous resin, and elution of the humic substance by backflushing the column with dilute base. The humic substances were then freeze dried until reconstituted with purified water. The objective of this research was to investigate how selected characteristics of humic substances can be related to DBP formation. In addition to the aquatic humic substances, soil, sediment, and commercial humic substances were included to increase diversity.

## Reagents

All DBP calibration standards, internal standards, and surrogate were purchased from either AccuStandard Inc., Aldrich Chemical Co., Supelco Inc., or ULTRA Scientific. All other chemicals employed in this study were reagent or certified grade (Fisher Scientific Co.). The reagent water used was Type 1 water (*5*) produced by

Barnstead NANOpure ultrapure water system. Sodium hypochlorite solutions prepared from reagent grade NaOCl were used as a source of free chlorine.

## Characterization of Humic Substances

The seven humic substances were characterized by a variety of techniques: including elemental analysis, total acidity titration, UV/VIS spectroscopy, and $^{13}$C-NMR spectroscopy. The results of these characterizations are presented in Tables I and II.

**Table I. Results of characterization of the humic substances: elemental composition (percent), total acidity titration, and C/H, C/N and $C_{rest}/H_{rest}$ ratios (4)**

| Source | Mois-ture | Ash | Elements[a] | | | | C/H[b] | C/N[b] | Total Acidity[c] | $C_{rest}/H_{rest}$ |
| | | | C | H | N | O | | | | |
|---|---|---|---|---|---|---|---|---|---|---|
| Aldrich | 2.06 | 1.50 | 50.8 | 3.57 | 0.60 | 45.0 | 1.19 | 98.9 | 14.5 | 1.24 |
| NLS | 0.81 | 15.6 | 42.3 | 4.76 | 2.51 | 50.4 | 0.74 | 19.7 | 19.9 | 0.69 |
| OHS | 2.60 | 12.6 | 47.7 | 4.41 | 0.74 | 47.2 | 0.90 | 75.1 | 18.7 | 0.88 |
| SFR | 1.65 | 1.15 | 49.4 | 3.69 | 0.77 | 46.1 | 1.12 | 74.9 | 17.0 | 1.14 |
| St MR | 0.62 | 0.43 | 51.2 | 3.82 | 0.61 | 44.4 | 1.12 | 97.8 | 18.1 | 1.15 |
| OHDOC | 1.20 | 2.35 | 50.5 | 4.10 | 0.49 | 44.9 | 1.03 | 120.3 | 15.5 | 1.04 |
| NL DOC | 0.28 | 6.62 | 50.2 | 4.60 | 0.79 | 44.4 | 0.91 | 74.3 | 15.2 | 0.89 |

[a] Corrected for moisture and ash content.
[b] Molar ratios
[c] meq $H^+$/g humic carbon.

Elemental analyses of C, H, and N were carried out on a Carlo Erba Model 1106 CHN analyzer. The oxygen content was determined by difference based on the C, H, and N content, after these values had been corrected for moisture and ash content. The moisture and ash content of each humic substance were measured by drying and ignition, respectively. The elemental composition of all samples of humic substances presented in Table 1 were analyzed in duplicate; the C/H, C/N, and $C_{rest}/H_{rest}$ ratios will be discussed later.

The Ba(OH)$_2$ titration method (6) was used to measure the total acidity. The results of total acidity are presented in Table I. Each humic substance was titrated in duplicate and the difference between replicate titrations was generally less than 5% (average,

3.7%; maximum, 9.9%; minimum, 0.1%). The mean variation for four replicate blank titrations was 1.0%.

UV/VIS absorbance measurements were recorded at 254 (UV254) and 272 (UV272 nm using a Perkin Elmer Model 552 double beam grating spectrophotometer with 1.0 cm cells. Solutions were prepared by dissolving each humic substance in buffered water and adjusting to pH 7 and 8.5 using 0.01 M NaOH or 0.01 M HCl. Measurements were performed with buffered water as a reference. The results are presented in Table II

Table II. Results of characterization of the humic substances: UV/VIS spectral analysis, and functional group percent compositions by $^{13}$C-NMR

| Source | UV[a] Absorptivity | | Composition % by $^{13}$C-NMR | | | |
|---|---|---|---|---|---|---|
| | 254 nm | 272 nm | 110-160 ppm aromatic C | 145-160 ppm phenolic C | 160–190 ppm carboxylic C | 190-220 ppm carbonyl C |
| Aldrich | 90.4 | 67.8 | 29.8 | 10.7 | 16.2 | 9.3 |
| NLS | 39.6 | 31.8 | 20.5 | 5.2 | 15.9 | 6.8 |
| OHS | 73.8 | 55.6 | 23.2 | 8.2 | 19.4 | 8.4 |
| SFR | 51.0 | 39.9 | 22.5 | 6.1 | 16.8 | 8.2 |
| St MR | 53.0 | 46.5 | 23.3 | 9.2 | 17.4 | 9.3 |
| OH DOC | 53.2 | 40.8 | 26.8 | 8.3 | 20.6 | 9.4 |
| NL DOC | 48.2 | 40.0 | 24.1 | 8.2 | 19.9 | 8.1 |

[a] L/(g humic carbon•cm)

The application of cross-polarization/magic angle spinning solid state $^{13}$C-NMR spectroscopy was used for functional group analysis of the humic substances studied in this research using a Brucker WP-200 SY spectrometer. The spectrometer was operated at 50 MHz for carbon and externally tuned to the known chemical shifts of *t*-butyl benzene. Samples were spun at the magic angle (54.7°) at about 4.5 KHz. A cross-polarization contact time of one ms and a repetition rate of 2 s were used. For each spectrum, 5000 scans were obtained and averaged before Fourier transformation. An internal standard (Delrin) and a primary standard of para-di-*t*-butyl benzene $^{13}$C-NMR technique was practiced to quantify the $^{13}$C response based on per unit carbon (*4*). The data obtained from the $^{13}$C-NMR analysis are shown in Table II.

## Chlorination Conditions

The water samples prepared from the characterized humic substances were chlorinated, stored in the absence of light, and analyzed for DBP formation in a series of experiments performed under the following conditions:

- Temperature = 25 ± 2 °C
- Br⁻ concentration = zero and 250 μg/L
- Chlorine dose = 12.5 mg/L (Cl₂/NPOC = 2.5)
- Contact time = 24 h
- pH = 7.0 ± 0.1 and 8.5 ± 0.1
- NPOC = 5.0 mg/L

The chlorine dose was chosen to achieve chlorine residual concentrations of at least 0.6 mg/L at the end of the 24-hour reaction period for all samples based on the results of preliminary disinfectant demand experiments. Chlorinated waters were quenched with different dechlorinating agents depending on the subsequent analytical method. The Br⁻ concentration of 250 μg/L was chosen in this research to give a sufficiently high impact on species distribution between chlorinated and brominated DBPs, while keeping the Br⁻ concentration within a reasonable range. Seasonal averages of Br⁻ for 100 utilities in the USA were reported to range from 2 to 429 μg/L (7).

## Experimental Design and Methods

For DBP formation investigation, chlorination was conducted in borosilicate glass vials (nominally 40 mL) sealed with teflon-faced septa. Each experiment was conducted using five vials. Two vials were used for chlorine residual analysis and for UV254 measurement at the end of the 24-hour reaction period, respectively. The chlorine residual concentrations were analyzed by Standard Method 4500-Cl G DPD Colorimetric Method (5) using a Milton Roy Spectronic 21 UV spectrophotometer. UV254 absorbances were measured by Standard Method 5910 UV-Absorbing Organic Constituents (5) for UV254 absorbance. The decrease in UV254 (ΔUV254) of each sample was determined by the difference of two measurements, one before chlorination, the other after the 24-hour reaction period. At least 20% of all samples taken for chlorine residual and UV254 measurements were analyzed concurrently as duplicate samples.

The remaining vials were used to determine the DBPs formed in chlorination using USEPA Method 551.1, USEPA Method 552.2 with a Perkin Elmer AutoSystem Gas Chromatograph (GC) equipped with an electron capture detector (ECD). Analytical separation was achieved with a 0.53 mm i.d., 30-meter fused silica capillary column (DB-5ms) with 1.5 μm film thickness (J & W Scientific). Selected samples were analyzed using GC/mass spectrometers (GC/MS) for confirmation. The QA/QC procedures described in the respective methods were followed.

## Results and Discussion of Structural Effects

The DBP formation results for chlorination under the aforementioned chlorination conditions are summarized in Table III, where total THMs and total HAAs are abbreviated as THM4 and HAA9, respectively. When pH was increased from 7.0 to 8.5, THM4 yield increased and HAA9 yield decreased. Although an increase in the initial bromide ion concentration changed the THM4 and HAA9 speciation, the THM4 and HAA9 concentrations were not appreciably affected with an increase in initial bromide ion concentration except for the NLS and SFR humic solutions at pH 8.5.

**Table III. Disinfection by-product formation as a function of bromide ion concentration and pH**

| Sites | THM4 ($\mu g/L$) | | | | HAA9 ($\mu g/L$) | | | |
|---|---|---|---|---|---|---|---|---|
| | Br⁻ = 0.0 $\mu g/L$ | | Br⁻ = 250 $\mu g/L$ | | Br⁻ = 0.0 $\mu g/L$ | | Br⁻ = 250 $\mu g/L$ | |
| | pH 7.0 | pH 8.5 | pH 7.0 | pH 8.5 | pH 7.0 | pH 8.5 | pH 7.0 | pH 8.5 |
| Aldrich | 441 | 556 | 447 | 565 | 520 | 414 | 606 | 419 |
| NLS | 264 | 354 | 218 | 361 | 371 | 201 | 315 | 197 |
| OHS | 369 | 426 | 361 | 445 | 454 | 339 | 474 | 213 |
| SFR | 397 | 420 | 368 | 436 | 455 | 331 | 472 | 254 |
| St MR | 374 | 476 | 407 | 472 | 512 | 374 | 571 | 366 |
| OH DOC | 392 | 464 | 413 | 502 | 483 | 370 | 597 | 345 |
| NL DOC | 374 | 442 | 359 | 480 | 407 | 205 | 459 | 216 |

The $\Delta UV254$ measurements associated with the DBP formation experiments are listed in Table IV.

Correlation of the observed DBP yields with the structural characteristics of elemental analysis, UV/VIS spectroscopy, and [13]C-NMR are discussed in the following sections.

### Elemental Analysis (C/H)

The activated aromatic structures such as phenolics have been reported to be especially reactive with chlorine producing large amounts of chlorinated by-products from model compound studies (8-10) and from humic substances (3, 11). The C/H ratio has been used to indicate the degree of aromaticity (a large value) or aliphaticity (a small value) of humic substances (12, 13). However, the use of C/H ratio has not been reported in water treatment settings with respect to DBP formation. Therefore the relationship between C/H ratio and the degree of aromaticity of humic substances was investigated in terms of DBP formation.

**Table IV. UV254 measurements and ΔUV254 values of the humic substances**

| Source | UV254 (cm⁻¹) | | ΔUV254 (cm⁻¹) | | | |
|---|---|---|---|---|---|---|
| | Before Chlorination | | 24 h After Chlorination Br⁻ = 0 µg/L | | 24 h After Chlorination Br⁻ = 250 µg/L | |
| | pH 7 | pH 8.5 | pH 7 | pH 8.5 | pH 7 | pH 8.5 |
| Aldrich | 0.445 | 0.452 | 0.145 | 0.136 | 0.146 | 0.133 |
| NLS | 0.195 | 0.198 | 0.051 | 0.040 | 0.050 | 0.041 |
| OHS | 0.364 | 0.369 | 0.076 | 0.067 | 0.075 | 0.060 |
| SFR | 0.246 | 0.255 | 0.068 | 0.065 | 0.068 | 0.067 |
| St MR | 0.262 | 0.265 | 0.082 | 0.072 | 0.087 | 0.076 |
| OH DOC | 0.263 | 0.266 | 0.078 | 0.068 | 0.083 | 0.073 |
| NL DOC | 0.240 | 0.241 | 0.059 | 0.041 | 0.059 | 0.053 |

The correlation of THM and HAA formation with C/H ratio is presented in Figure 1 and Table V, with the results of Br⁻ concentration of zero and 250 µg/L pooled together. All the correlations were remarkably good (significant at $\alpha = 0.005$) considering the qualitative relationship between C/H ratio and aromaticity and the wide variation in source characteristics.

**Elemental Analysis (C/N)**

The correlation of the C/N ratio and DBP formation was also significant ($\alpha = 0.005$). Amino acids, the building blocks of proteins, make up the largest known reservoir of organic nitrogen in most living organisms; amino sugars are important components of microbial cell walls, and nucleic acids are ubiquitous in living cells. Humic substances originate, in part, from some pool of these materials. Therefore, amino acids, amino groups, amino sugars, and nucleic acid derivatives usually account for more than 95% of the identifiable organic nitrogen in humic substances in soils, sediments, and water (14). Since most of the microbial sources are somewhat aliphatic in nature, humic substances derived from algae tend to contain less aromaticity, and accordingly, have lower C/H values. On the other hand, humic substances from streams and lakes in watersheds with lignin (aromatic organic polymer)-rich sources, such as the Suwannee River, are more aromatic in nature with higher C/H values. McKnight et al. (15), as well as Aiken and Cotsaris (16), in comparing the ratio of aromatic C/alphatic C

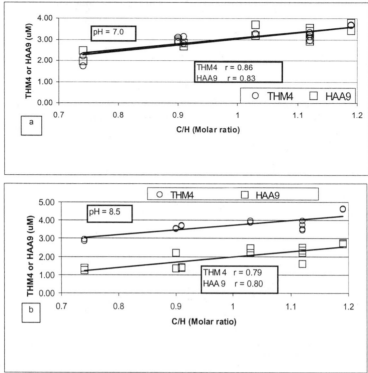

*Figure 1. Relationship between C/H ratio and DBP formation. (a) pH = 7.0*
*(b) pH= 8.5 All bromide data pooled*

content and the C/N ratio of aquatic fulvic acids, found that algal-derived fulvic acids clustered at one end of the range, having low values for both ratios while fulvic acids from lignin-rich sources clustered at the other end of the range, having high values for the same ratios, with intermediates between these extremes representing a combination of source diversity. In light of the above findings, it may be reasonable to assume that the C/N ratio is related to aromaticity and therefore DBP formation. The relationship between C/H and C/N gave a correlation coefficient of 0.76, but correlation with UV254 provided poorer results. On the other hand, the correlation between C/N ratio and THM as well as HAA formation gave high correlation coefficients generally comparable to those between C/H ratio and DBP formation (Table V).

**Table V. Correlation coefficients of DBP yields with humic structural
characteristics (n = 14; p ≤ 0.005 unless otherwise noted)**

(a) pH 7.0.

| Correlation Coefficients (r) | C/H | C/N | UV254 (cm⁻¹) | ΔUV254 (cm⁻¹) | $^{13}$C-NMR 110-160 ppm | $^{13}$C-NMR 145-160 ppm |
|---|---|---|---|---|---|---|
| THM4 (μM) | 0.86 | 0.86 | 0.67 | 0.74 | 0.79 | 0.79 |
| HAA 9 (μM) | 0.83 | 0.89 | 0.57[a] | 0.74 | 0.74 | 0.79 |

(b) pH 8.5.

| Correlation Coefficients (r) | C/H | C/N | UV254 (cm⁻¹) | ΔUV254 (cm⁻¹) | $^{13}$C-NMR 110-160 ppm | $^{13}$C-NMR 145-160 ppm |
|---|---|---|---|---|---|---|
| THM4 (μM) | 0.79 | 0.79 | 0.76 | 0.88 | 0.94 | 0.92 |
| HAA9 (μM) | 0.80 | 0.73 | 0.59[a] | 0.80 | 0.71 | 0.70 |

[a] p = 0.025.

**Total Acidity**

The C/H ratio has often been used to indicate the degree of aromaticity as discussed previously; however, this reasoning has be questioned because it does not consider the contribution of carbon-to-oxygen double bonds to the C/H ratio (both COOH and -HC=CH- have the same C/H ratio). The presence of a high concentration of COOH groups would adversely affect the use of C/H as a measure of -HC=CH- (13, 17, 18). To help clarify the relationship between aromaticity (as expressed as C/H) and DBP formation, a total acidity titration value was employed to calculate the contribution of COOH groups to the C/H ratio.

The acidic nature of humic substances has been attributed to both COOH and acidic OH groups (presumed to be phenolic OH), with COOH being the most predominant (19). Assuming total acidity is entirely attributable to COOH, a new C/H ratio with exclusion of COOH group can be derived by solving the following two equations with all terms expressed in units of mmole/L.

$$C_{rest} + C_{COOH} = C = (mg/L \text{ C in sample})/(mol. \text{ wt. C}) = 5/12 \quad (1)$$

$$H/C = (H_{rest} + H_{COOH})/C = (H_{rest} + H_{COOH})/(5/12) \quad (2)$$

Where $C_{COOH}$ is the amount of carbon existing in the COOH group; and $C_{rest}$ represents the total carbon with exclusion of $C_{COOH}$. Similarly, $H_{COOH}$ is the amount of hydrogen existing in the COOH group; and $H_{rest}$ represents the total hydrogen with exclusion of

$H_{COOH}$. Recall that the NPOC in solutions containing the studied humic substances was 5 mg/L, and 5/12 represents the carbon concentration expressed in mmole/L.

There is an equal number of carbon and hydrogen atoms in the COOH group; therefore, the eq. 1 can be rearranged to:

$$C_{rest} = 5/12 - H_{COOH} \tag{3}$$

By rearranging eq. 2, $H_{rest}$ can be determined:

$$H_{rest} = H/C * 5/12 - H_{COOH} \tag{4}$$

Assuming the extreme case of total acidity consisting of only the COOH group contribution, $H_{COOH}$ can be derived from the total acidity reported in Table I.

Because the difference between C/H ratio and $C_{rest}/H_{rest}$ ratio was very small (Table I), the contribution of the COOH group to the C/H ratio was not as significant in this study as reported in previously cited research (*13, 17, 18*). Note that if acidity consists of components other than COOH, the difference between the C/H and the $C_{rest}/H_{rest}$ ratios is even smaller.

## UV/VIS

UV254 is often a good surrogate parameter for DBP formation potential (*2, 20*). The results in Table V associated with UV254 illustrate the correlation between THM and HAA formation and UV254. The regression results were not as impressive as the correlations reported by Edzwald et al. (*2*) (THMs vs. UV254 for raw waters, r = 0.91-0.97) as well as Singer and Chang (*20*) (THMs vs. UV254 for raw waters, r = 0.89). However, the correlations reported in this research are certainly consistent with the claim that THM formation correlates well with UV254, and the regressions reported here were all significant at $\alpha = 0.025$. The inferior regression coefficients may be attributed to diversity of source (aqueous and solid) and the limited data (n = 14) included in this research compared to the two aforementioned studies (n = 42-54 by Edzwald et al.; n = 42 by Singer and Chang).

Initial UV254 measured before disinfection may only characterize the possible amount (viewed as potential) of activated aromatic rings in a particular organic matrix to form DBPs; but it does not represent the actual estimation of activated aromatic ring destruction which leads to DBP formation (viewed as reactivity). DBP formation can also be correlated with the change in UV254 absorption ($\Delta$UV254) induced by chlorination representing the extent of the destruction of activated aromatic rings. Korshin et al. (*21*) reported a strong correlation (r = 0.97) between chloroform formation and the $\Delta$UV254 after chlorination of a reservoir water for a range of reaction times, DOC concentrations, and chlorine doses. The same research group later reported that the change in UV272 absorption and $\Delta$UV254 correlated linearly with the amount of TOX formed, for a range of water quality conditions and reaction times (*22, 23*). Table V lists the results of THM and HAA formation versus $\Delta$UV254; this approach consistently gives better correlations compared to those associated with UV254.

## $^{13}$C-NMR

The chemical shift assignments in the $^{13}$C-NMR spectra of the humic substances listed in Table 2 were based on those used by Harrington et al. (*11*). The correlations of THM and HAA formation as well as chlorine consumption with aromatic carbon content and phenolic carbon content from $^{13}$C-NMR data are presented in Table V. Correlations of aromatic and phenolic character were substantially better for THMs than for HAAs. All the correlations were significant at $\alpha = 0.005$.

THM and HAA formation have been reported to be better correlated with phenolic carbon content than aromatic carbon content (*3, 11*); however, the research described in this paper does not support this finding as indicated by the similar correlation coefficients for aromaticity and phenolic carbon shown in Table V. This inconsistency may be addressed in two ways. First, although the absorption in the $^{13}$C-NMR 145 to 160 ppm region is associated with phenolic content, aromatic carbon substituted by oxygen and nitrogen, which may not be reactive in DBP formation, also can cause absorption in this band (*24*). Leenheer et al. proposed three structures of humic molecules from the Suwannee River, all with aromatic rings substituted with oxygen, not OH (*25*). Second, when small concentrations of chlorine react with humic substances, chlorine reacts with the most reactive sites (such as those associated with phenolic content). However, when chlorine concentrations are higher, as they were in this research and in many water treatment systems, chlorine remaining, after satisfying the demand of reactive sites associated phenolic content, will likely react with the humic aromatic structures which are substituted by a variety of other functional groups to produce DBPs. Accordingly, with higher chlorine doses, aromaticity may be as well correlated with DBP formation as phenolic character.

### Aromaticity

Aromaticity of humic substances is an important chemical property that can be used to explain formation, source, and potential interactions of the humic substances with pesticides and other contaminant organics, in addition to DBP formation. This important structural characteristic was quantified by a number of methods: elemental analysis, UV/VIS spectroscopy, and $^{13}$C-NMR. These methods were compared to $^{13}$C-NMR by correlational analysis. The results of this analysis (the correlation coefficients) are provided in Table VI. Either UV254 or UV272 absorptivity correlates well with the percent aromatic carbon ($r = 0.74$ and $0.73$, respectively). Thus, determinations of UV absorptivity at either wavelength provide an equal estimate of the aromatic carbon content in the humic substances studied. A poorer correlation ($r = 0.65$) was found for aromatic carbon and the C/H ratio, but better correlation was observed between percent aromatic carbon and the C/N ratio ($r = 0.74$). Only a slightly better correlation ($0.66$) was found between $C_{rest}/H_{rest}$ ratio and the percent aromatic carbon.

**Table VI. Correlation coefficients (r) from correlational analysis of specified methods with ¹³C-NMR (110-160 ppm) for determining aromatic carbon contents (n = 7; p = 0.05 unless otherwise noted)**

| Correlation Coefficients (r) | C/H | $C_{rest}/H_{rest}$ | C/N | UV254 (cm⁻¹) | UV272 (cm⁻¹) |
|---|---|---|---|---|---|
| ¹³C-NMR Aromatic C | 0.65[a] | 0.66 | 0.74 | 0.74 | 0.73 |

[a] p= 0.1.

The results in this research suggest that aromaticity is a good surrogate for predicting the reactivity of humic substances with chlorine and DBP formation. The ¹³C-NMR spectroscopy provides aromatic content information with definitive nature, but it requires sample sizes ≥ 100 mg with 6-12 h acquisition time (26). However, the amount of sample required for other characterization analyses is considerably smaller than that required with NMR spectrometers. Elemental analysis and ash content can be accomplished on 20 mg, and titration analysis for carboxyl and phenolic functional groups on 10 mg, of sample (27). UV/VIS analysis can be performed on aqueous samples with less than a mg of humic material. This small sample size is an advantage since the amount of extracted humic substances usually is limited, and UV/VIS analysis can be performed without extraction of the humic substance from aqueous solution. Therefore, from a practical perspective, techniques other than NMR spectrometry may be more attractive alternatives. This is particularly true if a qualitative understanding of DBP formation is sufficient. Therefore, in the interest of time and cost, it is advantageous to be able to obtain aromaticity data by simple methods that are independent of NMR spectroscopy.

**Acknowledgments**

The authors thank Dr. William T. Cooper and Jennifer Llewelyn, Department of Chemistry, Florida State University, for providing the NMR analyses.

**Literature Cited**

1. Krasner, S.W.; McGuire, M.J.; Jacangelo, J.G.; Patania, N.L.; Reagan, K.M.; Aieta, E.M. *Jour. Amer. Water Works Assoc.* **1989**, *81*, 41-53.

2. Edzwald, J.K.; Becker, W.C.; Wattier, K.L. *Jour. Amer. Water Works Assoc.* **1985**, *77*, 122-132.

3. Reckhow, D.A.; Singer, P.C.; Malcolm, R.L. *Envir. Sci. & Technol.* **1990**, *24*, 1655-1664.

4. Davis, W. M. Ph.D thesis, University of Florida, Gainesville, FL, 1993.

5. *Standard Methods for Examination of Water and Wastewater*, Clesceri, L.S.; Greenberg, A.E.; Trussell, R.R., Eds.; 19th Ed., American Public Health Assoc.: Washington, DC, 1995.

6. Perdue, E.M. In *Humic Substances in Soil, Sediment, and Water: Geochemistry, Isolation, and Characterization*; Aiken, G.R.; McKnight, D.M.; Wershaw, R.L.; MacCarthy, P., Eds.; John Wiley & Sons, Inc., New York, NY, 1985, pp 493-526.

7. Amy, G.L.; Siddiqui, M.; Zhai, W.; DeBroux, J.; Odem, W. *Survey of Bromide in Drinking Water and Impacts on DBP Formation*, Amer. Water Works Reseach Found., Denver, CO. 1994; pp 19-35

8. Norwood, D.L.; Johnson, J.D.; Christman, R.F; Hass, J.R.; Bobenrieth, M.J. *Envir. Sci. & Technol.* **1990**, *14*, 187-190.

9. De Laat, J.; Merlet, N.; Dore, M. *Wat. Res.* **1982**, *16*, 1437-1450

10. Reckhow, D.A.; Singer, P.C. In *Water Chlorination: Chemistry, Environmental Impact and Health Effects*; Jolley, R.L., Ed.; Vol. 5, Lewis Publ., Chelsea, MI, 1985, pp 1229-1257.

11. Harrington, G.W.; Bruchet, A.; Rybacki, D.; Singer, P.C. In *Water Disinfection and Natural Organic Matter: Characterization and Control*; Minear, R.A.; Amy, G.L., Eds.; American Chemical Society, New York, NY, 1996; pp 138-158.

12. Thurman, E.M. *Organic Geochemistry of Natural Waters*; Kluwer Academic Publ., Hingham, MA (1985).

13. Rice, J.A.; MacCarthy, P. *Org. Geochem.* **1991**, *17*, 635-648.

14. Anderson, H.A.; Bick, W.; Hepburn, A.; Stewart, M. In *Humic Substances II*; Hayes, M.H.B., MacCarthy, P., Malcolm, R.L. Swift, R.S., Eds.; John Wiley & Sons, Inc., New York, NY, 1989; pp. 223-253.

15. McKnight, D.M.; Andrews, E.D.; Spaulding, S.A.; Aiken, G.R. *Limnol. Oceanography* **1994**, *39*, 1972-1979.

16. Aiken, G.R.; Cotsaris, E. *Jour. AWWA*, **1995**, *87*, 36-45.

17. Perdue, E.M. *Geochim. Cosmochim. Acta* **1984**, *48*, 1435-1442.

18. Gauthier, T.D.; Seitz, W.R.; Grant, C.L. *Envir. Sci. & Technol.* **1987**, *21*, 243-248.

19. Stevenson, F.J. In *Humic Substances in Soil, Sediment, and Water: Geochemistry, Isolation, and Characterization;* Aiken, G.R., McKnight, D.M., Wershaw, R.L. & MacCarthy, P., Eds.; John Wiley & Sons, Inc.: New York, NY, 1985; pp. 13-52.

20. Singer, P.C.; Chang, S.D. *Jour. Amer. Water Works* **1989**, *81*, 61-65.

21. Korshin, G.V.; Li, C.W.;Benjamin, M.M. In *Water Disinfection and Natural Organic Matter: Characterization and Control;* Minear, R.A.; Amy, G.L., Eds., American Chemical Society, New York, NY, 1996; pp. 182-195.

22. Korshin, G.V.; Li, C.W.; Benjamin, M.M. *Water Res.* **1997**, *31*, 946-949

23. Korshin, G.V.; Li, C.W.; Benjamin, M.M. *Water Res.* **1997**, *31*, 1787-1795.

24. Malcolm, R.L. In *Humic Substances II*; Hayes, M.H.B., MacCarthy, P., Malcolm, R.L. Swift, R.S., Eds.; John Wiley & Sons, Inc., New York, NY, 1989; pp. 339-372.

25. Leenheer, J.A.; Wershaw, R.L.; Reddy, M.M. *Envir. Sci. & Technol.* **1995**, *29*, 399-405.

26. Stevenson, F.J. *Humus Chemistry: Genesis, Composition, Reactions*, 2nd ed., John Wiley & Sons, Inc., New York, NY, 1994.

27. Malcolm, R.L. In *Humic Substances in the Aquatic and Terrestrial Environment*; Allard, B. Ed.; Springer-Verlag Publ., Berlin Heidelberg, Germany, 1991; pp. 9-36.

Chapter 9

# Natural Organic Matter Characterization of Clarified Waters Subjected to Advanced Bench-Scale Treatment Processes

Eric R. V. Dickenson and Gary L. Amy

Department of Civil, Environmental, and Architectural Engineering, University of Colorado, Boulder, CO 80309-0421

Natural organic matter (NOM) from clarified-treated waters was removed or transformed via bench-scale activated carbon adsorption, an ultrafiltration membrane and ozonation/ biodegradation. Based on the results of NOM characterizations such as $SUVA_{254}$, molecular weight distribution by size exclusion chromatography (SEC), and XAD-8/4 resin adsorption chromatography, it was shown that activated carbon adsorption preferentially removed nonpolar dissolved organic carbon (DOC), ultrafiltration with a Desal GH membrane removed both polar and nonpolar DOC and biotreatment preferentially removed polar DOC that did not absorb UV light at 254 nm. For clarified Seine River water, PAC adsorbs nonpolar DOC compounds that are likely to form both THMs and HAAs, the Desal GH membrane did not preferentially remove DOC compounds that are likely to form THMs and HAAs, ozonation transforms DOC compounds that are likely to form THMs and biotreatment removes polar DOC compounds that are likely to form HAAs.

## Introduction

In water treatment, natural organic matter (NOM) is of concern because it serves as a precursor to the formation of disinfection by-products (DBPs), some of which have serious health effects (*1*). The NOM typically found in water supplies used for drinking water consists of humic (nonpolar) substances, generally from terrestrial origin, and nonhumic (polar) substances of biological origin (*2*). Note "nonpolar" refers to humic substances or hydrophobic NOM even though these substances have polar character. Analytical characterization of both the physical and chemical properties (structure, functionality and size) of NOM in water is central toward

122

understanding organic matter's fate and reactivity. The nonpolar NOM has been well characterized and is more readily removed by coagulation than polar NOM (*3*). By contrast, the polar NOM has not been well characterized and polar NOM and DBP relationships are not well understood. Utilities would like to be able to utilize advanced processes to optimize their treatment for the removal of polar and nonpolar NOM and thus the control of DBP formation in drinking water.

## Materials and Methods

### Characterization Tests

NOM characterization (specific ultraviolet absorption [SUVA], molecular weight [MW] distribution and XAD-8/4 resin adsorption chromatography) has been determined for clarified Colorado River (CRW), Seine River (SRW) and Sacramento-San Joaquin Delta (SDW) that have been subjected to advanced bench-scale treatment processes. The ability to remove (or transform) NOM has been investigated via activated carbon adsorption, membrane filtration, and ozonation/biodegradation,

NOM properties that were measured include SUVA, molecular weight as determined by size exclusion chromatography (SEC), and NOM fractionation into hydrophobic, transphilic and hydrophilic fractions as carbon. Note all samples were filtered through a 0.45 μm filter before any tests were performed, representing an operational definition of "dissolved" organic carbon (DOC). Conditions of the analyses are described in the following sections.

SUVA was determined by (1) measuring DOC of water samples using a Sievers 800 Total Organic Carbon (TOC) Analyzer and (2) measuring UV absorbance at 254 nm ($UVA_{254}$) using a Shimadzu UV-160U UV-Visible Scanning Spectrophotometer. The Sievers 800 is very accurate at low concentration ranges and utilizes a persulfate-ultraviolet oxidation method to determine TOC/DOC. The calibration curve for the TOC Analyzer was developed using potassium hydrogenphthalate as the standard and the standard concentrations ranged between 0 and 10 mg/L of DOC. All samples were acidified to a pH of 2 before analysis. $UVA_{254}$ measurement was performed with a 1 cm quartz cell, detecting only the presence of aliphatic alkenes and aromatic compounds.

Molecular weight as determined by SEC was measured using a high performance liquid chromatograph (HPLC) with UVA detection, a sodium phosphate buffer eluant (pH = 6.8), a Water Proteins Pak 125 column, and a 20 μL sample loop. The Water Proteins Pak column is a porous media with low residual hydrophobicity and minimal ion-exchange capacity (*4*). A calibration curve was developed with varying molecular weight polystyrene sulfonates (PSS) and acetone (MW = 58 g/mol) at a wavelength of 224 nm. Molecular weights of these standards ranged from 58 to 8,000 daltons. The PSS standards have been used to measure the molecular weight of Aldrich humic acid, which agrees with other literature values measured by other techniques (e.g. ultracentrifugation) (4). By integrating the resulting chromatogram from samples detected at 254 nm, the number-average molecular weight (Mn) and weight-average molecular weight (Mw; apparent molecular weight [AMW]) can be determined from the following equations (5):

$$Mn = \frac{\sum n_i M_i}{\sum n_i} \qquad\qquad Mw = \frac{\sum n_i M_i^2}{\sum n_i M_i}$$

$i =$     incrementing index over all molecular weights present
$M_i =$   molecular weight of $i$th molecule
$n_i =$   number of molecules of molecular weight $M_i$
$n_i M_i =$  weight of molecules of molecular weight $M_i$

These equations are based on a Gaussian distribution. The number-average molecular weight depends upon the number of particles in a test solution and in the case of a polydisperse system, Mn emphasizes the lower molecular weights. The weight-average molecular weight depends on the weight of molecules and favors the heavier-molecular-weight species of a mixture, resulting in higher-molecular-weight values than Mn. Note SEC uses a $UVA_{254}$ detector to measure the aliphatic alkenes and aromatic compounds, generally higher molecular weight fractions of NOM. DOC substances that do not absorb UV at 254 nm are not detected, generally lower molecular weight fractions.

Fractionation by XAD resin adsorption chromatography was used to determine the DOC distribution of operationally defined hydrophobic, transphilic and hydrophilic DOC fractions. Two sequential columns containing XAD-8 and XAD-4 resins operated at k' = 50 were used to isolate (adsorb) hydrophobic and transphilic DOC, respectively (6). Pre-acidified (pH 2) samples were first passed through a column containing XAD-8 resin at a flow rate of 2 mL/min, and then subsequently passed through an additional column containing XAD-4 resin at a flow rate of 1.5 mL/min. DOC measurements of influents and effluents of columns were used to perform a carbon mass balance, which yielded hydrophobic, transphilic and hydrophilic NOM fractions.

Note SUVA and molecular weight distributions are based on UV measurements, which does not take in account the entire DOC, and the distribution of NOM among hydrophobic, transphilic and hydrophilic groups is based on DOC measurements. These differing measurements may have an influence on the interpretation of results.

**Bench-Scale Experiments**

Advanced bench-scale treatment processes included activated carbon adsorption, membrane filtration, and ozonation/biodegradation.

DOC characterization was performed on clarified CRW, SRW and SDW subjected to powdered activated carbon (PAC) treatment. The PAC treatment involved the formation of equilibrium adsorption isotherms, followed by batch-mode experiments. The adsorbent used was Calgon's Filtrasorb 400 powdered activated carbon, which was granular activated carbon that was pulverized. Activated carbon was studied in powdered form to enhance kinetics and the ability to collect large quantities of treated water for characterization and DBP tests. Erlenmeyer flasks (125 mL) containing various amounts of carbon ranging from 0 to 32 mg were filled with 100 mL of clarified water sample. Next, the bottles were capped and placed in a rotational mechanical shaker for at least two days. The isotherms were performed at room temperature ($\approx 23°C$). The supernatent was then carefully decanted and filtered through a 0.45 µm filter and then the DOC was measured using the Sievers 800 Total Organic Carbon Analyzer. Based on the isotherms, batch-mode experiments then were performed on clarified water samples to produce larger volumes of PAC-treated waters corresponding to approximately 25%, 50% and 75% DOC removals; PAC and clarified water sample were mixed in 1 liter amber bottles on a rotational mechanical mixer for at least two days. Note, different doses of PAC were observed to achieve the same removal of DOC for clarified CRW, SRW and SDW. The 25%, 50% and 75% DOC removals approximate the initial, middle and later stages of a breakthrough curve for the adsorbent used in a column mode of operation.

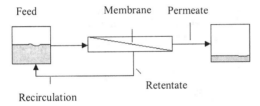

*Figure 1. Schematic of membrane apparatus.*

DOC characterization was performed on membrane-treated clarified CRW, SRW and SDW. These waters were tested with a rapid small-scale membrane test (RSSMT) using a *tight* ultrafiltration (UF) membrane. This membrane module is an Osmonics SEPA cell, which is a cross-flow unit operated with a flat-sheet specimen. A Desal GH membrane was used, which is a negatively charged UF membrane with a molecular weight cut off (MWCO) of 2,500. This membrane elucidates both size (steric) and charge (electrostatic) effects on NOM rejection. Selection of this membrane is purely pragmatic, since this membrane provides about 50% DOC removal under appropriate operating conditions, which allows enough DOC in the permeate to undergo characterization and DBP tests. The pressure, temperature, feed flow rate and permeate flow rate were monitored during the experiment. Table I lists the operating conditions of the experiment. Six liters of clarified CRW and ten liters each of clarified SRW and SDW were processes through the membrane system for approximately 4 days. The permeate was collected and the retentate recirculated back into the feed reservoir (Figure 1). The permeate volume was smaller than the final-feed recirculation volume.

## Table I. Average membrane operating conditions of experiments

| | | | Clarified Water | |
|---|---|---|---|---|
| Operating Condition | Unit | CRW | SRW | SDW |
| Pressure | psi | 15 | 15 | 15 |
| Temperature | °C | 24 | 23 | 23 |
| Permeate flow | mL/min | 0.5 | 0.5 | 0.3 |
| Feed flow | mL/min | 400 | 390 | 400 |
| Recovery | % | 0.13 | 0.13 | 0.08 |
| $UVA_{254}$ of permeate | $m^{-1}$ | 1.9 | 1.0 | 0.8 |
| Initial feed volume | L | 6 | 10 | 10 |
| Final permeate volume | L | 2.5-3 | 2.5-3 | 2.5-3 |

DOC characterization was performed on biotreated clarified CRW, SRW and SDW. Sample waters were first ozonated under ambient pH and a 1 mg $O_3$/mg DOC transferred dose and then biotreated for 24 hr. An OREC O3V5-0 ozone generator was used to produce ozone stock solution concentrations that ranged from 25-32 mg/L of $O_3$. The experimental setup, a true batch reactor, entailed the use of a 1 L graduated cylinder with a Teflon plunger, and ozonated samples were mixed until complete utilization of ozone (Figure 2).

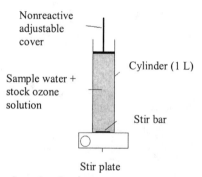

Nonreactive adjustable cover

Cylinder (1 L)

Sample water + stock ozone solution

Stir bar

Stir plate

*Figure 2. Schematic of true-batch reactor.*

The dilution of sample waters by the ozone stock solution ranged from 4 to 9%. Ozone was measured using the Hach indigo method to ensure completion of reaction. Ozonated samples were placed into biodegradable dissolved organic carbon (BDOC) reactors (recirculating biofilm acclimated sand type) (7). A vacuum was used to recirculate ($\cong$ 15 mL/min) and aerate the water sample (Figure 3). The BDOC apparatus has all glass, Teflon and stainless steel parts. The acclimated sand is rinsed three times with fresh sample (last rinse involves recirculation). The biotreated sample reflects the presence of residual non-biodegradable DOC, as well as the possible presence of soluble microbial products (SMPs). BDOC samples, after 24 hours of biotreatment, were filtered through a 0.45 μm filter for DOC analysis.

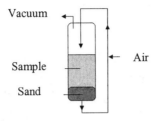

Vacuum

Sample

Sand

Air

*Figure 3. Recirculating biofilm acclimated sand type reactor.*

# Results

## Activated Carbon Treatment

Table II shows the actual DOC removal percentages obtained after the addition of PAC to clarified CRW, SRW and SDW waters in batch mode experiments. These low, mid and high DOC removals were close to the 25%, 50%, and 75% DOC removals. Figure 4 shows the clarified waters exhibited, in general, low SUVA values. Clarified SDW had the highest SUVA value of 1.8 L/(mg-m), while clarified CRW had the lowest SUVA value of 1.4 L/(mg-m). As more PAC is added, the SUVA values decreased. As the SUVA decreases, the degree of unsaturation and aromaticity decreases. Figure 5 shows that hydrophobic, transphilic and hydrophilic DOC were all removed after the addition of PAC. However, the hydrophobic DOC fraction was preferentially removed over the transphilic and hydrophilic DOC fractions for the three water samples. After the removal of about 75% of the DOC by the addition of PAC, nearly all the hydrophobic DOC adsorbed to the activated carbon. Also, the transphilic DOC fractions were preferentially removed over the hydrophilic DOC fractions.

**Table II. Percentages of DOC removals by PAC**

| Clarified Water | DOC Removals (%) | | |
|---|---|---|---|
| | Low | Mid | High |
| Colorado River | 22 | 53 | 76 |
| Seine River | 26 | 51 | 78 |
| Sacramento-San Joaquin Delta | 22 | 47 | 71 |

Note: Ten percent of $SUVA_{254}$ values were triplicated. Standard deviations for $SUVA_{254}$ values were 0.2 L/(mg-m).

128

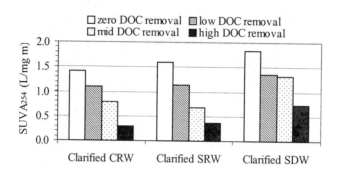

Figure 4. SUVA for PAC treated clarified waters.

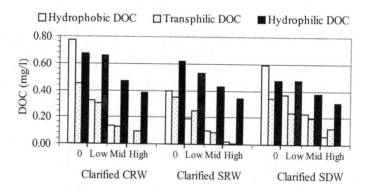

Figure 5. XAD 8/4 DOC profiles for PAC treated clarified waters.

The weight- and number-average molecular weight values decrease as more DOC was removed by the increased addition of PAC for clarified CRW and SDW (Figure 6). Intermediate molecular weight compounds are preferentially removed rather than higher molecular weight molecules. Mw and Mn decreased for these two waters because the compounds with the lowest molecular weights were poorly removed. However, for clarified SRW the weight and number-average molecular weight values increased as more DOC was removed. Note the SEC chromatograms in Figure 6 are nonGaussian, so the Mn and Mw results were calculated from a Gaussian model that may not be appropriate for NOM. Also, the SEC chromatograms for SDW that had been treated with activated carbon show a peak in the 58-250 Dalton range, a feature that was not observed for the clarified water. This peak may bias the results (especially Mn) that were obtained following treatment with activated carbon. The coefficient of variance (CV) for the SEC analysis is approximately 10%. The CV was based upon reproducibility of data when 10% of the samples were analyzed in triplicate.

Note, not only hydrophobicity, but also molecular size exclusion is an important factor in the preferential adsorption of NOM on activated carbon. The more hydrophobic a compound, the higher the adsorbability onto activated carbon, whereas the hydrophilic (more polar) compounds prefer to remain in solution. The hydrophobic DOC, which is generally larger molecular weight compounds, will adsorb more readily than smaller molecular weight compounds (hydrophilic-polar NOM). This is in agreement with the hydrophobic DOC fractions and MW trends presented. As more hydrophobic DOC is removed by activated carbon adsorption, larger molecular weight compounds are removed, except for the clarified SRW sample. If molecular weight distributions were based on DOC measurements and DOC fractions not being quantified by UV measurements are relatively more hydrophilic (polar) and have lower molecular weights, then the predicted MW trends would be similar to MW trends presented here. The adsorption of large MW hydrophobic compounds, however, maybe inhibited by size exclusion, whereby large compounds are unable to reach adsorption sites within the micropores of the activated carbon.

As more PAC was added, the SUVA values decreased for all three waters. Also, the hydrophobic DOC was preferentially removed in all three waters. Lastly, larger molecular weight DOC compounds were preferentially removed for clarified SDW and CRW. Powdered activated carbon adsorption has a propensity to preferentially remove nonpolar DOC.

**Membrane Treatment**

Table III shows the actual DOC removal percentages obtained after membrane treatment of clarified CRW, SRW and SDW in RSSMT experiments. The DOC removals were between 53 and 59%, which is near the targeted goal of 50% DOC removal. A 50 % DOC removal means that the permeate DOC concentration is half the untreated DOC concentration. Figure 7 indicates there is no preferential removal of unsaturated and aromatic carbon compounds for clarified CRW and SRW.

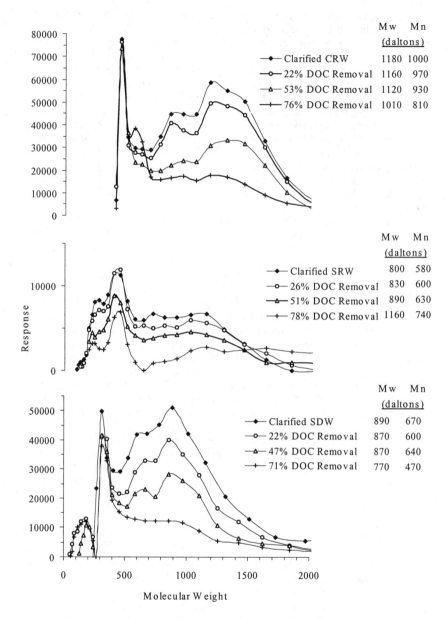

*Figure 6. SEC chromatographs for PAC treated clarified waters.*

**Table III. Percentages of DOC removals by membrane treatment**

| Clarified Water | DOC Removals (%) |
|---|---|
| Colorado River | 53 |
| Seine River | 57 |
| Sacramento-San Joaquin Delta | 59 |

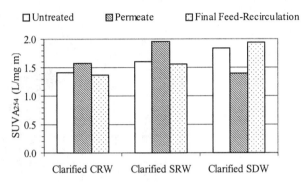

Figure 7. SUVA for membrane treated clarified waters.

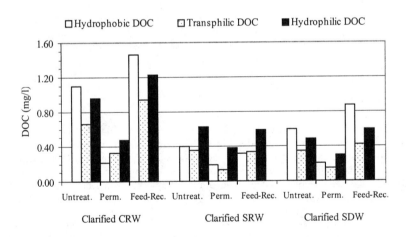

Figure 8. XAD 8/4 DOC profiles for membrane treated clarified waters.

The SUVA values for these waters are higher in the permeate than the untreated clarified (feed water) samples. However, for clarified SDW, the SUVA value is lower in the permeate than in the feed, suggesting a preferential removal of unsaturated and aromatic carbon compounds. The $UVA_{254}$ values of the membrane permeates were lower than the untreated clarified waters.

For clarified CRW and SDW the hydrophobic DOC was preferentially removed after membrane treatment (Figure 8). However, for clarified SRW there is no preferential removal of the fractions; following membrane treatment, equal amounts of DOC were removed from each of the fractions.

The weight- and number-average molecular weight values decrease as the membrane removes DOC for clarified CRW, SRW and SDW (Figure 9). This indicates that higher molecular weight compounds are preferentially removed. UV absorbing moieties are of higher molecular weight, but SUVA did not decrease because DOC was preferentially removed over $UVA_{254}$ for clarified CRW and SRW. Utilization of a DOC detector may have observed the removal of low molecular weight compounds.

After membrane treatment SUVA values increased for clarified CRW and SRW, but decreased for clarified SDW. Also, the hydrophobic DOC was preferentially removed for clarified CRW and SDW, while there was no preferential removal for SRW. Lastly, larger molecular weight DOC compounds were preferentially removed for clarified CRW, SRW and SDW. There appears to be no clear trend between SUVA, XAD fractions and MW values for the three waters. This is maybe due to the membrane's ability to remove DOC by steric and electrostatic means. The membrane did remove considerable amounts of hydrophobic, transphilic and hydrophilic DOC. Generally, hydrophobic DOC has a higher SUVA than hydrophilic DOC. High molecular weight compounds in the hydrophobic acid fraction (nonpolar) that exhibit a high SUVA may be removed by the membrane due to steric hindrance and electrostatic repulsion and also low molecular weight compounds in the hydrophilic fraction (polar) that exhibit a low SUVA, may be removed by the membrane due to electrostatic repulsion. Thus, the Desal GH membrane removes both polar and nonpolar DOC.

### Ozonation/Biotreatment

Table IV shows the actual BDOC removal percentages obtained after 24 hour biotreatment of ozonated-clarified CRW, SRW and SDW. Figure 10 shows that for the three clarified waters, the SUVA values decrease after ozonation. This indicates destruction by oxidation of carbon/carbon double and aromatic bonds. However, after the ozonated samples undergo biotreatment, the SUVA values increased slightly. This increase is explained by the DOC decreases due to biotreatment, while the $UVA_{254}$ does not change after biotreatment. This trend is shown in Table V for each of the waters. Biotreatment preferentially removes single carbon/carbon bond DOC compounds that do not absorb UV light. Note in Figure 10, there is shown an overall decrease in SUVA values due to the combination of ozone-biotreatment.

After ozonation, the hydrophobic DOC fractions decreased (Figure 11). This agrees with SUVA values decreasing after ozonation, since generally unsaturated and

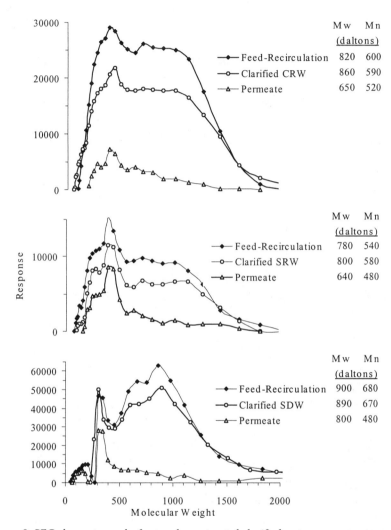

*Figure 9. SEC chromatographs for membrane treated clarified waters.*

aromatic carbon molecules tend to be hydrophobic DOC. After biotreatment the hydrophilic DOC fractions decreased. This fraction is the most likely to be removed by biotreatment. Note after biotreatment the hydrophobic DOC fraction increases for clarified SRW and SDW, which may indicate the presence (production) of soluble microbial products.

**Table IV. Percentages of DOC removals by biotreatment**

| Clarified Water | DOC Removals (%) |
|---|---|
| Colorado River | 18 |
| Seine River | 25 |
| Sacramento-San Joaquin Delta | 28 |

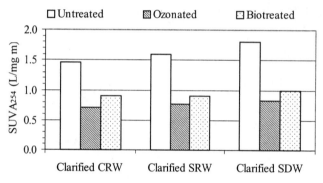

*Figure 10. SUVA for biotreated clarified waters*

**Table V.  DOC and UVA$_{254}$ data for biotreated clarified waters**

| | DOC (mg/L) Standard deviation = 0.02 | UVA$_{254}$ (1/m) Standard deviation = 0.2 | SUVA$_{254}$ (L/m-mg) Standard deviation = 0.2 |
|---|---|---|---|
| Clarified CRW | 2.47 | 3.6 | 1.5 |
| Ozonated Clarified CRW | 2.27 | 1.6 | 0.7 |
| Biotreated Clarified CRW | 1.87 | 1.7 | 0.9 |
| Clarified SRW | 1.19 | 1.9 | 1.6 |
| Ozonated Clarified SRW | 1.18 | 0.9 | 0.8 |
| Biotreated Clarified SRW | 0.89 | 0.8 | 0.9 |
| Clarified SDW | 1.50 | 2.7 | 1.8 |
| Ozonated Clarified SDW | 1.42 | 1.2 | 0.8 |
| Biotreated Clarified SDW | 1.02 | 1.0 | 1.0 |

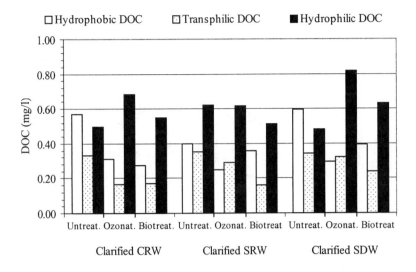

☐ Hydrophobic DOC  ☐ Transphilic DOC  ■ Hydrophilic DOC

*Figure 11. XAD 8/4 DOC profiles for biotreated clarified waters.*

The weight- and number-average molecular weight values decrease as DOC is transformed by oxidation via ozone for clarified CRW (Figure 12). However, for clarified SRW and SDW, the weight-average molecular weight values increased after ozonation, even though overall UV response decreased. Intermediate molecular weight compounds (~300-1000) are preferentially transformed and the compounds with higher molecular weights were essentially unaffected by ozonation for clarified SDW. For biotreated and ozonated CRW and SDW, the UV chromatograms (Figure 12) are nearly identical, even though biotreatment removed BDOC in both samples. This trend is consistent with the increase of SUVA values after biotreatment for all three waters. Bacteria consume low MW organics that do not absorb UV light, so a DOC detector, which has become recently available (8), is needed to determine the true change in MW distribution of the NOM.

After ozonation the SUVA values decreased for all three waters. Also, the hydrophobic DOC was preferentially transformed in all three waters. Large and intermediate molecular weight DOC compounds were preferentially transformed for clarified CRW and SDW, respectively. Oxidation by ozone preferentially transforms the nonpolar DOC into polar DOC. After biotreatment the SUVA values increased for all three waters. Also, the hydrophilic DOC was preferentially removed for all three waters. There was relatively no removal of unsaturated and aromatic carbon compounds for clarified CRW and SDW. Biotreatment has a propensity to remove polar DOC that is not detected by analytical techniques that rely on absorption of UV light at 254 nm.

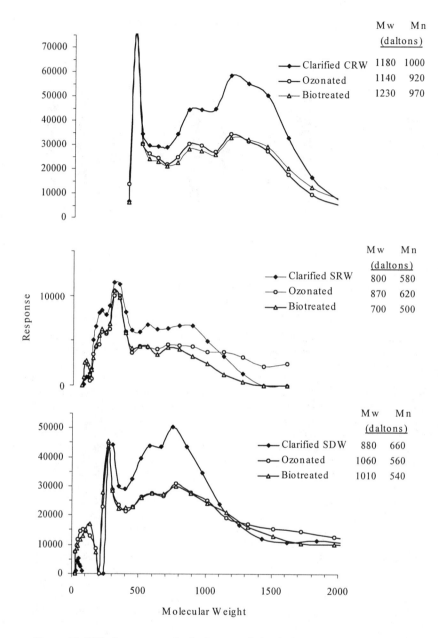

*Figure 12. SEC chromatographs for biotreated clarified waters.*

## Disinfection By-Products

Total trihalomethanes (TTHMs), total of nine haloacetic acids (HAA$_9$), and TTHM and HAA$_9$ formations normalized by DOC for bench-scale treated clarified SRW are reported in Table VI. The TTHM formation and the HAA$_9$ formation, both normalized by DOC, decreased after activated carbon. PAC adsorbs DOC compounds (nonpolar) that are likely to form both THMs and HAAs. After membrane treatment the TTHM and HAA$_9$ decreased, however, the TTHM and HAA$_9$, normalized by DOC, increased for TTHM and remained the same for HAA$_9$. The Desal GH membrane did not preferentially remove clarified SRW DOC compounds that are likely to form THMs and HAAs. After ozonation of clarified SRW the TTHM normalized by DOC decreased while the HAA$_9$ normalized by DOC stayed the same. Ozonation transforms DOC compounds (nonpolar) that are more likely to form THMs. After biotreatment of ozonated-clarified SRW the HAA$_9$ normalized by DOC decreased while the TTHM normalized by DOC remained the same. Biotreatment removes DOC compounds (polar) that are more likely to from HAAs.

**Table VI. TTHM and HAA$_9$ formations for bench-scale treated clarified SRW**

|  | DOC (mg/l) | SUVA$_{254}$ (L/mg-m) | TTHM ($\mu$mol/l) | HAA$_9$ ($\mu$mol/l) | TTHM ($\mu$mol/ $\mu$mol DOC) | HAA$_9$ ($\mu$mol/ $\mu$mol DOC) |
|---|---|---|---|---|---|---|
| Clarified SRW[a] | 1.19 | 1.6 | 0.14 | 0.06 | 0.0014 | 0.0006 |
| *PAC Treated* | | | | | | |
| Low DOC Removal | 0.88 | 1.1 | 0.10 | 0.04 | 0.0013 | 0.0005 |
| Mid DOC Removal | 0.59 | 0.7 | 0.06 | 0.01 | 0.0012 | 0.0002 |
| High DOC Removal | 0.27 | 0.4 | 0.02 | 0.00 | 0.0008 | 0.0000 |
| *Membrane Treated* | | | | | | |
| Permeate | 0.51 | 2.0 | 0.07 | 0.03 | 0.0017 | 0.0007 |
| *Ozonated-Biotreated* | | | | | | |
| Ozonated[b] | 1.18 | 0.9 | 0.09 | 0.07 | 0.0009 | 0.0007 |
| Biotreated | 0.87 | 1.0 | 0.08 | 0.03 | 0.0010 | 0.0004 |

Note:
[a] Ambient bromide = 40 $\mu$g/L and bromate is non-detectable.
[b] Following ozonation, 25 $\mu$g/L of bromate formed.
TTHM and HAA$_9$ results were obtained from uniform formation condition (UFC) tests (9). The UFC conditions for chlorine experiments, dosed with a buffered sodium hypochlorite solution, were pH = 8.0 (borate buffer), temperature = 20°C and incubation time = 24 hours. The target disinfectant residuals after incubation were 1.0 mg/L free chlorine.

138

## Acknowledgments

The authors thank AWWA Research Foundation (Project Manager: Jarka Popovicova) for their support. The authors also thank Cordelia Hwang, Stuart Krasner and Mike Sclimenti of the Metropolitan Water District of Southern California for their assistance in measuring DBP data.

## Literature Cited

1. International Life Science Institute, *Disinfection By-products in Drinking Water: Critical Issues in Health Effects Research, Workshop*, 1995, Chapel Hill, NC.
2. Krasner, S. W.; Croue, J. P.; Buffle, J.; Perdue, E. M. *Jour. AWWA* **1996**, *88*, 66-79.
3. Krasner, S. W.; Amy, G.; Zhu, H. W. *Proc. AWWA Enhanced Coagulation Research Workshop*, 1994, Charleston, SC.
4. Chin, Y.; Aiken, G.; O'Loughlin, E. *Environ. Sci. Technol.* **1994**, *28*, 1853-1858.
5. Yau, W. W.; Kirkland, J. J.; Bly, D. D. *Modern Size-Exclusion Liquid Chromatography*, 1979, John Wiley & Sons, New York.
6. Aiken, G. R.; McKnight, D. M.; Thorn, K. A.; Thurman, E. M. *Org. Geochem.* **1992**, *18*, 567-573.
7. Mogren, E. M.; Scarpino, P. V.; Summers, R. S. *Proc. Amer. Water Works Assoc. Conference*, 1990, Cincinnati, OH, 573-587.
8. Hesse, S.; Frimmel, F.H. *Acta. Hydrochim. Hydrobiol.* **1999**, *27*, 94-97.
9. Summers, R. S.; Hooper, S. M.; Shukairy, H. M.; Solarik, G.; Owen, D. *Jour. AWWA* **1996**, *88*, 80-93.

Chapter 10

# Disinfection By-Product Formation Potentials of Hydrophobic and Hydrophilic Natural Organic Matter Fractions: A Comparison Between a Low- and a High-Humic Water

Jean-Philippe Croué, David Violleau, and Laurence Labouyrie

Laboratoire de Chimie de l'Eau et de l'Environnement, UPRESA CNRS 6008, Université de Poitiers, 86022 Poitiers Cedex, France

The reactivities of hydrophobic and hydrophilic fractions of natural organic matter (NOM) isolated from two distinct rivers, the Suwannee River (Georgia) and the South Platte River (Colorado) have been studied. For both sources, the more hydrophobic and more acidic fractions provide the most active precursor sites; i.e., they have the largest formation potentials (FPs) for total organic halides (TOX), trihalomethanes (THMs), and haloacetic acids (HAAs). The formation of chlorinated disinfection by-products (DBPs) per gram of carbon was consistently greater for the Suwannee than for the South Platte, in corresponding fractions. Good correlations were obtained between specific ultraviolet absorbance at 254 nm (SUVA) and the FPs of THMs, TOX, or trichloroacetic acid (TCAA) for the isolates from both rivers, supporting the importance of aromaticity in the production of chlorinated DBPs regardless of the nature of the fraction. However, the relationship between DBPFPs and SUVA for the isolated fractions appears to differ between the two sites. THMs represented a substantially larger fraction of the TOX in the South Platte River NOM that might be supported by its stronger hydrophilic character while TCAA represented a larger fraction of the TOX in the Suwannee River possibly due to its stronger aromatic character. Results obtained with some neutral and base NOM fractions suggest that nitrogen rich structures are specific precursors of dichloroacetic acid.

## Background

The dissolved organic carbon (DOC) of natural waters can be fractionated into three NOM groups, based on their respective adsorption capability under acid pH conditions onto two XAD-8 and XAD-4 resin columns in series (*1-5*). The hydrophobic NOM adsorbs onto the XAD-8 resin, the transphilic NOM adsorbs onto the XAD-4 resin and the hydrophilic NOM adsorbs onto neither of the resins. Isolation of the NOM is then completed using sodium hydroxide elution of both resins, providing two major fractions, the humic fraction obtained from the XAD-8 resin (which can be further fractionated into humic and fulvic acids) and the transphilic acids—also called syn-fulvic acids (*6*) or XAD-4 acids (*2*)—eluted from the XAD-4 resin.

Considerable work has been done in the past on comparison of the reactivity with chlorine of humic and fulvic acids, the major constituents of the hydrophobic NOM. Because of the difficulty in isolating the more hydrophilic constituents of NOM, only few studies have been devoted to comparison of the reactivity of corresponding hydrophobic and hydrophilic NOM isolated from surface waters.

The major chlorination DBPs identified in both chlorinated surface waters and isolated humic materials are THMs, HAAs, haloketones (HKs), and halonitriles (HNs). HAAs represent a major portion of the non-THM halogenated organic compounds (*7-10*).

The work conducted by Reckhow, Singer, and Malcolm (*11*) with five surface waters (pH 7; 3-day contact time) showed humic acids to be higher chlorine consumers than the corresponding fulvic acids. In all cases, the DBP formation potentials—TOXFP, THMFP, TCAAFP, and dichloroacetic acid (DCAA) FP—were greater for humic acids than for the corresponding fulvic acids from the same source. Using similar experimental conditions, Martin (*12*) compared the reactivity of humic, fulvic and transphilic acids isolated from nine surface waters. As observed for all water sources, humic acids were more reactive with chlorine (i.e., chlorine demand and TOXFP and THMFP) than fulvic acids, which were more reactive than transphilic acids.

THMs (chloroform) represent an important fraction of the DBPs formed during the chlorination of humic materials. Reckhow et al. (*11*) and Martin (*12*) obtained good relationships between THM (chloroform) yield [expressed as micrograms of chlorine per milligrams of carbon ($\mu$g Cl/ mg C)] as a function of the TOX yield for a large variety of NOM. For Martin (*12*), who studied humic, fulvic and transphilic acids, chloroform accounted for about 25% of the TOX, while for Reckhow et al. (*11*), who studied only humic and fulvic acids, the ratio was close to 20%. It is interesting to note that whatever the origin and nature of the acid fraction of NOM, the proportion of TOX accounted for by chloroform remains almost constant.

Among the 10 fractions studied by Reckhow et al. (*11*) at neutral pH, only two fulvic acids produced more THM (chloroform) than TCAA (expressed as $\mu$g THM or TCAA/mg C), but for all 10 extracts, the sum of the yields of TCAA and DCAA were greater than the THM yield. When expressed as percent of TOX, the proportion of THM (average: 19.2%) was greater than the proportion of TCAA (average: 14.8%) for fulvic acids (with one exception), while the opposite was found for humic acids (19.1% and 20.4% respectively, as average values). DCAA accounted for 4.6 to 6.5% of the TOX.

Aromatic moieties contain $sp^2$-*hybridized* electron orbitals that are susceptible to electrophilic attack by chlorine. Moreover, oxygen atoms within hydroxyl groups locally donate electrons to the carbon atoms they are bound to, increasing the electron richness within the aromatic moieties. In humic acid structure, aromatic phenolic substituents are known to be associated with chlorine reactivity (*13, 14*). SUVA was found to be a good predictor of the aromatic carbon (C) content of hydrophobic organics. As expected, good relationships can also be established between this parameter and the TOXFP or THMFP of humic substances (*11, 15*). Previous work by Oliver and Thurman (*16*) showed that THMFP can be correlated with the specific absorbance of the NOM at 400 nm (color/mg C).

The results (which involved hydrophobic and transphilic acids) obtained by Martin (*12*) confirmed the good relationship between SUVA and THMFP or TOXFP. However, better relationships were obtained when each fraction was considered separately (similar observations were made for THMFP). This finding may indicate that the nature of the aromatic precursor sites differs with the characteristics of the NOM fraction. Reckhow et al. (*11*) indicated that the nature of the aromatic sites, and in particular the relative abundance of activated and non-activated rings, is one of the major parameters governing the reactivity of NOM with chlorine.

All of these results show that not only are chlorination conditions important in controlling the production and distribution of the DBPs, but also the origin and nature of the NOM plays an important role. However, this conclusion mainly relates to studies conducted with humic substances, which, in general, represent only one-half of the DOC of common surface waters. The objective of the present study was to compare the chlorine demand and DBP formation potentials of NOM fractions, including the most hydrophilic NOM fractions, isolated from two distinct rivers: the Suwannee River (Georgia), a low-salt/high-DOC water, and the South Platte River (Colorado), a moderate-salt/low-DOC water.

## Material and Methods

### Isolation and Fractionation of NOM

Two U.S. rivers with very different water quality and NOM characteristics were selected for fractionation case studies: the Suwannee River in southeastern Georgia and the South Platte River in Colorado. The Suwannee is a very soft, "black water" river with low salt content (30 to 60 µS/cm). It contains a high concentration of NOM (46.8 mg/L DOC) that is derived principally from terrestrial plants and that has been minimally fractionated by sorption onto soil mineral constituents. The river was sampled at its origin, at the outlet of the Okeefenokee Swamp. The South Platte River, which serves as a major source of drinking water for Denver, Colorado, was sampled in Waterton Canyon below Strontia Springs Reservoir. At the period of sampling, the river was almost completely covered with ice. In contrast to the NOM in the Suwannee, a substantial portion of the NOM in the South Platte is generated in the water itself, i.e., autochthonous NOM. The water in the South Platte is moderately hard with moderate salt content (400 μS/cm) and the NOM content is low (3.0 mg/L C). In addition, there are extensive mineral sediments and soils that act as solubility controls on the NOM content.

Both waters were fractionated and isolated into NOM fractions (5) based on their hydrophilic (hydrophobic, or HPO; transphilic, or TPH; and hydrophilic, or HPI) and acidic (acid, A; neutral, N; base, B) character. A total of 10 NOM fractions were isolated from the Suwannee (SW) River: hydrophobic acids and neutrals (HPOA and HPON); transphilic acids and neutrals (TPHA and TPHN); hydrophilic acids (HPIA and HPIA-2) and bases (HPIB); ultrahydrophilic humic acids (uHA); ultrahydrophilic acids and neutrals (uHPIA and uHPIN). The mass of hydrophilic NOM slightly exceeded the capacity of the $H^+$ exchange resin. A second filtration step was necessary to recover all the HPIA fraction under the acid form. Only a partial set of the fractions (six fractions: HPOA, HPON, TPHA, TPHN, HPIA, and HPIN) was obtained from the South Platte (SPL) River. See Leenheer (5) for details regarding the NOM fractionation.

All of these fractions of NOM—elemental analysis, total dissolved carbohydrates and amino acids, Fourier transform–infrared (FTIR) spectrometry, solid-state carbon 13 nuclear magnetic resonance ($^{13}$C-NMR) spectrometry, SUVA— were well characterized prior to chlorination (17).

## Chlorination Conditions

For the chlorination tests of the NOM isolates, the DOC concentration was 4 to 7 mg/L, the $(Cl_2)_{dose}$/DOC ratio was 4 mg/mg, pH was maintained at 8.0 using borate buffer, and samples were incubated for 72 hours in the dark at 20°C. At the end of the test, residual $Cl_2$ was quenched with sodium meta-arsenite. The only exception to these conditions was that the DOC concentration was only 1 mg/L in the test with the Suwannee River hydrophilic base fraction, due to a shortage of material.

## Analytical Methods

Free and total residual chlorine were analyzed (before the addition of sodium meta arsenite) using the colorimetric N,N-diethyl-p-phenylenediamine (DPD) method adapted from the AFNOR method (18).

TOX was analyzed in duplicate by the adsorption-pyrolysis-titrimetric standard method (19) using a TOX analyzer (Dorhmann DX 20 TOX analyzer). Analyses were conducted on either 25 mL or 50 mL of solution. The results were expressed as $\mu g$ Cl/mg C.

THMs were analyzed in duplicate using a headspace autosampler (Dani HSS 3950) coupled with a gas chromatograph (Varian 3300) equipped with a nickel 63 ($^{63}$Ni) electron capture detector and a split injector. Ten mL of sample was injected into a 20-mL flask sealed with a Teflon cap at 40°C. The injection loop volume was 100 $\mu L$. THMs were separated on a wide bore (0.53 mm internal diameter) column (DB-624, J&W Scientific) that was 30 m long and had a 3.0-$\mu m$ film, using a temperature program that increased temperature from 80°C to 120°C at 5°C/min. The temperatures of the injector and the detector were 200 and 300°C, respectively.

HAAs were analyzed in duplicate using a Fisons 8000 gas chromatograph equipped with a Fisons AS 800 autosampler, an on-column injector, and a $^{63}$Ni electron capture detector. After liquid-liquid extraction with methyl *tert*-butyl ether (MTBE) at pH 1 (50 mL of sample per 5 mL of solvent), the sample was derivatized with concentrated diazomethane (200 $\mu L$ per 3 mL of solvent). The derivatization reaction was stopped after a few minutes by adding two crystals of silica gel. HAAs were separated by injection of 1 $\mu L$ of sample onto a capillary column (DB-1701, J&W Scientific, 0.32 mm ID, 30 m long, 0.3-$\mu m$ film thickness) in the presence of dibromopropane (as an internal standard) using the following temperature program: 35°C (20 min) to 150°C (1 min) at 4°C/min, to 200°C (5°min) at 10°C/min.

## Results and Discussion

### DBP FPs

Normalized values of the chlorine demand, TOXFP, THMFP, TCAAFP and DCAAFP of the Suwannee and South Platte River isolates are listed in Tables 1 and 2, respectively. These two tables also include the SUVA and the C/N mass ratio of the NOM fractions. The yields of THMs (chloroform) and TCAA are similar (expressed as $\mu g$/mg C) for the Suwannee isolates, with the exception of the TPHA (TCAAFP>THMFP) and HPIA-2 (THMFP>TCAAFP) fractions. More TCAA was formed than DCAA in all fractions except for the HPIB. For the South Platte isolates, chloroform was the major chlorination by-product. More TCAA than DCAA was formed in the acid fractions, but the opposite trend was observed for two of the three neutral fractions (TPHN and HPIN).

The range of specific THM and TOX yields found with the HPOA and TPHA fractions of both the Suwannee and South Platte rivers fall within the ranges reported in the literature. However, the formation of chlorinated DBPs per gram of carbon was consistently greater in the Suwannee than in the South Platte, in corresponding fractions. It is widely reported that the relative production of THMs (chloroform) and TCAA is pH-dependent, so that production of chloroform increases when pH increases, while production of TCAA decreases (e.g., *7, 20*). The current study indicates that the nature of the NOM is also an important factor in this distribution.

**Table 1: Chlorine Demand and DBP Formation Potentials of the Suwannee River NOM Isolates**

| Fraction | C/N | Initial SUVA ($m^{-1} \bullet L/ mgC$) | Chlorine demand $\left(\dfrac{mgCl_2}{mgC}\right)$ | TOXFP $\left(\dfrac{\mu gCl}{mgC}\right)$ | THMFP $\left(\dfrac{\mu gCHCl_3}{mgC}\right)$ | TCAAFP $\left(\dfrac{\mu gTCAA}{mgC}\right)$ | DCAAFP $\left(\dfrac{\mu gDCAA}{mgC}\right)$ |
|---|---|---|---|---|---|---|---|
| HPON | 46.9 | 3.0 | 1.5 | 182 | 51 | 51 | 24 |
| HPOA | 69.5 | 4.6 | 1.4 | 268 | 55 | 59 | 25 |
| TPHA | 46.2 | 3.4 | 1.4 | 224 | 40 | 57 | 23 |
| TPHN | 29.7 | 3.0 | 1.2 | 204 | 40 | 44 | 22 |
| HPIA | 33.1 | 2.7 | 0.8 | 171 | 36 | 36 | 22 |
| HPIA-2 | 20.1 | 1.8 | 1.1 | 150 | 41 | 34 | 17 |
| UHPIAMe | 46.6 | 1.3 | 0.78 | 121 | 28 | 30 | 21 |
| HPIB | 7.75 | 2.3 | 2.5 | 175 | 29 | 31 | 39 |
| UHA | 14.8 | 5.0 | 2.8 | 285 | 63 | 66 | 45 |
| UHPIN | 14.7 | 0.7 | 0.8 | 117 | 23 | 26 | 22 |

**Table 2: Chlorine Demand and DBP Formation Potentials of the South Platte River NOM Isolates**

| Fraction | C/N | Initial SUVA ($m^{-1} \bullet L/ mgC$) | Chlorine demand $\left(\dfrac{mgCl_2}{mgC}\right)$ | TOXFP $\left(\dfrac{\mu gCl}{mgC}\right)$ | THMFP $\left(\dfrac{\mu gCHCl_3}{mgC}\right)$ | TCAAFP $\left(\dfrac{\mu gTCAA}{mgC}\right)$ | DCAAFP $\left(\dfrac{\mu gDCAA}{mgC}\right)$ |
|---|---|---|---|---|---|---|---|
| HPON | 8.92 | 1.6 | 0.83 | 114 | 29 | 16 | 12 |
| HPOA | 44.1 | 2.9 | 0.95 | 122 | 46 | 28 | 14 |
| TPHA | 17.7 | 1.8 | 0.81 | 100 | 39 | 21 | 14 |
| TPHN | 4.03 | 0.7 | 2.3 | 66 | 25 | 12 | 20 |
| HPIA | 14.9 | 1.7 | 0.86 | 98 | 35 | 24 | 16 |
| HPIN | 8.9 | 0.5 | 1 | 81 | 28 | 15 | 19 |

In Tables 3 and 4, THMFP, TCAAFP and DCAAFP are given in terms of µg Cl/mg C to facilitate a comparison with TOXFP. For some fractions, THMFP's expressed in this form include contributions from bromodichloromethane in addition to chloroform. In all cases, chloroform was the major THM; bromodichloromethane accounted for about 2% of the THMFP (in µg Cl/L) for the Suwannee isolates and around 10% for the South Platte isolates.

The sum of THMs, TCAA and DCAA accounted for 37 to 52% (average 42%) of the TOX for the Suwannee isolates, and 38 to 64% (average 55%) of the TOX for the South Platte isolates. The fact that these three compounds represent a larger proportion of the TOX in the South Platte isolates than in the Suwannee isolates primarily reflects differences in the production of chloroform. This representation identifies the South Platte TPHN and HPIN and the Suwannee HPIB fractions as having somewhat larger DCAA/TOX ratios than the other fractions. The structural characterization of the NOM fractions (17) showed that the South Platte TPHN and the Suwannee HPIB fractions are largely proteinaceous, while the South Platte HPIN fraction has a stronger carbohydrate and amino sugar character. Thus, it appears that nitrogenous units in the NOM structure (proteins and amino sugars) could be an important class of DCAA precursor sites.

However, it is also possible that the strong reactivity of the three fractions with chlorine (they had the highest chlorine demands of all the fractions studied) may have lowered the $Cl_2/DOC$ ratio during the course of the reaction more than in the other fractions, favoring the production of DCAA over more highly halogenated products. Reckhow (21) and Croué (9) demonstrated that for low $Cl_2/DOC$ ratios, DCAA appeared to be more abundant than TCAA.

**Table 3: DBP Formation Potentials of the Suwannee River NOM Isolates**

| Fraction | THMFP $\left(\dfrac{\mu g\, Cl}{mg\, C}\right)$ | THMFP/ TOXFP (%) | TCAAFP $\left(\dfrac{\mu g\, Cl}{mg\, C}\right)$ | CAAFP/ TOXFP (%) | DCAAFP $\left(\dfrac{\mu g\, Cl}{mg\, C}\right)$ | DCAAFP/ TOXFP (%) |
|---|---|---|---|---|---|---|
| HPON | 45 | 25 | 37 | 20 | 13 | 7 |
| HPOA | 49 | 18 | 38 | 14 | 14 | 5 |
| TPHA | 36 | 16 | 37 | 16 | 13 | 6 |
| TPHN | 36 | 18 | 29 | 14 | 12 | 6 |
| HPIA | 32 | 19 | 23 | 13 | 12 | 7 |
| HPIA-2 | 37 | 25 | 22 | 15 | 9 | 6 |
| UHPIAMe | 25 | 21 | 19 | 16 | 11 | 9 |
| HPIB | 26 | 15 | 20 | 11 | 21 | 12 |
| UHA | 57 | 20 | 43 | 15 | 25 | 9 |
| UHPIN | 21 | 18 | 17 | 14 | 12 | 10 |

**Table 4: DBP Formation Potentials of the South Platte River NOM Isolates**

| Fraction | THMFP $\left(\dfrac{\mu g\,Cl}{mg\,C}\right)$ | THMFP/ TOXFP (%) | TCAAFP $\left(\dfrac{\mu g\,Cl}{mg\,C}\right)$ | TCAAFP/ TOXFP (%) | DCAAFP $\left(\dfrac{\mu g\,Cl}{mg\,C}\right)$ | DCAAFP/ TOXFP (%) |
|---|---|---|---|---|---|---|
| HPON | 27 | 24 | 10 | 9 | 6 | 5 |
| HPOA | 43 | 35 | 18 | 15 | 8 | 7 |
| TPHA | 37 | 37 | 14 | 14 | 8 | 8 |
| TPHN | 23 | 35 | 8 | 12 | 11 | 17 |
| HPIA | 34 | 35 | 16 | 16 | 9 | 9 |
| HPIN | 26 | 32 | 10 | 12 | 11 | 14 |

**Relationships Between DBPFP and Spectral Properties**

Figure 1 shows the FPs for various DBPs as a function of SUVA for the Suwannee and the South Platte isolates.

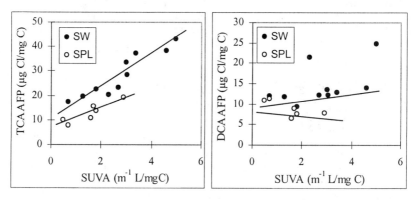

*Figure 1: Relationships between DBPFPs and SUVA ($m^{-1}$ •L/mg C) for NOM fractions isolated from the Suwannee and South Platte Rivers.*

Good correlations could be obtained between SUVA and THMFP, TOXFP, or TCAAFP for the isolates from both rivers, supporting the importance of aromaticity (a good relationship was obtained between SUVA and the aromatic C content as determined by $^{13}$C-NMR, Figure 2) for the production of chlorinated DBPs regardless of the nature of the fraction. Linear regression coefficients ($R^2$) ranged from 0.8 to 0.97 except for the THMFP data set of the South Platte isolates ($R^2 = 0.7$). The best linear regression coefficients were obtained for the correlation between TOXFP and SUVA, probably because TOX is a more general parameter for evaluating the reactivity of NOM with chlorine than are individual species such as THM (chloroform) or TCAA.

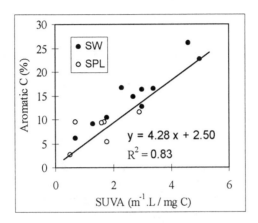

*Figure 2: Correlation between aromatic C and SUVA (254 nm) (m$^{-1}$•L/mg C).*

Correlations between DBPFP and the aromatic C content or the percent aromaticity determined from $^{13}$C-NMR spectra exhibited lower linear regression coefficients. For example, the $R^2$ calculations for the relationships between TOXFP and the phenolic content, the aromatic C content, the percent aromaticity, and SUVA in the Suwannee isolates were 0.86, 0.89, 0.91, and 0.97, respectively.

The correlation between DCAAFP and SUVA was poor for the two series of isolates, which might indicate that organic structures other than aromatics are responsible for the production of DCAA upon chlorination. DCAA has been identified as a chlorination by-product of several amino acids, methionine (*22*), aspartic acid (*23*), tyrosine (*24*), proline (*22, 23*) and cysteine (*25*). Consistent with the discussion above, these data further support the possibility that proteins may play a significant role in the production of DCAA.

For both sources, the more hydrophobic and more acidic fractions provide the most active precursor sites (i.e., they have the largest FPs for TOX, THMs, and

HAAs). Because the HPOA and TPHA fractions account for the largest part of the DOC, they comprise the main precursor sites for TOX, THM and TCAA in these waters.

## Comparison Between the NOM Source

The relationship between DBPFPs and SUVA for the isolated fractions appears to differ between the two sites. Reckhow et al. (11) indicated that the nature of the aromatic sites, and in particular the relative abundance of activated and non-activated rings, is one of the major parameters governing the reactivity of NOM with chlorine.

The plots of THMFP versus SUVA are linear with almost identical slopes for the two sources, but the y-intercept (the extrapolated estimate of THMFP corresponding to zero SUVA) is greater for the South Platte. The opposite trend is observed in the relationship between the HAAs and SUVA, i.e., at a given SUVA, the Suwannee fractions have a greater HAAFP than the South Platte fractions. These results suggest that the non-aromatic moieties in the South Platte NOM are somewhat better precursors for THMs and poorer precursors for TCAA than those in the Suwannee.

The best-fit lines in Figure 1 have y-intercepts that indicate that even in the absence of aromatic structures a significant amount of DBPs would be produced, and the intercept is significantly higher for THMs than for HAAs. Amino acid structures in the base fraction (HPIB for the Suwannee River) and β-hydroxyacids and β-ketones that might be predominant in neutral and hydrophilic acid fractions are known to be THM precursors. Because of the more pronounced hydrophilic character of the South Platte River NOM, β-hydroxyacids and β-ketones may account for a significant part of the THM precursors in NOM from this source.

Considering all DBPs together (as TOX), the Suwannee NOM has more Cl-reactive sites than the South Platte NOM. The relative TOXFPs of the two waters are consistent with the [13]C-NMR spectra, which illustrate that the Suwannee NOM has a greater aromatic C content than the South Platte NOM (Figure 2), and also contains a higher fraction of aromatic phenolic constituents in its structure. Reckhow et al. (*11*) reported that molecules having a high degree of conjugation preferentially lead to TCAA formation over chloroform formation. The production of TCAA in the two NOM sources is consistent with this observation (Figure 3a). Because the NOM in the South Platte has a higher THMFP:SUVA ratio and a lower yield of other DBPs than does the NOM in the Suwannee, THMs represent a substantially larger fraction of the TOX in the South Platte (Figure 3b).

A linear relationship was obtained between TCAAFP and DCAAFP for most of the isolated NOM fractions (Figure 4) but an amount of DCAA that far exceeded the value expected based on this relationship was produced in a few cases (i.e., the NOM fractions that exert the strongest chlorine demand). These results may indicate that two categories of DCAA sites exist in the NOM structure (Figure 5): nonspecific sites that can produce either DCAA or TCAA (e.g., activated aromatic rings) and specific sites such as amino acid type structures that lead to formation of DCAA but not TCAA.

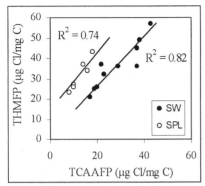 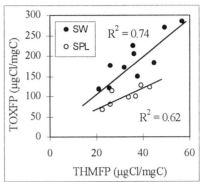

*Figure 3a:  THMFP vs TCAAFP.*          *Figure 3b:  TOXFP vs THMFP.*
*Figure 3:  Relationships between DPBs in Suwannee and South Platte River NOM fractions.*

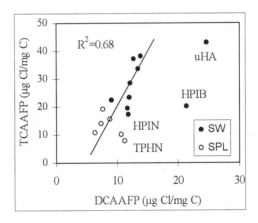

*Figure 4:  Relationship between TCAAFP and DCAFP for the Suwannee River and South Platte NOM*

One can note that the fractions of NOM that do not follow the DCAAFP vs TCAAFP correlation are the ones that show a lower C/N ratio. Based on pyrolysis GC/MS analysis, Labouyrie (26) found that the C/N ratio was a good surrogate for the protein plus aminosugar content.

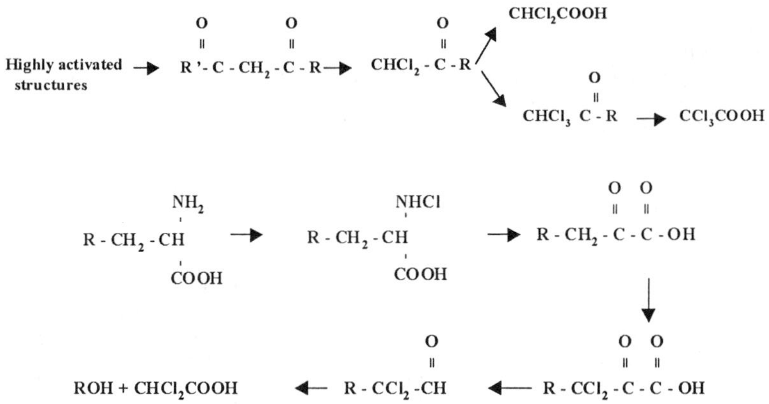

$$
\text{Highly activated structures} \rightarrow R' - \overset{O}{\overset{\|}{C}} - CH_2 - \overset{O}{\overset{\|}{C}} - R \rightarrow CHCl_2 - \overset{O}{\overset{\|}{C}} - R
$$

$$
\nearrow CHCl_2COOH
$$

$$
\searrow CHCl_3 \ \overset{O}{\overset{\|}{C}} - R \rightarrow CCl_3COOH
$$

$$
R - CH_2 - \underset{\underset{COOH}{|}}{\overset{\overset{NH_2}{|}}{CH}} \rightarrow R - CH_2 - \underset{\underset{COOH}{|}}{\overset{\overset{NHCl}{|}}{CH}} \rightarrow R - CH_2 - \overset{O}{\overset{\|}{C}} - \overset{O}{\overset{\|}{C}} - OH
$$

$$
ROH + CHCl_2COOH \leftarrow R - CCl_2 - \overset{O}{\overset{\|}{C}} H \leftarrow R - CCl_2 - \overset{O}{\overset{\|}{C}} - \overset{O}{\overset{\|}{C}} - OH
$$

Figure 5:  *Two distinct reaction pathways for the production of dichloroacetic acid during the chlorination of NOM.*

## Conclusions

A significant number of studies have shown that humic and fulvic acids (i.e., hydrophobic acids) isolated from natural waters are highly reactive with chlorine, and reactivity (chlorine demand and production DBPs) can be correlated with SUVA, a good surrogate parameter for the aromatic character of NOM. The present study supports the importance of aromaticity in the production of chlorinated DBPs regardless of the nature (hydrophobic and hydrophilic NOM) of the fraction. However, the relationship established between DBPFPs and SUVA for the isolated fractions appears to differ between the two sites, the Suwannee and South Platte rivers, two rivers with "end member" NOM characteristics. The chlorination of South Platte River NOM, hydrophilic in character, led to a higher relative production of THM (THMFP/TOXFP ratio) as compared to the Suwannee River NOM (more hydrophobic), a finding which may support the important role of β-hydroxyacids and β-ketones as THM precursor sites. Hydrophilic NOM fractions enriched in nitrogeneous type structures were found to exert a strong chlorine demand associated with high dichloroacetic acid production. These results demonstrate that not only humic substances but also the more hydrophilic NOM (i.e., nonhumic substances) play an important role for the production of Dbps upon chlorination. More work needs to be done on this second portion of NOM in order to better understand DBP speciation.

## Acknowledgments

The authors want to acknowledge the major role of Dr. Jerry Leenheer and Dr. George Aiken (U.S. Geological Survey, Denver, Colorado, 80225) in the isolation and characterization of the samples. This work was funded by AWWA Research Foundation (grant # 159-94). Support from the engineering companies HDR Engineering, Inc. (Omaha, NE), SAUR (Maurepas, France) and DYNAMCO (West Sussex, UK) is greatly appreciated.

## References

1. Aiken, G.R.; McKnight, D.M.; Thorn, K.A.; Thurman, E.M. Isolation of hydrophilic organic acids from water using nonionic macroporous resins. *Org. Geochem.* **1992**, *18*, 567-573.
2. Malcolm, R.L.; McCarthy, P. Quantitative evaluation of XAD-8 and XAD-4 resins used in tandem for removing organic solutes from water. *Environment Int.* **1992**, *18*, 597-607.
3. Croué, J.-P.; Martin, B.; Deguin, A.; Legube, B. Isolation and characterization of dissolved hydrophobic and hydrophilic organic substances of a reservoir water. In *Natural Organic Matter in Drinking Water: Origin, Characterization, and Removal,* Workshop Proceedings, Chamonix, France, Sept 19-22, 1993. AWWARF and AWWA: Denver, CO, 1994.
4. Andrews, S.A.; Huck, P.M. Using fractionated natural organic matter to quantitate organic byproducts of ozonation. *Ozone Sci. Eng.* **1994**, *16*, 1-12.
5. Leenheer, J.A. Fractionation, isolation and characterization of hydrophilic constituents of dissolved organic matter in water. In *Influence of Natural Organic Matter Characteristics on Drinking Water Treatment and Quality,* Workshop Proceedings, Université de Poitiers, Poitiers, France, Sept 18-19, 1996.
6. Aiken, G.; McKnight, D.; Harnish, R.; Wershaw, R. Geochemistry of aquatic humic substances in the Lake Fryxell Basin, Antartica. *Biogeochemistry* **1996**, *34*, 157-188.
7. Miller, J.W.; Uden, P.C. Characterization of non-volatile aqueous chlorination products of humic substances. *Environ. Sci. Technol.* **1983**, *17* (3), 150-157.
8. Reckhow, D.A.; Singer, P.C. Mechanisms of organic halide formation and implications with respect to pre-ozonation. In *Water Chlorination: Chemistry,*

*Environmental Impact and Health Effects*; Jolley, R.L., et al., Eds. Lewis Publishers: Chelsea, MI, 1985; Vol 5.

9. Croué, J.-P. Contribution à l'étude de l'oxydation par le chlore et l'ozone d'acides fulviques naturels extraits d'eaux de surface. Thèse de Doctorat, Université de Poitiers, Poitiers, France, 1987 ($n^0$ d'ordre 89).

10. Stevens, A.A.; Moore, L.A.; Miltner, R.J. Formation and control of non-trihalomethane disinfection by-products. *J.—Am. Water Works Assoc.* **1989,** *81* (8), 54-60.

11. Reckhow, D.A.; Singer, P.C.; Malcolm, R.L. Chlorination of humic materials: byproduct formation and chemical interpretations. *Environ. Sci. Technol.* **1990,** *24,* 1655-1664.

12. Martin, B. La matière organique naturelle dissoute des eaux de surface: fractionnement, caractérisation et réactivité. Thèse de Doctorat, Université de Poitiers, Poitiers, France, 1995.

13. Hanna, J.V.; Johnson, W.D.; Quezeda, R.A.; Wilson, M.A.; Xia-Qiao, L. Characterization of aqueous humic substances before and after chlorination. *Environ. Sci. Technol.* **1991,** *25* (6), 1160-1164.

14. Harrington, G.W.; Bruchet, A.; Rybacki, D.; Singer, P. Characterization of natural organic matter and its reactivity with chlorine. In *Water Disinfection and Natural Organic Matter: Characterization and Control*; Minear, R.A., and Amy, G.L., Eds. American Chemical Society: Washington, DC, 1996; Chapter 10.

15. Legube, B.; Xiong, F.; Croué, J.-P.; Doré, M. Etude sur les acides fulviques extraits d'eaux superficielles françaises: extraction, caractérisation et réactivité avec le chlore. *Revue des Sciences de l'Eau* **1990,** *4,* 399-424.

16. Oliver, B.G.; Thurman, E.M. Influence of aquatic humic substance properties on trihalomethane potential. In *Water Chlorination: Chemistry, Environmental Impact and Health Effects.* Ann Arbor Science Publishers: Ann Arbor, MI, 1983; Vol. 4, pp 231-252.

17. Croué, J.-P.; Korshin, G.V.; Leenheer, J.A.; Benjamin, M.M. Isolation, fractionation and characterization of natural organic matter in drinking water. AWWARF report, in preparation. (1999).

18. Jadas-Hécart, A.; El Morer, A.; Stitou, M.; Bouillot, P.; Legube, B. Modélisation de la demande en chlore d'une eau traitée. *Wat. Res.* **1992,** *26,* 1073-1084.

19. Eaton, A.D., Clesceri, L.S., Greenberg, A.E., Eds. *Standard Methods for the Examination of Water and Wastewater.* American Public Health Association, American Water Works Association, Water Environment Federation: Washington, DC, 1995; 19th Ed.

20. Arora, H.; LeChevallier, M.; Dixon, K.L. DBP occurrence survey. *J.—Am. Water Works Assoc.* **1997,** *89* (6), 60-68.

21. Reckhow, D.A. Organic halide formation and the use of pre-ozonation and alum coagulation to control organic halide precursors. Ph.D. Thesis,

Department of Environmental Sciences and Engineering, University of North Carolina, Chapel Hill, NC, 1984.

22. Hureiki, L.; Croué, J.-P. Identification par couplage CG/SM des sous-produits de chloration de deux acides aminés libres, la proline et la méthionine. *Revue des Sciences de l'Eau* **1997,** *2,* 249-264.

23. de Leer, W.B.; Erkelens, C.; de Galan, L. The influence of organic nitrogen compounds on the production of organochlorine compounds in the chlorination of humic material. In *Water Chlorination: Environmental Impact and Health Effects*; Jolley, R.L., Condie, L.W., Johnson, J.D., Katz, S., Minear, R.A., Mattice, J.S., Jacobs, V.A.,Eds. Ann Arbor Science Publishers: Ann Arbor, MI, 1990; Vol. 6, pp 763-781.

24. Horth, H. Identification of mutagens in drinking water. *JWSRT—Aqua,* **1989,** *38,* 80-100.

25. Hureiki, L. Etude de la chloration et de l'ozonation d'acides aminés libres et combinés en milieu aqueux dilué. Thèse de Doctorat, Université de Poitiers, Poitiers, France, 1993 (n$^0$ d'ordre 21).

26. Labouyrie, L. Extraction et caractérisation des matières organiques naturelles dissoutes d'eaux de surface: etude comparative des techniques de filtration membranaire et d'adsorption sur résines macroporeuses non ioniques. Thèse de Doctorat, Université de Poitiers, Poitiers, France, 1997.

Chapter 11

# Characterization and Disinfection By-Product Formation Potential of Natural Organic Matter in Surface and Ground Waters from Northern Florida

Colleen E. Rostad[1], Jerry A. Leenheer[1], Brian Katz[2], Barbara S. Martin[1], and Ted I. Noyes[1]

[1]U.S. Geological Survey, Box 25046, Mail Stop 408, Denver Federal Center, Denver, CO 80225
[2]U.S. Geological Survey, 227 North Bronough Street, Tallahassee, FL 32301

Streamwaters in northern Florida have large concentrations of natural organic matter (NOM), and commonly flow directly into the ground water system through karst features, such as sinkholes. In this study NOM from northern Florida stream and ground waters was fractionated, the fractions characterized by infrared (IR) and nuclear magnetic resonance (NMR), and then chlorinated to investigate their disinfection by-product (DBP) formation potential (FP). As the NOM character changed (as quantified by changes in NOM distribution in various fractions, such as hydrophilic acids or hydrophobic neutrals) due to migration through the aquifer, the total organic halide (TOX)-FP and trihalomethane (THM)-FP yield of each of these fractions varied also. In surface waters, the greatest DBP yields were produced by the colloid fraction. In ground waters, DBP yield of the hydrophobic acid fraction (the greatest in terms of mass) decreased during infiltration.

## Introduction

Chlorination of natural organic matter (NOM) during drinking water treatment for the purpose of disinfection produces a variety of chlorinated and brominated disinfection by-products (DBPs) that are of potential concern as health hazards (1). Surface waters found on the Atlantic and Gulf coastal plains in northern Florida frequently have large concentrations of NOM that produce DBPs during drinking water treatment. Certain of these "black waters" infiltrate into the Upper Floridan limestone aquifer where NOM concentrations are significantly reduced by adsorption, dilution, and precipitation processes. Understanding the dominant geochemical factors involved in degradation or alteration of NOM in the environment is critical to accurate prediction of potential DBP formation. One water quality characteristic of concern is total organic halogen or halide (TOX), a collective term encompassing halogenated organic compounds in the water, many of which may be toxic, carcinogenic, or mutagenic, and including trihalomethanes (THMs; chloroform, bromodichloromethane, dibromochloromethane, and bromoform) (2). Non-purgeable TOX (NPTOX) excludes primarily THMs. NPTOX and THMs result from the disinfection of natural waters with chlorine due to reaction with dissolved organic carbon (DOC). The types and concentrations of the NPTOX and THMs so generated vary with the nature and reactivity of the DOC. Even though ground water in the area contains high levels of natural DOC, no systematic studies have been conducted to determine the extent to which chlorination of these waters might produce DBPs.

The purpose of this study is to fractionate NOM in certain surface and ground waters into more homogeneous analytical classes, characterize the fractions, chlorinate the fractions and analyze any DBPs formed. This study seeks to identify NPTOX-FP and THM-FP associated with specific fractions of the NOM in surface and ground waters. As the NOM character changes (as quantified by changes in DOC distribution in various fractions (such as hydrophilic acids, hydrophobic neutrals) and their IR and NMR spectra), the TOX-FP and THM-FP of each of these fractions also may be altered. The chemistry of the chlorination reactions can then be deduced from the characterization data, and the biogeochemistry involved in causing changes in NOM concentrations and characteristics can be deduced by the changes that occur during infiltration of surface water into the limestone aquifer. This study may aid water-treatment utilities in the selection and treatment of drinking water sources to minimize DBP formation.

# Hydrogeologic Setting

The study sites are located in the Woodville Karst Plain and the Apalachicola Coastal Lowlands of humid subtropical north central Florida. The karst plain has a flat or gently rolling surface of highly porous quartz sand overlying limestones (St. Marks Formation) that comprise the Upper Floridan aquifer, from which the Donahue and Trout Pond well samples were taken. Lost Creek and Fisher Creek originate in the Apalachicola Coastal Lowlands, a flat sandy area that is underlain by thick clay and sand deposits. The lowlands area is characterized by shallow "bays" (densely wooded, swamp-like areas) and creeks with poorly defined channels that drain the bays (3). Both Fisher and Lost Creeks receive water from smaller creeks flowing out of bays and swamps, flow through narrow meandering valleys, and disappear into sinkholes that are connected to the Upper Floridan aquifer. Lost Creek and Fisher Creek are referred to as black-water streams due to high concentrations of tannins and lignins dissolved in the water. The ultimate source of recharge to the Upper Floridan aquifer in the study area is about 58 cm of annual rainfall. Recharge to the aquifer ranges from 20 to 46 cm/yr in the northern and southern parts of the study area, respectively (4).

# Methods

Water samples from streams and wells were collected in September 1997 during low-flow conditions using standard USGS techniques (*5, 6, 7*). Samples for analysis of tannin and lignin were analyzed in the field. This method detects hydroxylated aromatic compounds, including tannin, lignin, phenol and cresol (*8*).

Stable isotopes of oxygen ($^{18}O/^{16}O$), hydrogen (D/H), and carbon ($^{13}C/^{12}C$) were used to determine the origin of waters and to identify and quantify biogeochemical processes that control the chemical composition of ground water (*9*). Tritium ($^3H$) was used to estimate ground-water recharge time by comparing measured $^3H$ concentrations in ground water with the long-term $^3H$ input function of rainfall measured in Ocala, Florida (*10*). Water samples for $^3H$ were collected and analyzed according to methods presented by Michel (*10*).

Approximately 20 liters of the two surface-water samples and 80 liters of the two ground-water samples were collected for the NOM fractionation, characterization, and chlorination studies. These samples were field-filtered through 25-μm porosity and 0.9-μm porosity glass fiber cartridge filters in series.

## NOM Fractionation

Filtered water samples were physically and chemically fractionated to produce characteristic NOM fractions. Dissolved organic carbon concentrations were determined by high-temperature oxidation to carbon dioxide on a total organic carbon analyzer before fractionation and on subsequent samples. The analytical error on the two ground-water samples is about ±1.0 mg/L because of the high levels of inorganic carbon relative to organic carbon in these samples. Physical fractionation involved tangential-flow ultrafiltration through 30,000-dalton (0.005 μm) porosity regenerated cellulose flat-plate membranes (*11*). Suspensions of the colloids recovered from the membranes were then vacuum-rotary evaporated to 100 mL, placed inside 12,000-14,000 dalton dialysis membranes, and the remainder of the inorganic salts removed by dialysis against distilled water. The colloids were then lyophilized and weighed.

Dissolved organic matter (DOM) was chemically fractionated by the procedure shown in Figure 1, which is described in detail elsewhere (*12, 13, 14*). Use of trade names in this report is for identification purposes only and does not constitute endorsement by the U.S. Geological Survey. Hydrophobic and hydrophilic acids were separated by acidifying to pH 2 and passing through XAD-8. About 10% of the colored DOM in the two surface-water samples passed through all four columns listed in Figure 1. This colored DOM fraction, designated hydrophobic acids 2, was isolated by acidifying to pH 2 and passing it through the 1-liter XAD-8 column. The colored DOM was quantitatively sorbed and eluted with 750 mL of 75% acetonitrile/25% water. Elution solvents were removed by vacuum evaporation and lyophilization. Compared to the more traditional initially acidified XAD-8 resin isolation of NOM, the procedure shown here should isolate humic acids in the colloid and hydrophobic acids 2 fraction, whereas fulvic acids should be isolated in the hydrophobic acid fraction.

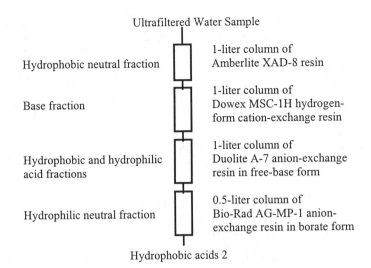

Figure 1. Flow chart for preparative dissolved organic matter fractionation procedure.

## NOM Spectral Characterizations

Infrared spectra of 2 to 5 mg of NOM fraction isolates and standard compounds in potassium bromide pellets were determined on a Perkin Elmer System 2000 Fourier Transform Infrared (FTIR) Spectrometer. All spectra were normalized after acquisition to a maximum absorbance of 1.0 for comparative purposes.

Solid-state, cross-polarization, magic-angle-spinning (CP/MAS) [13]C-NMR spectra were measured on a 200-MHz (megahertz) Chemagnetics CMX spectrometer with a 7.5 mm-diameter probe. The spinning rate was 5000 Hz. The acquisition parameters for the freeze-dried samples included a contact time of 1 ms, pulse delay of 1 s, and a pulse width of 4.5 μs for the 90° pulse. All spectra were normalized to the largest peak.

Ultraviolet/visible spectra from 190-900 nm were obtained on field-filtered and ultrafiltered water samples on a Perkin-Elmer Lambda 4B spectrometer using quartz cells with a 10 cm light path.

## NOM Disinfection By-Product Formation Potentials

Each fraction, isolated as a dry powder, was redissolved in distilled, deionized water at about 10 mg DOC/L for further analysis. Samples were taken for DOC (standard method 5310B (15)) and UV analyses. Ultimate DBP FPs were determined using standard method 5710 (15), which involves pH 7.0 buffered solutions incubated in the dark at 25°C for 168.0 hours, using initial free chlorine:DOC concentration ratios of 2:1 to 3:1, with a residual of at least 3.5 mg free chlorine/L.

Non-purgeable TOX (NPTOX) resulting from the FP process was determined by standard method 5320 (15). Samples with an exceptionally high (greater than 30%) relative percent difference between duplicate measurements were reanalyzed.

Trihalomethanes (THM), haloacetonitriles (HAN), haloketones (HK), chloropicrin (trichloronitro-methane) and carbon tetrachloride resulting from the FP process were determined by EPA method 551 (16) using capillary gas chromatography with electron-capture detection. Compounds were identified according to their retention times, and compound retention times were confirmed using GC/MS under similar conditions.

## Results and Discussion

### Initial Field and Laboratory Measurements

Initial field and laboratory measurements for the four water samples are listed in Table I. The increase in pH and specific conductance between the Lost Creek and Fisher Creek samples reflects the influence of underlying limestone with which only Fisher Creek water is in contact. The pH and specific conductance values of the ground-water samples are much greater than the surface-water samples because of the inputs of carbonate minerals from the limestone aquifer. Total organic carbon values in the two surface waters are typical for "black waters," similar to the nearby Okefenokee Swamp and Suwannee River (17), whereas TOC values of the ground-water samples are much lower, suggesting significant biogeochemical attenuation of DOC by biodegradation, sorption, precipitation, and/or dilution processes. The values for tannic acid may be important for chlorination DBPs because condensed tannins of the proanthocyanidin and flavonoid types are important precursors (18).

Table I. Initial Field and Laboratory Measurements (NA, not available)

| Measurement | Lost Creek | Fisher Creek | Donahue Well | Trout Pond Well |
|---|---|---|---|---|
| pH | 3.9 | 4.7 | 7.2 | 7.6 |
| Specific conductance ($\mu$S/L) | 38 | 52 | 251 | 265 |
| Dissolved oxygen (mg/L) | 7.0 | 6.9 | 4.9 | 0.3 |
| Temperature (°C) | 24.8 | 23.9 | 21.1 | 22.0 |
| Total Organic Carbon, before ultrafiltration (mg/L) | 35.8 | NA | NA | 1.3 |
| Total Organic Carbon, after ultrafiltration (mg/L) | 25.2 | 20.3 | 2.5 | 0.2 |
| Tannic Acid (mg/L) | 8.7 | 7.5 | NA | NA |

## Chemical and Isotopic Composition of Ground Water and Surface Water

Values of delta deuterium ($\delta$D) and delta 18-oxygen ($\delta^{18}$O) in ground and stream water were compared to concentrations in rainfall to determine the source(s) of water. Similar $\delta^{18}$O and $\delta$D values were observed for water from all four sites. Isotopic values for the two ground-water samples plot along the global meteoric water line, GMWL (*19*), as do samples of rainfall from the study area (*9*) indicating that they are probably little affected by evaporation. The isotopic composition of stream water also plots along the GMWL. Close agreement among tritium concentrations for ground water, surface water, and rainfall (*9*) indicates that recharge by rainfall to the Upper Floridan aquifer occurred during the past 30 to 40 years and most likely represents very recent recharge.

## NOM Fractionation and Organic Carbon Recovery

Table II presents NOM fraction mass and organic carbon recovery data. The mass of the hydrophilic neutral fraction was so small that this fraction was not spectrally characterized or reacted with chlorine. Recovery of organic carbon using the column isolation procedure in Figure 1 averaged near 50% for the surface-water samples, but the recovery data were erratic for the two ground-water samples because of known variability associated with low TOC determination. However, the NOM fractions obtained for the chlorination study should be reasonably representative of the whole NOM.

## FTIR Spectral Characterization of NOM Fractions

FTIR is especially useful for qualitative identification of oxygen and nitrogen functional groups in NOM. It can serve as an assay of the purity of NOM fractions because borate, bicarbonate, carbonate, nitrate, phosphate, silicate, and sulfate salts are readily detected. For interpretation of the spectra of pure compounds, the reader is referred to Pouchert (*20*), and for analysis of complex biomolecular structures and humic substances, to Bellamy (*21*) and Stevenson (*22*), respectively.

FTIR characterization of isolates is useful for identifying the proteinaceous component of NOM, which is difficult to identify using [13]C-NMR. The hydrophobic and hydrophilic nature of NOM isolates can also be assessed by the relative abundances of hydrocarbon (hydrophobic) and carbohydrate (hydrophilic) moieties indicated by the IR spectrum. Inorganic salts (with the exception of chloride salts) can be a major interference, so purification requirements are substantial in order for FTIR spectrometry to be a useful tool for NOM characterization.

FTIR for the NOM fractions isolated in this study are shown in Figures 2a for Lost Creek (Fisher Creek was similar, unless noted), and Figure 2b for Donahue Well (Trout Pond Well was similar, unless noted). The IR spectra for the two surface-water samples were almost identical with no direct indications of inorganic constituents in the NOM fractions.

**Table II   NOM fractionation mass and organic carbon recovery data**
**(NA, not available)**

| NOM Fraction | *Lost Creek* *Mass fraction* *Percent C* *Mass C* | *Fisher Creek* *Mass fraction* *Percent C* *Mass C* | *Donahue Well* *Mass fraction* *Percent C* *Mass C* | *Trout Pond Well* *Mass fraction* *Percent C* *Mass C* |
|---|---|---|---|---|
| Colloids | 388.1 mg 30.1 % 116.8 mg C | 239.9 mg 22.5 % 54.0 mg C | 60.6 mg 1.6 % 1.0 mg C | 15.3 mg 3.6 % 0.6 mg C |
| Hydrophobic Acids | 436.6 mg 41.8 % 182.5 mg C | 299.6 mg 45.3 % 135.7 mg C | 220.2 mg 45.7 % 100.6 mg C | 62.5 mg 44.0 % 27.5 mg C |
| Hydrophobic Acids 2 | 63.4 mg 37.1 % 23.5 mg C | 37.9 mg 50.1 % 19.0 mg C | NA | NA |
| Hydrophilic Acids | 190.6 mg 31.0 % 59.1 mg C | 103.3 mg 37.9 % 39.2 mg C | 35.5 mg 26.5 % 9.4 mg C | 16.2 mg 3.6 % 0.6 mg C |
| Bases | 78.8 mg 5.0 % 3.9 mg C | 29.7 mg 5.7 % 1.7 mg C | 42.2 mg 9.1 % 3.8 mg C | 32.1 mg 4.9 % 1.6 mg C |
| Hydrophobic Neutrals | 78.7 mg 50.5 % 39.7 mg C | 22.1 mg 42.0 % 9.3 mg C | 21.6 mg 25.7 % 5.6 mg C | 3.0 mg NA NA |
| Hydrophilic Neutrals | 10.2 mg NA NA | 6.3 mg NA NA | 4.3 mg NA NA | 1.6 mg NA NA |
| Mass C Recovered | 425.5 mg C | 258.9 mg C | 120.4 mg C | 30.3 mg C |
| Mass C in Filtered Sample | 841.3 mg C | NA | NA | 109.6 mg C |
| Percent C Recovery in Filtered Sample | 50.6 % | NA | NA | 27.6 % |
| Mass C Recovered in Ultra-Filtered Sample | 308.7 mg C | 204.9 mg C | 119.4 mg C | 29.7 mg C |
| Mass C in Ultra-filtered Sample | 592.2 mg C | 408.4 mg C | 330.2 mg C | 16.9 mg C |
| Percent C Recovered in Ultra-filtered Sample | 52.1 % | 50.2 % | 36.2 % | 175.8 % |

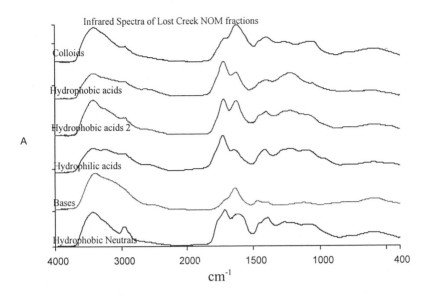

*Figure 2a. Infrared spectra of Lost Creek NOM fractions.*

Peaks near 1600 and 1400 $cm^{-1}$, indicating carboxylate groups for the colloid and hydrophobic neutral fractions in all these samples, suggest these fractions were isolated as the neutralized salt form of the acids. The acid fractions all have peaks near 1720 $cm^{-1}$ that are indicative of free carboxylic acid groups. The hydrophobic neutral fractions in all these samples are richest in (hydrophobic) aliphatic hydrocarbons as indicated by the peak near 2940 $cm^{-1}$. The colloid and hydrophilic acid fractions have a broad peak near 1100 $cm^{-1}$ indicative of alcohol groups in carbohydrates and hydroxy acids, especially in the ground waters. The broad peak near 1620 $cm^{-1}$ in the hydrophobic acid and hydrophobic acid 2 fractions from the surface waters are indicative of aromatic rings and aromatic ketonic groups such as quinones in fulvic acid that predominates in these black surface waters. This 1620 $cm^{-1}$ peak is smaller in the hydrophobic acids from the ground-water samples and indicates that fulvic acid has been removed by biogeochemical processes during ground-water infiltration. For the ground-water samples, peaks near 800 and 470 $cm^{-1}$ in the colloid fractions are indicative of silica (probably clay minerals) in these fractions.

### [13]C-NMR Spectral Characterization of NOM Fractions

Application of solution and solid-state NMR spectrometry to obtain quantitative and qualitative information on organic carbon structural distributions in NOM has been reviewed by Nanny et al. (*23*), Wershaw and Mikita (*24*), and Wilson (*25*). [13]C-NMR spectra of the NOM fractions isolated from Lost Creek are shown in Figure

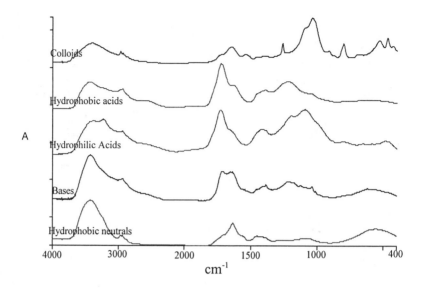

*Figure 2b. Infrared spectra of Donahue Well NOM fractions.*

3a (Fisher Creek spectra were similar, unless noted). For both surface-water samples, the colloid composition is predominately carbohydrate (peaks at 72 and 103 ppm) with lessor amounts of aliphatic carbon (peak at 31 ppm), phenols (152 ppm), and carboxylic acids (174 ppm). Generally, this composition corresponds to tannins and carbohydrates freshly leached from plants. The hydrophobic acid fractions in the two surface-water samples have spectral profiles very similar to fulvic acid isolated from the Suwannee River that has been extensively characterized (*17*). These black-water fulvic acids are likely derived from oxidized tannins.

The hydrophobic acid 2 (HPO-A2) fractions in the surface-water samples have a composition intermediate between the colloid and hydrophobic acid (HPO-A) fractions. Based on the ultrafiltration and resin sorption behavior, their molecular size is smaller than the colloids, but larger than the hydrophobic acid fraction because they were initially size-excluded from sorption on the resins. When this hydrophobic acid 2 fraction was acidified, the size must have decreased because it subsequently adsorbed on the XAD-8 resin. The Fisher Creek hydrophobic acid 2 fraction had more aliphatic carbon (peak at 30 ppm, not shown) component than the corresponding fraction from Lost Creek.

The hydrophilic acid (HPI-A) fractions from surface waters are predominately aliphatic hydroxy acids; the aromatic peak characteristic of tannins and lignins at 130 ppm is now a minor peak. The sharp peak at 20 ppm is likely the methyl group of acetate esters formed during desalting of this fraction with acetic acid. Although the base fractions show small peaks near 50 ppm that may be due to C-N bonds of

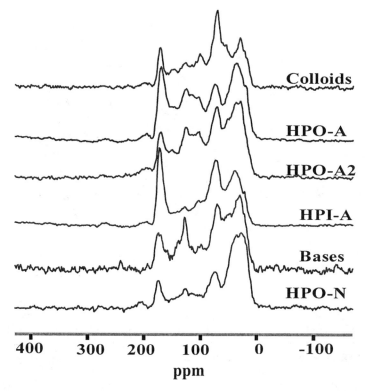

*Figure 3a. [13]C- NMR spectra of the NOM fractions isolated from Lost Creek.*

amines, these fractions consist predominately of very hydrophobic (peaks at 30 and 130 ppm) acids that adsorb to the matrix of the cation exchange resin during the NOM fractionation procedure. Finally, the hydrophobic neutral (HPO-N) fraction of the two surface-water samples are predominately complex plant lipids as evidenced by the broad aliphatic hydrocarbon peak from 0-60 ppm.

[13]C-NMR spectra of the NOM fractions isolated from Donahue Well are shown in Figure 3b (Trout Pond Well spectra were similar, unless noted). The ground-water NOM fractions show markedly different composition, as indicated by [13]C-NMR spectral profiling than do the surface-water fractions (Figure 3a). The colloid fraction, despite the poor signal to noise ratio because of its very small mass, is predominantly aliphatic hydrocarbon in nature in the ground-water samples.

The hydrophobic acid fraction that constitutes the majority of the mass in these ground waters (Table II) is substantially different than this corresponding fraction in the surface waters. Based on overall abundance, the aliphatic carbons (0-60 ppm) are greater; hydroxyl groups (75 ppm) are less; aromatic carbon content is similar, but the

164

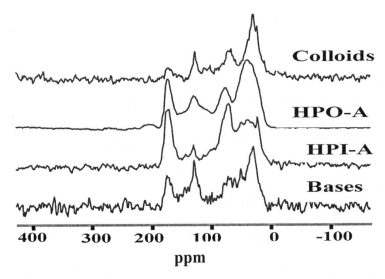

**Colloids**

**HPO-A**

**HPI-A**

**Bases**

400　　300　　200　　100　　0　　-100
ppm

*Figure 3b. ¹³C- NMR spectra of the NOM fractions isolated from Donahue Well.*

aromatic peak profile is more symmetrical than the skewed aromatic profile in the surface-water samples that is caused by phenolic constituents in the tannins. Comparing the surface- and ground-water spectra, the very short chain aliphatic acids have decreased during infiltration through limestone due to their complexation with available calcium, as reflected in the hydrophilic acids NMR spectra. Lastly, the small ketone peak at 205 ppm in the ground-water samples is indicative of aliphatic ketones, whereas this peak at 195 ppm in the surface-water sample indicates aromatic ketones. These ground-water hydrophobic acids appear to be more degraded and fractionated (loss of aromatic tannins) than these acid fractions from the surface-water samples.

The hydrophilic acid fraction from the Donahue Well is predominately hydroxy acids, but moderately well-defined peaks at 38, 49, 130, and 143 are suggestive of specific hydroxy-acid metabolites formed during biodegradation of the NOM. The base fractions in the ground water have peaks near 50 ppm suggestive of amine C-N linkages, but little significance should be given to this fraction because of its small mass.

The relatively large percentage of colloid carbon for the surface-water samples (22-30% carbon, Table II) is unusual. Colloidal carbon percentages for Mississippi River samples isolated similarly averaged about 15% of the total organic carbon, and its composition is high in N-acetyl amino sugars characteristic of bacterial cells (*26*). The unusual size behavior of the hydrophobic acid 2 fraction upon acidification led to an experiment whereby half of the Lost Creek colloid fraction was resuspended in 10 liters of water, acidified with HCl to pH 2, and successively ultrafiltered through 30,000 dalton (the lower size limit of the initial colloids before acidification) and

10,000 dalton membranes to determine composition changes with size changes upon acidification. Of the total colloid mass after acidification, 52% was greater than 30,000 daltons, 19% was between 10,000 and 30,000 daltons, and 29% was less than 10,000 daltons. The [13]C-NMR spectra of these three colloid size fractions is shown in Figure 4.

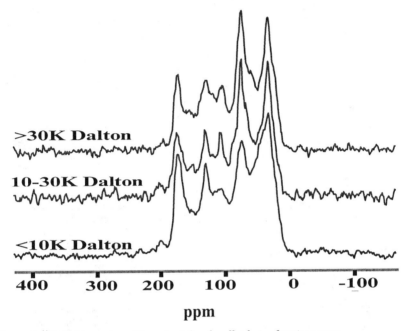

*Figure 4. [13]C-NMR spectra of three Lost Creek colloid-size fractions upon acidification.*

The two colloid fractions retained on the membrane are enriched in aliphatic carbon (30 ppm) and carbohydrate (75 ppm and 104 ppm), whereas the low molecular-weight fraction is enriched in aromatic (90-160 ppm) and carboxylic acids (174 ppm). The broad aromatic carbon hump from 90 to 115 ppm may be related to aromatic carbons between the meta-dihydroxy benzene structures that are common in condensed tannins (*27*). These structures are very important for the present study because meta-dihydroxybenzene structures have been found to produce THMs upon reaction with chlorine (*1*). The release of these lower molecular weight tannins upon acidification may be related to depolymerization reactions with acid (*28*). Alternatively, metal-NOM complexes might be disrupted by acidification. Released NOM macromolecules then might coil in acid due to increased intramolecular hydrogen bonding resulting in smaller size.

Indications of phloroglucinol A rings in condensed tannins were suggested by the aromatic carbon intensity from 90-115 ppm in the colloid, hydrophobic acid, and hydrophobic acid 2 fractions, but anomeric carbons of carbohydrates significantly interfere in this region. Dipolar dephased NMR measurements that detect only methyl and quaternary substituted carbons were developed as a more specific probe to detect residues of this phloroglucinol A ring in NOM fractions based upon known chemistry of tannin degradation. Flavonoid pigments, such as quercetin, released from tannins upon plant senescence are very unstable in the presence of oxygen and degrade in a matter of hours according to the following reaction (29):

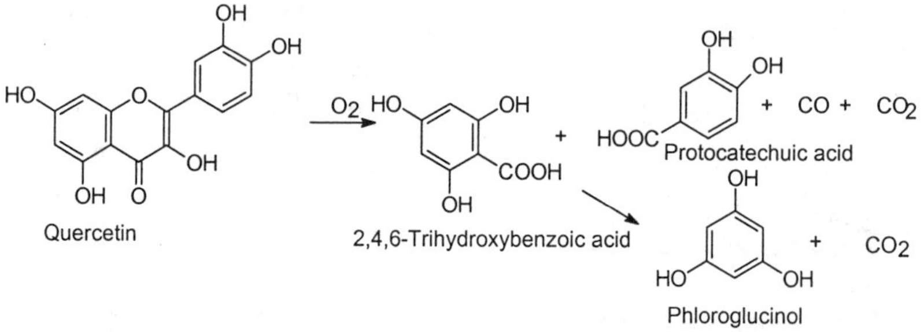

Quercetin

2,4,6-Trihydroxybenzoic acid

Protocatechuic acid

+ CO + CO₂

Phloroglucinol

+ CO₂

The phloroglucinol product of this reaction is very reactive and condenses with itself, with amines to form browning reaction condensation products, and with aldehydes and ketones. The condensation reaction with an aldehyde is illustrated by the following reaction (30):

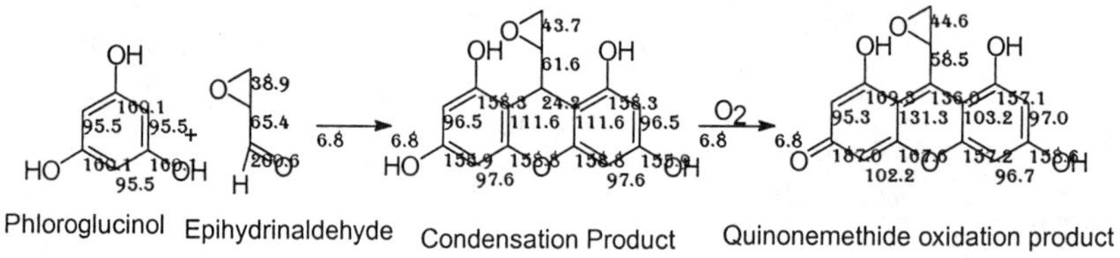

Phloroglucinol  Epihydrinaldehyde  Condensation Product  Quinonemethide oxidation product

The numbers on the chemical structures of the above reaction are calculated [13]C-NMR chemical shifts. The unsubstituted positions in the phloroglucinol ring before and after condensation reactions give shifts from 95 to 98 ppm, whereas the carbon-substituted positions produce shifts from 103 to 112 ppm. It is postulated that almost all of the unsubstituted positions in the phloroglucinol rings will become substituted when humic substances form from these complex condensation reactions during plant senescence. The carbon-substituted phloroglucinol ring should be detected as a peak between 100 and 115 ppm by the dipolar dephasing pulse sequence, and most of the anomeric carbons of carbohydrates should not be detected because these carbons are protonated.

A broad peak from 95 to 120 ppm that is indicative of the condensed tannin hypothesis for NOM was found in most of the spectra of Figure 3, but it is by far the most intense for the hydrophobic acid fractions of the two surface-water samples.

This peak would not be as intense for the colloid faction because the condensation reactions have not proceeded to completion. Also, it would not be as intense for the Donahue Well hydrophobic acid fraction because of the removal of humic substances during infiltration. Lastly, it should not be intense in the Lost Creek hydrophilic acid fraction because of its low aromatic carbon content. These spectra provide strong evidence for flavone-type tannin pigments as precursors for NOM in the two surface-water samples.

## Chlorination and DBP Production from NOM Fractions

Because DBPs result from the reaction of chlorine with DOC, the DOC of each reconstituted FP sample (close but not exactly as in the original source water) are listed in Table III. Hydrophobic acids were the most abundant fraction based on mass and organic carbon (Table II). Second most abundant was the colloid fraction. DOC concentration decreases dramatically between the surface- and ground-water sites. The colloid fraction is effectively filtered out by the aquifer, leaving only minute amounts in the ground waters. The colloid fraction, as isolated here by ultrafiltration, would probably be partially removed by initial sand filtration treatment in drinking water treatment plants, but provide an important contribution to the chemistry of the whole water.

Specific ultraviolet absorbance (SUVA), the absorbance per meter at 254 nm per mg organic carbon per liter (L/mg-m), is shown for each fraction in Table III. Absorbance at 254 nm is due predominantly to aromatic constituents in the NOM (*31*). Absorbance and fluorescence analyses indicate that chlorination of NOM is accompanied by selective oxidation of activated aromatic rings (*32*). SUVA is indicative of aromatic content, which is considered to be a surrogate parameter for the THM-formation reactive constituent of the DOC (*33, 34*). In these surface and ground waters, the colloid, hydrophobic acid, and hydrophobic acid 2 fractions had the highest SUVA, except for the Trout Pond Well. If only aromatic components contribute to the formation potential, the colloids, hydrophobic acids, and hydrophobic acids 2, based on their SUVA, would be the most reactive with chlorine, and the most DBP productive. SUVA for these NOM fractions are higher, in general, than SUVA reported in the literature for whole waters (*35*). SUVA of five surface-water fulvic acids, the most similar to the hydrophilic acids isolated here, ranged from 2.86 to 4.28 L/mg-m, whereas SUVA of surface-water humic acids, the most similar to the colloids isolated here, ranged from 4.86 to 7.36 L/mg-m (*36*).

Samples of each fraction were analyzed for FP of DBPs. In order to compare these data with other studies, NPTOX concentrations were converted to NPTOX yield, that is, microgram NPTOX formed per milligram organic carbon (Table III). These yields are comparable to yields in the literature from a variety of water sources. Typical TOX yields from FP studies on whole waters range from about 80 to 500 µg/mg (*37*). NPTOX yields from whole surface waters ranged from 174 to 348 µg/mg (*37*). TOX yields of raw drinking waters from seven cities ranged from 170 to 298 µg/mg (*34*). TOX yields of more specific fractions from five surface waters ranged for fulvic acids from 136 to 232 and averaged 191 µg/mg, and for humic acids from 230 to 288 and averaged 256 µg/mg (*36*). FPs of drinking waters

**Table III. Dissolved organic carbon (DOC) of reconstituted fractions, specific ultraviolet absorbance (SUVA), and disinfection by-product (NPTOX=nonpurgeable total organic halide, THM=trihalomethane, HAN=haloacetonitriles, HK=haloketones) formation potential yields by NOM fraction of four Florida surface- and ground-water samples (Fisher Creek=FC, Lost Creek=LC, Trout Pond Well=TPW, Donahue Well=DW, HPOA-2=Hydrophobic acids 2)**

| | Colloids | | | | Hydrophobic Acids | | | | Hydrophobic Neutrals | | |
|---|---|---|---|---|---|---|---|---|---|---|---|
| | FC | LC | TPW | DW | FC | LC | TPW | DW | FC | LC | DW |
| DOC, mg/L | 3.8 | 5.1 | 0.1 | 0.2 | 5.3 | 4.4 | 5.5 | 3.2 | 2.3 | 6.6 | 1.9 |
| SUVA, L/mg-m | 5.5 | 6.1 | 13 | 9.7 | 5.2 | 5.8 | 4.4 | 8.1 | 4.1 | 3.1 | 2.9 |
| NPTOX yield, ug NPTOX/mg DOC | 440 | 370 | 210 | 800 | 220 | 250 | 140 | 170 | 120 | 150 | 100 |
| THM yield, ug THM/mg DOC | 130 | 120 | 110 | 300 | 80 | 90 | 46 | 56 | 44 | 46 | 41 |
| HAN yield, ug HAN/mg DOC | 1.5 | 1.3 | 5.9 | 5.6 | 1.1 | 1.1 | 0.53 | 0.87 | 0.44 | 0.43 | 0.40 |
| HK yield, ug HK/mg DOC | 0.34 | 0.30 | 2.00 | 1.60 | 0.29 | 0.42 | 0.27 | 0.40 | 0.48 | 0.31 | 0.43 |

| | Bases | | | | Hydrophilic Acids | | | | HPOA-2 | |
|---|---|---|---|---|---|---|---|---|---|---|
| | FC | LC | TPW | DW | FC | LC | TPW | DW | FC | LC |
| DOC, mg/L | 0.29 | 0.73 | 0.40 | 0.65 | 3.5 | 4.3 | 0.12 | 2.4 | 3.1 | 3.9 |
| SUVA, L/mg-m | 4.1 | 4.7 | 3.7 | 2.8 | 3.2 | 3.4 | 5.0 | 1.9 | 6.0 | 6.5 |
| NPTOX yield, ug NPTOX/mg DOC | 240 | 630 | 220 | 200 | 170 | 210 | 350 | 100 | 280 | 390 |
| THM yield, ug THM/mg DOC | 93 | 140 | 89 | 72 | 79 | 83 | 140 | 81 | 78 | 110 |
| HAN yield, ug HAN/mg DOC | 4.2 | 3.0 | 2.4 | 2.3 | 0.94 | 0.83 | 6.5 | 1.3 | 1.8 | 1.7 |
| HK yield, ug HK/mg DOC | 1.1 | 1.0 | 0.84 | 0.75 | 0.37 | 0.39 | 1.8 | 0.29 | 0.31 | 0.33 |

are usually much lower because steps are sometimes taken to minimize DOC before chlorination takes place. We found NPTOX yields ranged from 13 to 50 μg/mg in reclaimed water, where low reactivity would be expected because of chlorination and dechlorination prior to release (unpublished data).

Some of the THM-FP yields in μg THM/mg organic carbon, are high (Table III) compared to other studies (usually on whole waters). Only traces of brominated DBPs were found, except from the Donahue Well hydrophilic acids at slightly elevated amounts. THM yields from the literature range from 32 to 190 μg/mg for various conditions (37). THM yields of surface (river) waters ranged from 47.5 to 188 μg/mg under the same conditions used here (37). THM yields of raw drinking waters from 7 cities ranged from 41 to 68 μg/mg (34). Chloroform yields from five surface-water fulvic acids (3-day reaction time at 20°C, only chloroform reported) ranged from 30.8 to 50.8 μg/mg (36), compared to 90 μg/mg from Lost Creek and 80 μg/mg from Fisher Creek hydrophobic acids. Chloroform yields from five humic acids ranged from 47.2 to 68.2 μg/mg (36), compared to 120 and 130 μg/mg from these surface-water colloids. THM yields from 50 Kansas ground waters, whose TOC ranged from 0.21 to 2.14 mg/L, ranged from about 25 to 83 μg/mg (38).

Minor components formed during chlorination and quantified include the HANs (trichloroacetonitrile, dichloroacetonitrile, bromochloroacetonitrile, and dibromo-acetonitrile), HKs (1,1-dichloropropanone and 1,1,1-trichloropropanone) (Table III), chloropicrin (trichloronitromethane), and carbon tetrachloride (data not shown). These compounds were found at only trace concentrations, amounts typical for the chlorination disinfection process, such as those reported for Utah drinking waters (39). Again, during drinking water treatment some DBP precursors may be removed from the surface waters prior to or during chlorination. Concentrations of these compounds in drinking water have decreased substantially over the last 20 years, as water treatment processes have been improved and refined.

Most importantly, the different NOM fractions had different chlorination reactivity (some variability in yield is due to dividing by the DOC value, which is highly variable at the very low concentrations found in the ground waters). The colloid fraction typically had the greatest reactivity or yield for producing NPTOX and THMs. This colloid fraction resembles degraded tannins, based on the NMR. In surface waters, the colloid fraction contributes a significant portion of the DOC, and that portion is very reactive. In general, the second most reactive fraction was the base fraction (although it is a minor portion of the DOC) especially for HAN formation. This was not surprising due to the higher nitrogen content usually present in the base fraction due to amino acids (seen in the NMR spectra as C-N linkages), a possible precursor for the HANs. The hydrophobic neutrals had the lowest yields, due to lack of reactive functional groups, as seen in the NMR spectra.

For average NPTOX yield, the colloids, bases and hydrophobic acids 2 were the most reactive. For THM, the colloids were the most reactive, followed by bases, hydrophilic acids and hydrophobic acids 2, which all had similar average yields. The hydrophobic acids were the most abundant fraction by mass, but had lower yields than most other fractions, on average. This would be expected based on the FTIR and NMR, which indicated the presence of more aliphatic carbon (less DBP reactive) and less hydroxyl groups (more DBP reactive).

During infiltration, the character of the DOC changes, as reflected in the changes in yield. The mass of the most reactive colloid fraction decreased during infiltration. Colloid yield of NPTOX and THM remained high, whereas the colloid yield of HAN and HK increased after infiltration. For the hydrophobic acid fraction, the most dominant in terms of mass, the yield of NPTOX and THM decreased during infiltration. As seen in the NMR, this reflects a decrease in degraded tannins, that produce higher yields.

Major technologies for removal of DBP precursors are enhanced coagulation, granular activated carbon adsorption, and membrane filtration (40). The effectiveness of TOC removal by coagulation depends on the TOC content and alkalinity of the raw water, the hydrophobic/hydrophilic distribution of the TOC and pH of coagulation (40). Infiltration through limestone karst did not change the hydrophobic/hydrophilic distribution of the DOC. It was surprising that the drastic decrease in DOC during infiltration produced only minor differences in the distributions of the DOC fractions, except for the colloids. However, the character of each DOC fraction changed during infiltration, as indicated by FTIR and NMR, and reflected by the NPTOX-FP and THM-FP yields. Based on the reactions discussed earlier, condensations fill the reactive sites on the tannins and the DBP yields decrease.

## Literature Cited

1. Larson, R.A.; Weber, E.J. *Reaction Mechanisms in Environmental Organic Chemistry;* Lewis Publishers, Ann Arbor, MI, 1994.

2. Kronberg, L.; Christman, R.F. *Sci. Total Environ.* **1989**, *81-82*, 219-230.

3. Rupert, F.R.; Spencer, S. *Florida Geological Survey Bulletin No. 60*, 46 pp., **1988**.

4. Davis, J.H. *Water Resour. Invest. (U.S. Geol. Surv.) 95-4296*, 55 pp., **1996**.

5. Brown, E.; Skougstad, M.W.; Fishman, M.J. *Techn. Water Resour. Invest. (U.S. Geol. Surv.)*, Bk 5, Ch. C1, 160 pp. **1970**.

6. Wood, W.W. *Techn. Water Resour. Invest. (U.S. Geol. Surv.)*, Bk 1, Ch. D2, 24 pp., **1976**.

7. Skougstad, M.W.; Fishman, M.J.; Friedman, L.C.; Erdman, D.E.; Duncan, S.S. *Open-File Rep. (U.S. Geol. Surv.), 78-679*, 159 pp., **1979**.

8. Kloster, M.B. *J. Am. Water Works Assoc.* **1974**, *66*, 44-51.

9. Katz, B.G.; Coplen, T.B.; Bullen, T.D.; Davis, J.H. *Ground Water* **1997**, *35* (6), 1014-1028.

10. Michel, R.M. *Water Resour. Invest. (U.S. Geol. Surv.) 89-4072*, 46 pp., **1989**.

11. Leenheer, J.A.; Meade, R.H.; Taylor, H.E.; Pereira, W.E. *Water Resour. Invest. (U.S. Geol. Surv.) 88-4220*, 501-511, **1989**.

12. Leenheer, J.A.; Noyes, T.I. *U.S. Geol. Surv. Water-Supply Paper 2230*, 16 pp., **1984**.

13. Leenheer, J.A. Natural Organic Matter Workshop, Sept. 18-19, 1996, Poitiers, France; pp. 4-1 to 4-5, 1996.

**171**

14. Leenheer, J.A., Croue, J-P., Benjamin, M., Korshin, G.V., Hwang, C.J., Bruchet, A., Aiken, G.R. Comprehensive isolation of natural organic matter from water for spectral characterizations and reactivity testing. In *Natural Organic Matter and Disinfection By-Products: Characterization and Control in Drinking Water*; Barrett, S.E., Krasner, S.W., Amy, G.L., Eds.; Am. Chem. Soc., Washington, DC (in press).

15. Greenberg, A.E.; Clesceri, L.S.; Eaton, A.D., Eds. *Standard Methods for the Examination of Water and Wastewater, 19th Ed.*; Am. Public Health Assoc., Am. Water Works Assoc., Water Environ. Fed.; pp. 5-17 to 5-19, 5-22 to 5-27, 5-50 to 5-59; 1995.

16. U.S. Environmental Protection Agency. In *Methods for the Determination of Organic Compounds in Drinking Water, Suppl. 1*; PB91-146027, EPA-600/4-90/020. USEPA; pp. 169-200, 1990.

17. Averett, R. C.; Leenheer, J. A.; McKnight, D. M.; Thorn, K. A. *U.S. Geol. Surv. Water-Supply Paper 2373*, **1994**.

18. Fam, S.; Stenstrom, M.K. *J. Wat. Pollut. Control Fed.*, **1987**, *59*, 969-978.

19. Craig, H. *Science*, **1961**, *133*, 1702-1703.

20. Pouchert, C.J. *Aldrich Library of FT-IR Spectra, Vol. 1 and 2*; Aldrich Chem. Co., Milwaukee, WI; 958 pp., 1985.

21. Bellamy, L.J. *Infrared Spectra of Complex Molecules, 3rd Ed.;* John Wiley and Sons, London; 1975.

22. Stevenson, F.J. In *Humus Chemistry, Genesis, Composition, Reactions*; John Wiley and Sons, New York; pp. 264-284, 1982.

23. Nanny, M.A.; Minear, R.A.; Leenheer, J.A., Eds. *Nuclear Magnetic Resonance in Environmental Chemistry*; Oxford University Press, New York; 326 pp., 1997.

24. Wershaw, R.L.; Mikita, M.A. *NMR of Humic Substances and Coal*; Lewis Publishers, Chelsea, MI; 1987.

25. Wilson, M.A. *NMR Techniques and Applications in Geochemistry and Soil Chemistry*; Pergamon Press, Sydney, Australia; 1987.

26. Leenheer, J.A.; Barber, L.B.; Rostad, C.E.; Noyes, T.I. *Water Resour. Invest. (U.S. Geol. Surv.) 94-4191*, 47 pp., **1995**.

27. Porter, L.J.; Newman, R.H.; Foo, L.Y.; Wong, H. *J. Chem. Soc., Perkin Transactions*, **1982**, *1*, 1217-1221.

28. Steelink, C. In *Humic Substances in Soil, Sediment, and Water, Geochemistry, Isolation, and Characterization*, Aiken, G.R.; McKnight, D.M.; Wershaw, R. L., Eds.; John Wiley and Sons, New York; pp. 457-476, 1985.

29. Nordstrom, C.G. *Suomen Kemistilehti* **1968**, *B 41*, 351-353.

30. Fiegl, F. *Spot Tests in Organic Analyses*; Elsevier Press, Amsterdam; 1960.

31. Novak, J.M.; Mills, G.L.; Bertsch, P.M. *J. Environ. Qual.* **1992**, *1*, 144-147.

32. Korshin, G.V.; Li, C-W.; Benjamin, M.M. In *Water Disinfection and Natural Organic Matter: Characterization and Control*, Minear, R.A., Amy, G.L., Eds.; Am. Chem. Soc., Washington, DC; pp. 182-195, 1996.

33. Singer, P.C.; Chang, S.D. *J. Am. Water Works Assoc.* **1989**, *81*, 61-65.

34. Reckhow, D.A.; Singer, P.C. *J. Am. Water Works Assoc.* **1990**, *82*, 173-180.

35. Rathbun R.E.; Bishop, L.M. *Open-File Rep. (U.S. Geol. Surv.) 93-158*, 57 pp., **1993**.

36. Reckhow, D.A.; Singer, P.C.; Malcolm, R.L. *Environ. Sci. Technol.* **1990**, *24,* 1655-1664.
37. Rathbun, R.E. *Arch. Environ. Contam. Toxicol.* **1996**, *31* (3), 420-425.
38. Miller, R.E.; Randtke, S.J.; Hathaway, L.R.; Denne, J.E. *J. Am. Water Works Assoc.* **1990**, *82,* 49-62.
39. Nieminski, E.C.; Chaudhuri, S.; Lamoreaux, T. *J. Am. Water Works Assoc.* **1993**, *85,* 98-105.
40. Singer, P.C. *J. Environ. Eng.* **1994**, *120,* 727-744.

Chapter 12

# Disinfection By-Product Formation Reactivities of Natural Organic Matter Fractions of a Low-Humic Water

Cordelia J. Hwang, Michael J. Sclimenti, and Stuart W. Krasner

Metropolitan Water District of Southern California, 700 Moreno Avenue, La Verne, CA 91750–3399

Colorado River water (CRW), a low-humic water (specific ultraviolet absorbance of 1.0 to 1.6 L/mg·m), was fractionated and isolated into five natural organic matter (NOM) fractions—hydrophobic, transphilic, hydrophilic acid plus neutral (A+N), hydrophilic bases, and colloids—and then chlorinated or chloraminated. The disinfection by-product (DBP) yields of the major NOM fractions in CRW (on the basis of µmol DBP/µmol dissolved organic carbon) contrasted strongly with those of high-humic waters. The highest yield of trihalomethanes came from the hydrophilic A+N, not the hydrophobic NOM. Production of dihaloacetic acids (DXAAs) was significantly higher than that of trihaloacetic acids (TXAAs) after both chlorination and chloramination, whereas TXAAs are typically higher than DXAAs in high-humic waters after chlorination. The greatest decrease in haloacetic acid yield after ozonation was in the hydrophilic A+N fraction. The haloacetonitrile yields were highest for the hydrophilic A+N and hydrophilic base fractions, which were highest in nitrogen content. In a water low in humic NOM, the more polar fractions can impact DBP type and level almost as much as the hydrophobic NOM. Thus, treatments to minimize DBP formation in CRW should address removal of the hydrophilic A+N fraction.

Natural organic matter (NOM) in drinking-water sources typically consists of humic (hydrophobic) substances, such as humic and fulvic acids, and nonhumic (hydrophilic) material, which is often of biological origin. These materials include precursors for disinfection by-products (DBPs), which form upon reaction with disinfectants such as chlorine and chloramines. Some of these DBPs are of health concern. Stage 1 of the Disinfectants/DBP Rule lowers the allowable level of trihalomethanes (THMs) and adds regulations for haloacetic acids (HAAs) and bromate (*1*).

In addition, most surface-water utilities are required to remove DBP precursors—as measured by total organic carbon (TOC)—by enhanced coagulation *(1)*. TOC or dissolved organic carbon (DOC), ultraviolet (UV) absorbance at 254 nm, and specific UV absorbance (SUVA)—UV $(m^{-1})$/DOC (mg/L)—have been used as indicators of the DBP-forming capability of a source water. In U.S. surface waters, typical ranges for TOC are 2 to 20 mg/L and for SUVA are 1.4 to 5.5 L/mg·m *(2)*. Typically, SUVA at <3 L/mg·m contains largely nonhumic material, whereas SUVA in the range of 4-5 L/mg·m contains mainly humic material *(3)*. Humic materials have been more extensively studied and are more easily removed through coagulation than nonhumic substances *(4)*.

Colorado River water (CRW) is a low-DOC water (~2.5 mg/L), containing at least 50 percent nonhumic DOC, with a low SUVA value (1.0–1.6 L/mg·m). Such a low-humic water is not readily amenable to NOM removal by enhanced coagulation *(5)*. The NOM present in CRW may pose difficulties as DBP regulations become more stringent. Understanding the character of various components of the NOM and their relation to DBP formation should provide guidance in selection of treatment options for such low-humic waters.

## Experimental Conditions and Analytical Methods

### Experimental Approach

NOM from CRW was isolated and fractionated for two treatment schemes with two final disinfectants. Each water sample (raw and treated) was separated and isolated into five fractions generally based on polarity, but operationally on behavior on various resins. Isolation of XAD-8 and XAD-4 resin-retained NOM has been reported in the literature *(6)*; however, isolation of hydrophilic and colloidal fractions free of inorganic salts has required development of new techniques *(7)*. The two treatment scenarios evaluated were conventional coagulation with dual-media (anthracite/ sand) filtration (currently used at Metropolitan) and ozonation with biofiltration (may be used to meet future, more stringent DBP regulations). Final disinfection with chorine was evaluated because it is the most common disinfectant and is currently used for primary disinfection at plants operated by the Metropolitan Water District of Southern California (Metropolitan). Chloramines were evaluated because they are a common alternative disinfectant used in the distribution system to minimize DBP formation. For each treatment/disinfection scenario evaluated, the reactivities of the various NOM fractions with respect to the formation of 20 halogenated DBPs were determined.

### Treatment Plant Schemes and Sample Locations

CRW was collected from Metropolitan's 6-million-gallon-per-day Oxidation Demonstration Plant during two modes of operation: (1) preozonation (for virus and *Giardia* inactivation), followed by coagulation with ferric chloride (2 mg/L) and cationic polymer (2 mg/L), sedimentation, and biologically active filtration, and (2) conventional treatment (no ozone, not biologically active filters) with coagulant dose optimized for turbidity (not NOM) removal. Under the first mode, samples were collected at the plant influent (PI#1), at the ozone contactor effluent (OE), and at the effluent of the biologically active filters (BF). Under the second mode, samples were collected at the plant influent (PI#2) and filter effluent (FE).

## Sample Collection, NOM Isolation and Fractionation, and Profiling Procedure

Approximately 1000 L of each water sample was concentrated by reverse osmosis (RO) (Filmtech TW30-4040 membrane) to 20 L. The NOM was then isolated and fractionated by a series of resin separations, evaporations, selective precipitations, and freeze-drying techniques (7). Hydrophobic NOM, transphilic NOM, and the hydrophilic bases were separated on Amberlite XAD-8, XAD-4, and MSC-1H (cationic) resins, respectively, and the hydrophilic acid plus neutral (A+N) fraction was isolated by freeze drying after extensive desalting (7). The colloids were recovered by ultrafiltration from the 20-L recirculated sodium hydroxide (0.02 N) rinse of the RO membrane. Bulk water samples were also collected and characterized for comparison with isolated fractions.

In addition, the NOM of bulk water was profiled by sequentially passing the bulk water through XAD-8 and XAD-4 resins. Rather than isolating the NOM, estimates of the NOM content were based on the difference in DOC between the influent and effluent of a resin column. The hydrophobic NOM is retained on the XAD-8 resin, and the transphilic NOM is retained on the XAD-4 resin. The DOC not retained by the XAD-8 or XAD-4 resin was called the hydrophilic NOM.

## Chlorination and Chloramination

The NOM fractions were individually reconstituted in laboratory reagent water (Milli-Q), with sonication, to NOM concentrations of approximately 10 mg/L (by weight) of dry isolate, resulting in DOC concentrations of 3.3–5.7 mg/L. The bulk water samples and the reconstituted NOM fractions were chlorinated under uniform formation conditions (UFC) (8): pH 8.0 with 2 mM borate buffer, 20°C temperature, 24-h incubation time, and addition of buffered sodium hypochlorite to yield a free chlorine residual of 1.0 mg/L after incubation. The major NOM fractions (hydrophobic, transphilic, hydrophilic A+N) of the PI#1 were also analyzed with a bromide spike (100 µg/L). Chloramination conditions paralleled UFC chlorination but used preformed chloramines (pH 8), at a 5:1 chlorine:ammonia-nitrogen weight ratio, to yield a total chlorine residual of 2.0 mg/L after incubation. After incubation, pH and disinfectant residual were measured, the aliquots separated, the disinfectant residuals quenched, and the aliquots preserved for DBP analysis.

## DBP Analyses

Eleven neutral extractable DBPs were analyzed by liquid/liquid extraction (LLE) with *t*-butyl methyl ether and by gas chromatography with electron-capture detection (GC/ECD) according to U.S. Environmental Protection Agency (USEPA) Method 551.1 (*9*). These DBPs consisted of four THMs (chloroform, bromodichloromethane, dibromochloromethane, and bromoform); four haloacetonitriles, or HANs (trichloro-, dichloro-, bromochloro-, and dibromoacetonitrile); two haloketones (1,1-di- and 1,1,1-trichloropropanone); and chloropicrin. In addition, nine HAAs (bromochloro-, bromodichloro-, dibromochloro-, dibromo-, dichloro-, monobromo-, monochloro-, tribromo-, and trichloroacetic acid) were analyzed by LLE followed by derivatization with acidic methanol and by GC/ECD according to USEPA Method 552.2 (*10*).

# Results and Discussion

General findings based on DOC will be presented first. DBP formation will be discussed for the bulk waters and then the fractions of treated waters by major DBP classes, followed by the minor DBPs in the plant influent NOM fractions. An approximate mass balance was calculated by summation of the relative contributions of each NOM fraction and was compared to the bulk water DBP yields.

The observed DBP concentrations ($\mu$g/L) were converted to molar concentrations, and the yields were normalized to the carbon concentration on a molar percent basis (i.e., 100 × $\mu$mol of DBPs produced per $\mu$mol of DOC). The yields were normalized on a carbon basis to facilitate comparisons among NOM fractions because the concentration of DOC was different in each reconstituted NOM solution. Expressing the DBP yields on a molar basis also minimized the effect of variations in speciation (for example, bromo- versus chloro-) within the classes of DBPs.

All NOM fractions were free of bromide except the hydrophilic A+N fractions, which resulted in 75-159 $\mu$g/L bromide in the reconstituted fractions. Studies of DBP formation in CRW NOM fractions in the presence of a bromide spike (100 $\mu$g/L) showed that although a shift to increased bromine incorporation was observed, in general, the presence of bromide did not significantly increase the molar yield of THMs or HAAs. However, other researchers have observed a significant increase in the molar yield of THMs with increasing bromide concentration (*11*).

### Bulk Water Characterization

Raw CRW had the following water quality: total dissolved solids = 636 mg/L; alkalinity = 132 mg/L as calcium carbonate; pH = 8.2; DOC = 2.5 mg/L; UV = 0.038 cm$^{-1}$; bromide = 64-71 $\mu$g/L. The water quality of the two PI samples was consistent within experimental error. Ozonation coupled with biofiltration and conventional treatment of CRW resulted in small DOC reductions in the water (Table I).

**Table I. Effect of Treatment on DOC and SUVA of CRW**

| Parameter | PI#1 | OE | BF | PI#2 | FE |
|---|---|---|---|---|---|
| DOC, mg/L | 2.53 | 2.51 | 2.15 | 2.49 | 2.28 |
| SUVA, L/mg·m | 1.5 | 1.2 | 1.2 | 1.7 | 1.4 |

**NOM Distribution**

The DOC distribution among the CRW NOM fractions is given in Table II. The three major fractions—the hydrophobic, transphilic, and hydrophilic A+N fractions—constituted 96-99 percent of the isolated NOM. The sum of the DOC of the NOM fractions corresponded to 59 to 77 percent of the bulk water DOC. The DOC of the hydrophobic and transphilic fractions was comparable by both isolation and profiling procedures (as shown in Figure 1 for the PI samples. Portions of the hydrophilic and colloidal NOM were probably retained on the membrane and precipitated silica during RO concentration. The "not recovered" portion of the bulk water from the isolation procedure is probably a combination of hydrophilic A+N, hydrophilic base, and colloidal NOM *(7)*.

**Table II. NOM Distribution for CRW Samples**

| NOM Fraction | PI#1 | OE | BF | PI#2 | FE |
|---|---|---|---|---|---|
| Hydrophobic | 42% | 38% | 36% | 44% | 52% |
| Transphilic | 15% | 15% | 15% | 17% | 17% |
| Hydrophilic A+N | 13% | 9% | 7.5% | 13% | 7.6% |
| Hydrophilic bases | 0.4% | 0.7% | 0.2% | 1.0% | 0.5% |
| Colloids | 0.8% | 1.2% | 0.5% | 1.7% | 1.0% |
| Not recovered | 28% | 35% | 41% | 23% | 23% |

NOTE: Units are DOC w/w percent of bulk water DOC.

**NOM Reactivity with Respect to Chlorination**

*Effect of Treatment on DBP Yields in Bulk Water Samples*

Figure 2 shows the effects of ozonation and conventional treatment on the yields of THMs and HAAs in the bulk water samples. Ozonation resulted in 33- and 49-

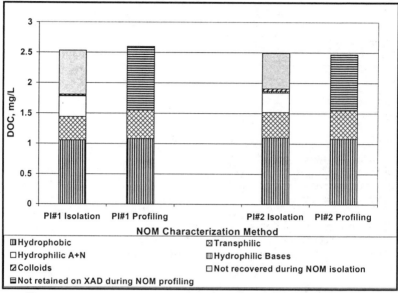

Figure 1. DOC of NOM fractions by isolation versus profiling methods.

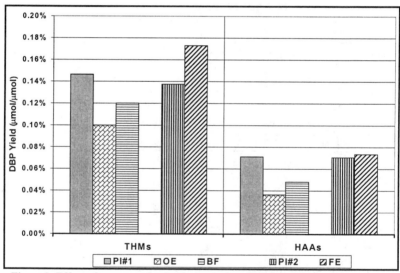

Figure 2. Effect of treatment (and chlorination) on THM and HAA yields for CRW bulk water samples.

percent decreases in the molar yield of THMs and HAAs, respectively. Ozonation can destroy a portion of the THM and HAA precursors (*12*) even though the NOM is not mineralized. In contrast, conventional treatment of CRW did not lower the DBP yields and resulted in a slight increase in the THM yield. After biofiltration, there was an increase in the DBP yields (20 percent for THMs and 33 percent for HAAs) as compared to the ozonated water. However, the DBP yields in the ozonated/biofiltered water were lower than in the raw water.

*Effect of Treatment on THM Yields in NOM Fractions*

Figure 3 shows the yields of THMs from chlorination of the five NOM fractions and illustrates how they were affected by ozonation and biofiltration. The trends for the NOM fractions followed those of the bulk water, with some exceptions. After ozonation, the THM yields of the three major NOM fractions were significantly lowered (40-48 percent), but the THM yields of the hydrophilic bases and colloids remained relatively constant. The highest THM yields were in the hydrophilic A+N fractions. This finding for CRW differs from the results for eight other waters, where the hydrophobic fraction had the highest THM yields (*13*). CRW is a water that is relatively low in hydrophobic NOM (*4*) (Table II), as exhibited by a relatively low raw water SUVA value, 1.54 L/mg·m. In reality, hydrophobic NOM fractions are a complex mixture of organic substances. These results suggest that the particular hydrophobic substances in CRW may be on the low end of the spectrum (as compared to waters higher in hydrophobic NOM) in terms of their reactivity to form THMs.

The NOM fractions isolated after final treatment with ozonation, coagulation, and biofiltration were more reactive toward THM production than the NOM isolated after ozonation alone. The colloid yield increased 86 percent, and the yield of the other four NOM fractions increased approximately 30 percent. These increases may be caused by a combination of factors, such as the removal of less reactive but biodegradable NOM or the sloughing-off of bacteria (or bacterial exudates) from the biofilters. This question will be examined further by pyrolysis GC/mass spectrometry (MS) and other characterizations of the NOM fractions (*13*).

The effect of conventional treatment on the reactivity of the five NOM fractions ranged from no change for the hydrophobic, transphilic, and colloidal fractions to a 28 percent increase in the THM yield for the hydrophilic A+N fraction and a 120 percent increase for the hydrophilic bases. In contrast to biofiltration, the conventional filters did not increase the THM yield of the colloids.

*Effect of Treatment on HAA Yields of NOM Fractions*

Figure 4 shows the effects of ozonation, biofiltration, and chlorination on the yield of HAAs in the ambient-bromide NOM fractions. As was observed with the THM yields (Figure 3), ozonation reduced the HAA yield of each of the three major NOM fractions and had only minimal impact on the hydrophilic bases and colloids. Biofiltration resulted in an even greater increase in reactivity of the hydrophilic A+N fraction and the colloids for HAA formation than for THM formation.

The highest yields of HAAs in the CRW samples were in the more hydrophilic fractions—that is, hydrophilic A+N and bases for all five sample locations (ozonation

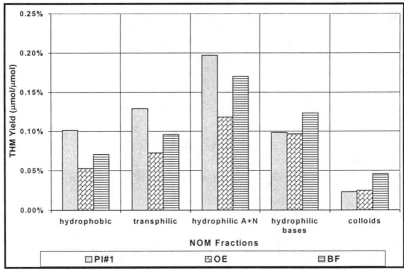

Figure 3. Effect of ozonation and biofiltration on THM yields of NOM fractions.

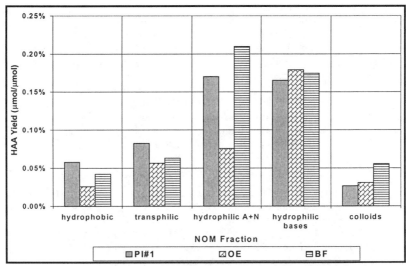

Figure 4. Effect of ozonation and biofiltration on HAA yields of NOM fractions.

and conventional treatment scenarios) and the colloids from the biofilter effluent. In contrast, Sinha and colleagues (*13*) found that both the hydrophobic and hydrophilic fractions of NOM produce significant concentrations of HAAs. This again suggests that CRW contains an atypical mix of DBP precursors. The study of CRW, however, represents an opportunity to characterize the reactivity of source waters that are more nonhumic.

The HAA evaluations were based on the degree of halogenation—that is, the percentage of dihalo- and trihaloacetic acids (DXAAs and TXAAs)—that could be done in this study because all nine HAAs were measured. Cowman and Singer (*15*) reported HAA formation in the hydrophobic extracts of two waters with SUVA values of 4.2 and 2.8 L/mg·m. The formation of TXAAs represented over 60 percent of the HAAs on a molar basis, whereas the formation of DXAAs represented over 30 percent of the HAAs on a molar basis. In contrast, the formation of DXAAs was significantly higher than the formation of TXAAs in the CRW NOM fractions. The mean percent of DXAAs for all fractions of the two plant influent samples, the ozone contactor effluent, and the conventional filter effluent was 61 percent of the HAAs, with a range of 54-79 percent. The biofiltered sample differed, with the hydrophobic and transphilic NOM having DXAA values of 47-48 percent and the hydrophilic bases having 82 percent. In CRW, the average percentage of HAAs represented by TXAAs was 29 percent—highest (37 percent) for the hydrophobic fractions and lowest (22 percent) for the hydrophilic base fractions—with little change caused by biofiltration.

Reckhow and Singer (*12*) developed a simplified conceptual model for the formation of major organic halide products from fulvic acid, in which there is initially the formation of a ß-diketone moiety (R'–CO–CH$_2$–CO–R). In this model, the middle (activated) carbon atom will quickly become fully substituted with chlorine. Hydrolysis then occurs rapidly, yielding a monoketone group. If the remaining "R" group is a hydroxyl group, the reaction will stop, yielding DCAA. Otherwise, the structure will be further chlorinated to a trichloromethyl species. This intermediate species is base-hydrolyzable to chloroform. At neutral pH, if the "R" group is an oxidizable functional group capable of readily donating an electron pair to the rest of the molecule, TCAA is expected to form. NOM with higher values of SUVA probably has a higher amount of oxidizable functional groups, which would result in a higher formation of TCAA (*16*). The Reckhow and Singer model (*12*) suggests that the CRW NOM fractions may include—during chlorination—a higher proportion of ß-diketone moieties with "R" groups that are hydroxyl groups (i.e., ketoacids), whereas higher-SUVA waters may include a higher proportion of "R" groups that are oxidizable functional groups. These results suggest that DXAA and TXAA have different precursors and that CRW (a low-SUVA water) contains a higher proportion of DXAA precursors.

*Effect of Treatment on Yields of Other DBPs in NOM Fractions*

Figure 5 shows the yields of three other classes of DBPs—the HANs, halo-ketones, and chloropicrin for the CRW PI#1 NOM fractions. Although these DBPs occur at lower concentrations in water, these results can provide some insight into the formation of other DBPs containing other functionalities such as carbonyl, cyano, and nitro groups. The haloketones and chloropicrin do not include the brominated analogues, so conditions that affect the distribution within a DBP class will impact the results. Therefore, to facilitate comparisons, the results for the three major fractions are shown in Figure 5 for similar levels of bromide (~20:1 μg Br⁻/mg DOC ratio). However, the DBP yields for the hydrophilic bases and colloids were only determined in the absence of bromide.

The HAN yields are highest for the hydrophilic A+N and hydrophilic base fractions. Comparison of the ozonated and biofiltered CRW also showed that the hydrophilic fractions produced two or more times the yield of HANs than did the hydrophobic and transphilic fractions. These nitrogen-containing DBPs have been shown to form from chlorination of amino acids (17). Elemental analysis showed that the highest nitrogen content for the plant influent NOM fractions was in the hydrophilic bases; for the treated-water samples, it was in the hydrophilic A+N fraction. In contrast, the hydrophobic and colloid fractions had the lowest nitrogen contents and the lowest HAN yields.

The yields of haloketones ranged from 0.004 to 0.005 percent for all fractions except the colloidal NOM, which yielded only 0.001 percent. In contrast, the reactivity in terms of chloropicrin formation was highest in the hydrophilic A+N fraction.

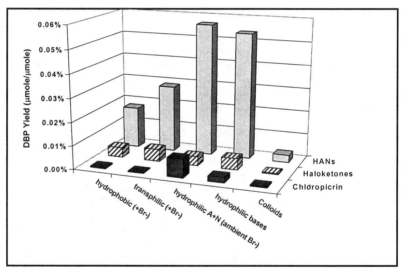

Figure 5. Yields of the HANs, haloketones, and chloropicrin for the CRW PI#1 NOM fractions.

**NOM Reactivity with Respect to Chloramines**

Figures 6 and 7 show the effects of chlorination versus chloramination on the yields of THMs and HAAs, respectively, in the PI#1 sample (the ambient-bromide NOM fractions). For the bulk water, the yield of THMs during chloramination was 3.6 percent of that formed during chlorination. For the hydrophobic and transphilic fractions, the yields of THMs during chloramination were 1.0 and 1.5 percent, respectively, of those formed during chlorination, whereas the yields were 7.0, 5.7, and 7.4 percent of those formed during chlorination for the hydrophilic A+N, hydrophilic base, and colloid fractions, respectively.

For the bulk water, the yield of HAAs during chloramination was 6.2 percent of that formed during chlorination. The general trend seen in THM yield reduction as a result of chloramination is also seen in HAAs yields, although it is not as decisive. For the hydrophobic and transphilic fractions, the yields of HAAs during chloramination were approximately 10 percent of those formed during chlorination. The yields for the polar fractions, however, ranged from 16 to 27 percent of those formed during chlorination. It is important to note that in contrast to chlorination, where TXAAs account for approximately 30 percent of the HAAs in CRW, no TXAAs were formed during chloramination in the CRW samples. DXAAs were the major component in the HAAs after chloramination. These findings are consistent with the observations of Krasner and colleagues (*18*) that chloramines minimized the formation of THMs and TCAA, whereas DXAA could form to a significant extent during chlorination. These results continue to suggest different DBP precursors for the DXAA and the trihalogenated DBPs (i.e., THMs and TXAAs).

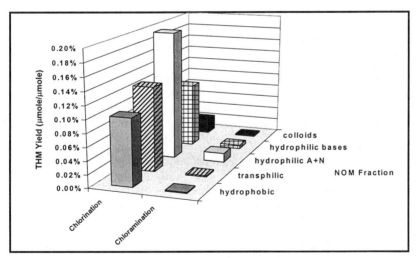

Figure 6. Effect of chlorination versus chloramination on the yields of THMs for CRW PI#1.

184

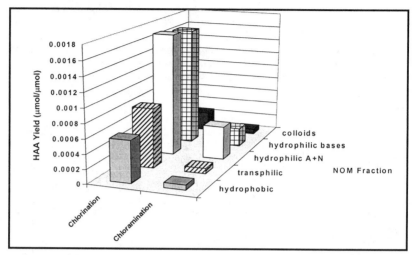

Figure 7. Effect of chlorination versus chloramination on the yields of HAAs for CRW PI#1.


## Estimated Relative Contribution of NOM Fractions to Bulk Water DBP Reactivity

To assess the contribution of the individual NOM fractions to the bulk water, a weighted DBP yield normalized to the DOC of the bulk water was calculated for the three major DBP classes. The sum of the three major NOM fractions accounted for over 95 percent of the NOM recovered. However, 28 percent of the NOM was not isolated for this sample (Table II). Figure 8 shows the relative contributions and mass balance for CRW PI#1. Data for the bromide-spiked hydrophobic and transphilic fractions were used, as that bromide level (100 μg/L) was comparable to the amount of bromide in the other major NOM fraction, the ambient-bromide hydrophilic A+N fraction. In general, the "mass balance" of HAA and HAN yields from the NOM fractions results in DBP yields similar to those of the bulk water. For the THM yield, the mass balance accounted for less of the bulk water yield, undoubtedly because of the unrecovered NOM missing from the mass balance.

On a mass-balance basis, the majority of the THM and HAA precursors were in the hydrophobic and hydrophilic A+N fractions. All three major NOM fractions in the CRW plant influent contributed significantly in terms of DBP yield. These results show how each of the major NOM fractions in the CRW plant influent contributed to the bulk water DBP formation, which can provide insight into precursor control requirements. Treatment processes that focus on the removal of hydrophobic NOM are not sufficient. A process that can reduce the amount of hydrophilic A+N NOM in CRW would contribute to a reduction in the overall amount of DBPs.

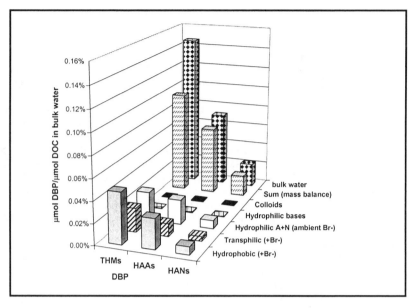

Figure 8. Comparison of estimated relative contributions of the NOM fractions to the bulk water DBP yields for the CRW PI#1.

## Summary and Conclusions

### NOM Reactivity

General trends were observed in DBP yields upon chlorination as a function of the NOM fraction in CRW. The highest yield of THMs was from the hydrophilic A+N fraction, whereas the THM yields were lowest in the colloid fraction (except for the biofilter effluent, in which the yield was relatively high). The hydrophilic A+N and hydrophilic base fractions had the highest HAA yields. The hydrophilic fractions also had the highest HAN yields, consistent with those fractions having the highest nitrogen content.

The reactivity of individual fractions to form DBPs provided some insight into the nature of NOM. For example, the predominance of DXAAs over TXAAs for all NOM fractions with free chlorine suggests that CRW NOM has a higher proportion of "R" groups that are hydroxyl groups rather than oxidizable functional groups *(12)*. Additional work (e.g., [13]C-nuclear magnetic resonance, Fourier transform infrared spectroscopy, pyrolysis GC/MS) on NOM fractions of this and other waters should make it possible to better link specific NOM characteristics with the reactivity to form specific DBPs.

## Impact of Treatment

Conventional treatment of CRW did not lower DBP yields. A decrease was seen in DBP reactivity after ozonation; however, the reactivity increased after subsequent biofiltration, but not to the same level as that of the initial untreated water.

Similar changes were observed upon treatment for most of the NOM fractions from the corresponding bulk water samples. The THM and HAA yields of the hydrophobic, transphilic, and hydrophilic A+N fractions were lower after ozonation, but the yields from the hydrophilic bases or colloids were not affected. Increased reactivity was observed for the NOM fractions present after biofiltration, especially the colloid fraction and the HAA yield for the hydrophilic A+N fraction.

Chloramination is an effective means of minimizing the formation of THMs and TXAAs and, to a lesser extent, the formation of DXAAs. During chloramination, as during chlorination, the hydrophilic A+N fraction had the highest yields of THMs and DXAAs. Future research should consider the reactivity of NOM fractions to form "chloramination-specific" DBPs (e.g., cyanogen chloride).

The hydrophobic, transphilic, and hydrophilic A+N fractions constituted more than 95 percent of the isolated NOM from all samples. All three major NOM fractions contributed significantly in terms of DBP yield. On a mass balance basis, the hydrophobic and hydrophilic A+N fractions were the major sources of THM and HAA precursors in this low-humic water. Thus, a treatment process that can reduce the amount of hydrophilic A+N NOM in CRW will significantly reduce the overall amount of DBPs. These results demonstrate the importance and benefits of studying the effects of treatments on the removal of different NOM fractions.

## Acknowledgments

The authors would like to thank Jerry Leenheer (U.S. Geological Survey) for the fractionation and isolation of the NOM fractions used in this work; Gary Amy (University of Colorado at Boulder), Auguste Bruchet (Lyonnaise des Eaux), and Jean-Philippe Croué (University of Poitiers) for their guidance and participation in the project; and Eric Dickenson (University of Colorado at Boulder) for analytical support. Appreciation is extended to the American Water Works Association Research Foundation for its financial, technical and administrative assistance in funding and managing the project through which this information was discovered.

## Literature Cited

1. U.S. Environmental Protection Agency. *Fed. Reg.* **1998,** *63* (241), 69390.
2. Krasner, S. W.; Chowdhury, Z. K.; Edwards, M. A.; Bell, K. A. Use of SUVA

**187**

in Developing Revised TOC Removal Requirements. In *Proceedings, AWWA Water Quality Technology Conference*, Denver, CO, Nov 9-12, 1997; American Water Works Association, Denver, CO, 1998.

3. Edzwald, J. K.; Van Benschoten, J. E. In *Chemical Water and Wastewater Treatment*; Hahn, H. H., Klute, R., Eds.; Springer-Verlag: Berlin, 1990; pp 341-359.
4. White, M. C.; Thompson, J. D.; Harrington, G. W.; Singer, P. S. *J.—Am. Water Works Assoc.* **1997,** *89* (5), 64.
5. Krasner, S. W.; Amy, G. L. *J.—Am. Water Works Assoc.* **1995,** *87* (10), 93.
6. Malcolm, R. L.; MacCarthy, P. *Environment Int.* **1992,** *18,* 597.
7. Leenheer, J. A.; Croué, J.-P; Benjamin, M.; Korshin, G. V.; Hwang, C. J.; Bruchet, A.; Aiken, G. R. In *Natural Organic Matter and Disinfection By-Products*; Barrett, S. E., Krasner, S. W., Amy, G. L., Eds.; American Chemical Society: Washington, DC, 2000 (in press).
8. Summers, R. S.; Hooper, S. M.; Shukairy, H. M.; Solarik, G.; Owen, D. *J.—Am. Water Works Assoc.* **1996,** *88* (6), 80.
9. Munch, D. J.; Hautman, D. P. *Method 551.1: Determination of Chlorination Disinfection Byproducts, Chlorinated Solvents, and Halogenated Pesticides/Herbicides in Drinking Water by Liquid-Liquid Extraction and Gas Chromatography with Electron-Capture Detection*, Revision 1.0; U.S. Environmental Protection Agency, National Exposure Research Laboratory: Cincinnati, OH, 1995.
10. Munch, D. J.; Munch, J. W.; Pawlecki, A. W. *Method 552.2: Determination of Haloacetic Acids and Dalapon in Drinking Water by Liquid-Liquid Extraction, Derivatization and Gas Chromatography with Electron Capture Detection*, Revision 1.0; U.S. Environmental Protection Agency, National Exposure Research Laboratory: Cincinnati, OH, 1995.
11. Trussell, R. R.; Umphres, M. D. *J.—Am. Water Works Assoc.* **1978,** *70* (11), 604.
12. Reckhow, D. A.; Singer, P. C. In *Water Chlorination: Chemistry, Environmental Impact and Health Effects;* Jolley, R. L., Bull, R. J., Davis, W. P., Katz, S., Roberts, M. H., Jacobs, V. A., Eds.; Lewis Publishers: Chelsea, MI, 1985; Vol. 5, p 1229.
13. Sinha, S.; Amy, G. L.; Sohn, J. In *Proceedings: Water Quality*, AWWA Annual Conference, Atlanta, GA, June 15-19, 1997; American Water Works Association: Denver, CO, 1998; pp 329-353.
14. Hwang, C. J.; Sclimenti, M. J.; Krasner, S. W.; Croué, J-P.; Bruchet, A.; Amy, G. L. In *Proceedings*, AWWA Water Quality Technology Conference, Tampa, FL, Oct 31—Nov 3, 1999; American Water Works Association, Denver, CO, 1999 (in press).
15. Cowman, G.; Singer, P. C. In *Proceedings*, AWWA Water Quality Technology Conference, New Orleans, LA, Nov 12-16, 1995; American Water Works Association, Denver, CO, 1996; pp 947-955.
16. Reckhow, D. A.; Singer; P. C; Malcolm, R. L. *Environ. Sci. Technol.* **1990,** *24* (11), 1655.
17. Trehy, M. L.; Yost, R. A.; Miles, C. J. *Environ. Sci. Technol.* **1986,** *20,* 1117.
18. Krasner, S. W.; Symons, J. M.; Speitel, G. E., Jr.; Diehl, A. C.; Hwang, C. J.; Xia, R.;. Barrett, S. E. In *Proceedings: Water Quality*, AWWA Annual Conference, Toronto, ON, Canada, June 23-27, 1996; American Water Works Association: Denver, CO, 1997; pp 601-628.

# FORMATION OF CHLORINATION DISINFECTION BY-PRODUCTS

# Chapter 13

# The Use of Granular Activated Carbon Adsorption for Natural Organic Matter Control and Its Reactivity to Disinfection By-Product Formation

Tanju Karanfil[1], Mehmet Kitis[1], James E. Kilduff[2], and Andrew Wigton[2]

[1]Department of Environmental Engineering and Science, Clemson University, Clemson, SC 29634–0919
[2]Department of Environment and Energy Engineering, Rensselaer Polytechnic Institute, Troy, NY 12180–3590

The roles of granular activated carbon (GAC) surface chemistry and pore structure on the adsorption of natural organic matter (NOM) and subsequent formation of disinfection by-products (DBPs) were investigated. It was found that NOM removal can be maximized by selecting GACs with minimal surface acidity and pores widths in the range of 20-100 Å (e.g., mesoporous). However, the optimum pore size distribution for a given NOM will depend on its molecular size distribution, a property that is system specific. The reactivity of carbon surfaces controlling NOM uptake was shown to depend on the raw material type, activation conditions and surface treatment. The surfaces of coal-based carbon showed more reactivity for NOM than the wood-based carbon.

The results also indicated that GAC fractionates NOM solutions by preferentially removing high specific UV absorbance (SUVA) components. NOM exhibits a heterogeneous reactivity to DBP formation, with the UV-absorbing fractions being the major contributors. The coal-based carbons with low surface acidity minimized subsequent DBP formation. It was found that low SUVA components of NOM are more efficient at incorporating bromine. SUVA removal may be used to select among GAC materials, and to monitor and control of DBP formation on-line in practical applications. Finally, a "DBP reactivity profile" was developed using GAC fractionation. This profile can be a potential tool for practitioners to assess the heterogeneity of NOM in natural waters, evaluate the effectiveness of different precursor control technologies, and to provide a fundamental approach for incorporating the effects of treatment processes in water treatment plants.

# Introduction and Objectives

The simultaneous control of microbial quality and disinfection by-product (DBP) formation is one of the greatest challenges currently faced by the water treatment industry. DBPs are comprised of several organic and inorganic compounds that are formed by reactions between naturally occurring organic matter (NOM) and oxidants employed for disinfection and/or oxidation (1). The presence of DBPs in drinking water is undesirable because they are suspected to be toxic, carcinogenic and mutagenic to humans if ingested in high concentrations (2-4). Because of potential risk to public health, the United States Environmental Protection Agency (USEPA) is planning to impose more stringent standards under the Disinfectants/Disinfection By-Products (D/DBP) Rule (5). To comply with forthcoming regulations, drinking water utilities must develop strategies to minimize DBP formation.

NOM is considered to be the primary precursor to DBP formation and it is present in all natural waters. Today, the majority of DBP control strategies (e.g., enhanced coagulation, carbon adsorption, and membrane filtration) focus on the removal of NOM from water prior to disinfectant/oxidant addition. In general, total organic carbon (TOC) removal is used as a criterion for effectiveness. However, it is known that NOM is a heterogeneous mixture of complex organic materials including humic substances, hydrophilic acids, proteins, lipids, carboxylic acids, polysaccharides, amino acids, and hydrocarbons. It is likely that different components exhibit different reactivity with disinfectants to produce DBPs. Therefore, the desired effect of any control technology is to remove the most reactive components.

Adsorption by granular activated carbon (GAC) has been designated as one of the best available technologies for the removal of NOM to achieve the requirements of the D/DBP rule (5,6). *Physical size exclusion, specific surface-chemical interactions, and solution interactions between adsorbates and water* are the major interactions that control the extent of NOM adsorption by GAC (7,8). Size exclusion determines accessible surface area for adsorption and is a function of NOM molecular and GAC pore size distributions. Chemical interactions between NOM and the GAC surface, and interactions between NOM and water then determine the overall extent of adsorption on the available surface area. Reviews of these interactions can be found elsewhere (7-14). In general, uptake is lower when solutes interact strongly with water; this occurs when the NOM molecular weight is small and its acidity is high. Despite voluminous literature on activated carbon adsorption, little is known about how GAC characteristics (i.e., surface chemistry and pore structure) affects the removal of DBP precursors and subsequent DBP formation. This is partly because it is not possible to control both the effects of pore structure and surface chemistry by using only as-received carbons, as documented in previous experimental investigations (9-12).

The objectives of the research presented here were (1) to investigate how activated carbon characteristics (i.e., surface chemistry and pore structure) influence adsorption of NOM from water; (2) to provide insight to the fractionation of NOM during adsorption by GAC; (3) to characterize the extent of DBP (e.g., TTHM, HAAs) formation subsequent to GAC adsorption; (4) to evaluate the reactivity of different NOM fractions to DBP formation; and (5) to determine the role of GAC characteristics on the DBP formation control.

## Experimental Materials and Methods

### Activated Carbons

To investigate the role of carbon surface chemistry on NOM adsorption and subsequent DBP formation, the surfaces of a thermally activated, coal-based carbon (Calgon F400) and an acid-activated, wood-based carbon (Westvaco WVB) were modified using liquid-phase oxidation ($HNO_3$) and heat treatment in an inert atmosphere ($N_2$). These carbons were selected, in part, because of the significant differences in their pore structures and raw materials. The surface modification experiments included several steps. Activated carbons were used as received from the manufacturers without any crushing. Initially, they were sonicated and rinsed with distilled and deionized water. Subsequently, ash components and alkaline impurities were removed using 2-N HCl in a Soxhlet extractor. Acid-washed (AW) carbon samples were then heat treated under nitrogen atmosphere at 1000°C for 24 hours (designated by HT 1000) to remove oxygen-containing functionality. A portion of the heat treated carbons were oxidized in aqueous solutions of 70% $HNO_3$ at different temperatures (50°C to 90°C) and for different reaction times (2 to 9 hours) to increase surface acidity (designated by OX followed by time in hours and temperature C, e.g., OX2/70) by creating new surface carboxyl, lactonic, and phenolic groups. A portion of oxidized carbons was subsequently heat-treated at 650°C for 24 hours (designated by HT 650) to remove strongly acidic and other $CO_2$ evolving groups. Wood is much less resistant to high pressure, temperature and oxidation conditions than coal material. Therefore, less vigorous treatment techniques were employed for the wood carbon. A total of 10 carbons with different surface characteristics were prepared.

Surface-treated activated carbons were characterized by: (i) elemental analysis; (ii) surface area and pore size distribution; (iii) water uptake and water vapor adsorption; (iv) acid-base adsorption characteristics, measured by the Boehm technique *(15)*; and (v) a mass titration/pH equilibration method to determine the $pH_{pzc}$, the pH at which the total net surface charge of GAC is zero. The details of the experimental protocols for carbon preparation and characterization can be found elsewhere *(13,16)*. The results showed that as-received coal-based carbon (i.e., F400) was more microporous and significantly less acidic than the wood-based carbon (i.e., WVB). It was found that the surface treatments successfully changed the surface chemistry while not significantly changing the pore structure *(13)*. Therefore, the impact on the NOM uptake can be systematically linked to the changes in the carbon surface chemistry alone. In general, heat treatment decreased the surface acidity selectively (depending on the temperature) whereas oxidation in $HNO_3$ solution increased the acidity by creating new carboxylic, lactonic and phenolic type functionality. Surface acidity was the surface property chosen for detailed investigation because it has been implicated as a factor controlling uptake in the adsorption of several compounds (17,20). No study to date has examined the impact of surface oxidation on the uptake of NOM and subsequent DBP formation. The role of carbon pore structure was evaluated by using activated carbons having significantly different pore size distributions (i.e., microporous coal and mesoporous wood) and by comparing the pore size distribution of treated carbons with the estimated average molecular size of the NOM.

## NOMs

Seven different natural and synthetic organic macromolecules were used during this study. Model NOMs included polymaleic acid (PMA), a fulvic acid surrogate; natural fulvic acid (LaFA) extracted from Laurentian soil; and Aldrich humic acid (AHA) purified to remove ash components. Physicochemical characteristics of these macromolecules have been reported in our previous publications (7,8) and a summary of the important characteristics with respect to this work is provided in Table I.

Surface water samples were collected from the influents of Charleston, SC, and Myrtle Beach, SC, drinking water treatment plants, from the Tomhannock reservoir, the water supply for the City of Troy, NY, and from a stream draining a rural agricultural watershed in Rensselaer County, NY. The compositional properties of these waters are given in Table II.

## Adsorption Isotherms

Isotherm experiments were conducted according to the bottle-point variable-dose (0.02-2.00 of g GAC/l) method as described in detail elsewhere (7,8). In brief, an NOM aqueous solution (i.e., water sample filtered through a 0.45 μm prewashed filter) was equilibrated with activated carbon for 30 days in well-mixed 250-ml batch reactors (CMBRs) under oxic conditions. After equilibration, an aliquot of solution was filtered and analyzed for UV absorbance and dissolved organic carbon (DOC) concentration. The wavelength of 280 nm was selected for UV measurements to minimize the interference of sodium azide that was added to the water samples during collection for the control of biological activity. All isotherms were conducted in the presence of a 0.001 M phosphate buffer, at pH 7 and room temperature of $21 \pm 3°C$, except isotherms with surface water samples, which were conducted after filtration (0.45 μm) without any pH or buffer adjustment. Specific UV absorbance (i.e., $SUVA_{280} = UV_{280}/DOC$) was used to characterize NOM fractionations before and after GAC adsorption.

Table I. Physicochemical Characteristics of Model NOMs

| NOMs | Molecular Weight (g/mol PSS) | Carboxylic Acidity (meq/g DOC) | Phenolic Acidity (meq/g DOC) | Aromatic Content (%) | Apparent Average Radius (Å) |
|------|------------------------------|--------------------------------|------------------------------|----------------------|------------------------------|
| PMA  | 2601 | 13.0 | 3.2 | 23.0 | 16 |
| LaFA | 2402 | 11.7 | 9.3 | 28.2 | 16 |
| AHA  | 4006 | 7.9  | 3.6 | 57.7 | 21 |

Table II. Compositional Characteristics of Surface Water Samples

| Sample | DOC (mg/l) | $UV_{280}$ (abs) | $SUVA_{280}$ (L/mg org-C.m) | Alkalinity (mg CaCO$_3$/l) | pH (-) |
|--------|-----------|------------------|------------------------------|----------------------------|--------|
| Charleston   | 3.9  | 0.120 | 3.08 | 66 | 7.8 |
| Myrtle Beach | 14.0 | 0.421 | 3.01 | 94 | 7.8 |
| Tomhannock   | 2.8  | 0.053 | 1.89 | 35 | 7.2 |
| Rensselaer   | 4.9  | 0.114 | 2.32 | 69 | 7.4 |

**DBP Formation and Analysis**

Chlorination of NOM supernatant solutions after adsorption were conducted according to the Uniform Formation Conditions (UFC) protocol described elsewhere *(21)*. Experiments were conducted at the pH of each natural water sample for a 1-day contact time. A residual of 0.5 to 1 mg Cl$_2$/l was always present in the reaction vials at the end of the experiments. The formation of several DBPs (THMs, HANs, HKs, CHP, CH and HAA9) was measured according to EPA method 551.1 and Standard Method 6251B. The change in the extent of NOM removal, NOM characteristics and subsequent DBP formation was related to the surface chemistry and pore size distribution of the activated carbons. UV measurements were also conducted before and after chlorination and the change in absorbance (i.e., differential UV spectroscopy) was correlated to DBP formation.

# Results and Discussion

**Adsorption of NOMs**

The results of isotherm experiments for model and the surface water NOMs along with the dose and surface-area normalized Freundlich isotherm parameters are reported and discussed elsewhere *(8)*. Because the pore structure of the coal-based carbon was not affected by surface treatment, the role of surface chemistry on the NOM adsorption was examined by comparing the isotherm results on a *mass* basis as exemplified in Figure 1. In general, the trends were consistent for both model and surface water NOMs: (1) no significant capacity difference was observed between untreated, acid-washed (AW) and heat treated (HT 1000) carbons samples; (2) oxidation of the carbon surface (OX 2/70, OX 9/70) significantly decreased the uptake; and (3) the capacity was partially restored by subsequent heat treatment (OX 9/70 H650) of the oxidized surfaces.

The results of coal-based carbons indicated that surface acidity appears to have a significant effect on the NOM uptake. Therefore, it was hypothesized that uptake by wood-based carbon (WVB) should increase upon reduction in its high level of initial surface acidity with heat treatment. Furthermore, given the mesoporous nature of wood carbon and its higher surface area, its overall NOM capacity was expected to exceed the lower surface area and more microporous coal-based carbon (F400). Heat treatment reduced the total surface acidity of the wood carbon by about 50%; however, the impact on the NOM adsorption was surprisingly minimal (e.g., PMA) or absent *(8)*. Subsequent oxidation of the heat-treated carbon doubled the total surface acidity and tripled the density of strongly carboxylic groups on the carbon surface. However, the impact on the NOM adsorption was still not significant.

The significant impact of surface oxidation on NOM uptake by the coal-based carbon and the absence of such an impact for the oxidized wood-based carbon suggests two possibilities: (1) The type of acidity created on the two carbon surfaces under similar surface treatment conditions and their reactivities toward NOM adsorption may be quite different and may depend on the type of raw material; this is consistent with the significant difference observed for the two low molecular weight priority pollutants, TCE and TCB on the same carbons previously *(13)*. (2) Other differences in surface chemistry exist and may contribute to the observed trends for wood-carbons, but such differences cannot be characterized using the methods employed in this study.

## Fractionation of NOMs during GAC Adsorption

Fractionation of NOM during GAC adsorption was investigated for four surface waters. Similar trends were observed for all waters; therefore, results for a single representative (i.e., Myrtle Beach) water will be discussed in detail. As shown in Figure 2, measurements of UV absorbance and DOC of NOM components remaining in solution after contacting different carbon doses demonstrate that carbon adsorption fractionates NOM solutions with regard to SUVA. Because it is generally accepted that the UV absorbance of natural waters at wavelengths over 250 nm is caused primarily by aromatic structures *(7,22-25)*, the changes in SUVA as a result of adsorption can be interpreted as changes in the aromatic carbon content of NOM remaining in solution. The SUVA values of supernatants from oxidized carbons (e.g., OX 9/70) were significantly higher than those of the heat treated and acid washed carbons, demonstrating that increasing carbon surface acidity inhibited adsorption of organic matter with aromatic character. This is probably a result of

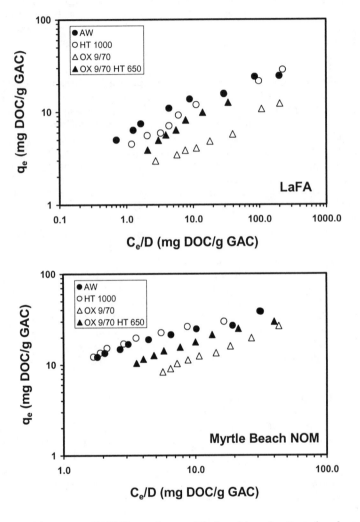

Figure 1. Adsorption of NOM by surface-modified coal-based activated carbons.

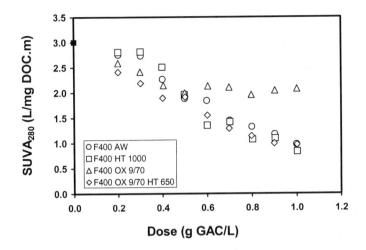

*Figure 2. Effect of carbon dose on SUVA of NOM components remaining in solution.*

reducing the affinity between the carbon surface and aromatic NOM components. It was found that increasing surface acidity increases surface polarity of the surface as indicated with increasing water vapor uptake *(13)*. Since aromatic components of NOM are hydrophobic in character, increasing surface acidity inhibits their uptake by GAC. This is undesirable from DBP formation perspective because aromatic (more correctly, *reactive* aromatic) NOM fractions are known to be more reactive for DBP formation. Subsequent heat treatment of the oxidized carbons (e.g., OX 9/70 HT 650) decreased surface acidity and polarity, thus recovering most of the NOM uptake, and decreasing the SUVA of the supernatant solutions. Experiments with wood carbons showed trends relatively similar to those observed with coal-carbons (data not shown).

For all the coal and wood based carbons tested, very good correlations have been obtained between $UV_{280}$ and DOC remaining after fractionation. Most of the correlations (i.e., $UV_{280}$ vs. DOC) have a positive x-intercept, for example ranging between 0.15 and 0.67 mg/L for the Charleston water. We interpret this to indicate the presence of a non-UV absorbing fraction(s) of NOM. Different values for the x-intercept suggest that different carbons removed the non-UV-absorbing fraction to different extents.

**Reactivity of NOM fractions to DBP formation**

Following GAC adsorption, all NOM fractions, having a wide range of $SUVA_{280}$ values, were chlorinated. After a 24-hr reaction time, the formation of DBPs was measured. To account for possible differences in DBP formation resulting from different levels of DOC (i.e., organic DBP precursors) in the different fractions, the DBP formation (e.g. µg/L TTHMs) was normalized by DOC and then plotted as a function of the $SUVA_{280}$ values of the fractions (Figures 3A and B). In general, an increase in the DBP formation was observed with increasing SUVA, relatively independent of the carbon type. A non-linear relationship between DBP (TTHM and HAA) formation and SUVA is clearly apparent – the reactivity significantly increases with increasing SUVA, most dramatically at values of SUVA above 1.5 to 2.0. The reactivity below a SUVA of about 1.5 is low, and while DBPs were identified in this region, concentrations were near detection limits for individual species ($\approx 1\mu g/L$). Therefore, these data were not shown; rather, the trend toward zero reactivity is indicated by a dashed line. For all waters tested, the TTHM and HAA yields per precursor concentration (i.e., TTHM/DOC or HAA9/DOC) were relatively similar but significantly higher than the formation of other DBPs (i.e., HANs, HKs, chloral hydrates and chloropicrin) that ranged around <1-3 $\mu g/mg$ DOC.

The results indicated that the reactivity of NOM fractions within a water sample fundamentally depends on the SUVA; the reactivity (i.e., TTHM or HAA9/DOC) for all the supernatant solutions (about 60) fell on a single line, independent of the carbon type used to fractionate the NOM solution. However, it is also evident that at the same SUVA values, waters from different sources have different reactivity, indicating that in addition to aromaticity, additional chemical structural characteristics of NOM may influence DBP formation. Since the supernatants tested were obtained *without any chemical treatment*, they are expected to represent the original reactivity of NOM fractions for DBP formation. Therefore, the single line observed between TTHM/DOC and $SUVA_{280}$, appears to represent a *natural DBP formation reactivity profile* of the surface water tested. For all four waters tested, it was found that there are two significantly different reactivity regions: in the low reactivity region, NOM fractions having SUVA smaller than 1.0 to 1.5 did not exhibit significant TTHM and HAA9 formation, while in the high reactivity region (SUVA>1.5), TTHM and HAA9 formation increased dramatically with increasing SUVA (Figures 4A and B). The differential UV spectroscopy data showed strong correlations between TTHM, HAAs and $\Delta UV_{280}$ (Figures 5A and B) confirming previous reports that UV absorbing sites are the major components of NOM molecules leading to DBP formation *(26)*.

Although normalization of the data in Figures 4 and 5 is useful in controlling for the effects of DOC, it does not fully account for all factors that may affect TTHM yield. Three factors play a major role in DBP formation: organic and inorganic precursor concentrations (DOC and bromide) and oxidant (chlorine) concentration. Bromide concentration is important because it substitutes more rapidly than chlorine and to a greater extent *(27)*. During fractionation, the DOC is lowered while the bromide concentration remains constant, because bromide is not removed significantly by carbon adsorption *(28)*. As a result, the $Br^-/DOC$ ratio increases at constant $[Br^-]$. To control for this effect, data points from different fractionation experiments having a narrow range of DOC were selected. In these samples, the reactivity of NOM as a function of SUVA alone could be examined, because DOC, chlorine, and bromide were constant. The results of such an analysis are shown in Figures 6A and B for selected DOC ranges. These data confirm that there is a non-linear increase in the formation of TTHMs and HAAs with increasing SUVA, and that the yield is independent of the fractionation technique used. Similar trends were observed for other DOC ranges.

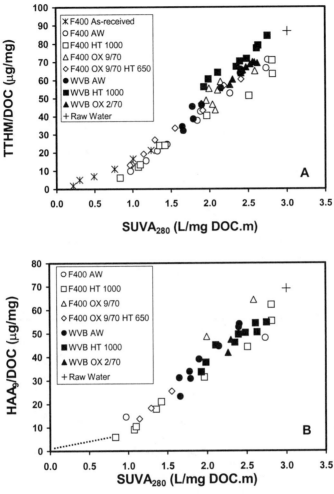

*Figure 3. For adsorption by carbon with widely varying characteristics, the NOM "reactivity profile" fundamentally depends on SUVA, and not on how a given SUVA level was achieved.*

Future D/DBP regulations are expected to focus more on individual instead of combined DBPs because recent toxicology studies indicate that individual THMs and HAAs may have different health effects. Of particular interest are the brominated species; for example, bromodichloro-methane has been reported to pose a higher cancer risk than chloroform *(29)*. Therefore, the degree of bromine substitution, as expressed by the bromine incorporation factor (*n*), was calculated. The bromine incorporation factor is the moles of bromine incorporated into a class of DBP species

200

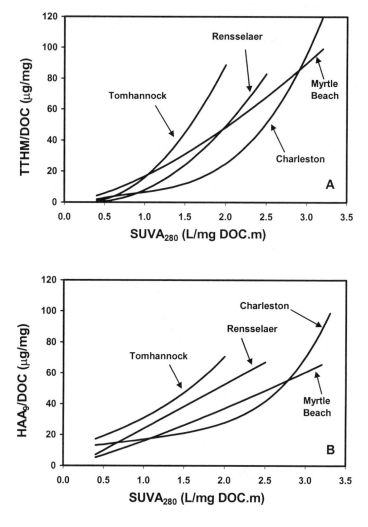

*Figure 4. Reactivity profiles of four surface water NOMs tested. The lines represent regression fit to the experimental data.*

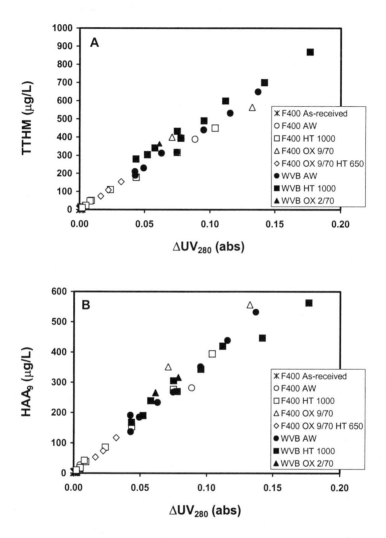

*Figure 5. Differential UV spectroscopy results indicating that aromatic structure are the primary sites responsible sites for the formation of THMs and HAAs.*

202

(e.g. THMs or HAAs) per number of moles of species formed. For example, the bromine incorporation factor defines an "average" THM species with the formula $CHBr_nCl_{3-n}$. To examine the affinity of different NOM components to bromine substitution, the bromine incorporation factor was computed and related to the SUVA of NOM fractions at constant DOC, chlorine, and bromide concentrations, as described above. The results of this analysis are shown in Figures 7A and B for selected DOC ranges; these results are representative of other DOC ranges. The data in Figure 7 show that the bromine incorporation factor decreases with increasing SUVA – it appears that bromine is more readily incorporated into less aromatic NOM constituents. This finding is consistent with those of Sinha et al. *(30)*, who studied bromine incorporation in fractions prepared using the XAD resin technique.

## Conclusions

The results of this study indicate that adsorbent capacity is expected to increase with decreasing NOM size as long as both the surface and sorbate macromolecules are chemically compatible. Repulsive forces between strongly acidic functionalities (such as carboxylic groups) on the GAC surface and within the NOM structure may significantly reduce adsorption capacity. Our results indicate that NOM removal can be maximized by selecting GACs with minimal surface acidity and pores widths in the range of 20-100 A. However, the optimum pore size distribution for a given NOM will depend on its molecular size distribution, a property that is system specific. It appears that the reactivity of carbon surfaces to NOM uptake depends on the raw material type, activation conditions and surface treatment. It was found that the surfaces of coal-based carbon showed more reactivity for NOM than the wood-based carbon.

The findings in this research also confirm that GAC fractionates NOM solutions by preferentially removing high-SUVA components. NOM has a heterogeneous reactivity for DBP formation, and the UV absorbing fractions were found to be the major contributors to DBP formation. Coal-based carbons with low surface acidity maximized NOM uptake by preferentially removing high-SUVA components and minimized subsequent DBP formation. However, it was also found that low-SUVA components of NOM are more efficient at incorporating bromine. SUVA is a promising parameter for prediction of NOM reactivity with chlorine in terms of DBP yield; SUVA removal may be a useful criterion to select among GAC materials. In addition, SUVA has promise as a parameter for on-line monitoring and control of DBP formation in practical applications. Understanding how reactivity is correlated to SUVA may allow utilities to optimize the degree of treatment required to comply with D/DBP regulations. The reactive components that require removal, and the degree of treatment necessary to accomplish this removal, may be directly obtained from the relationship between SUVA removal and the degree of treatment.

## Acknowledgments

This publication was funded in part by the Clemson University Research Grants (Grant: 1-30-0919-51-4455). Partial support for J. E. Kilduff and A. Wigton from the Kodak Corporation is gratefully acknowledged.

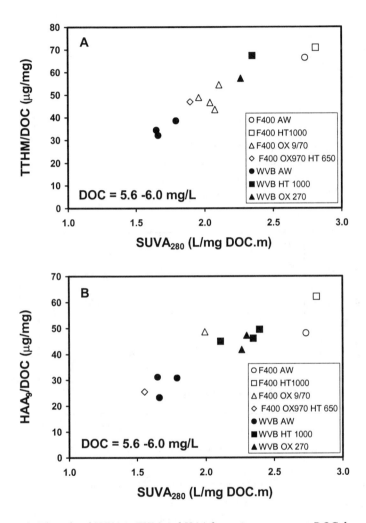

*Figure 6. The role of SUVA in THM and HAA formation at constant DOC, bromide and Cl₂ levels.*

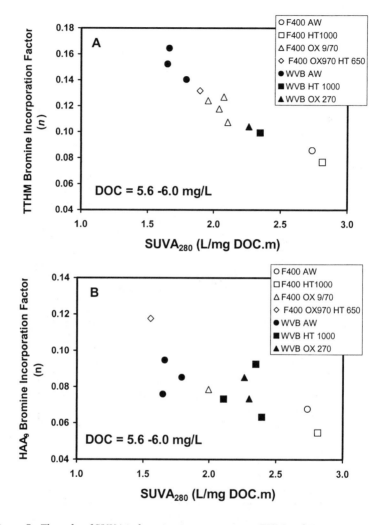

Figure 7. The role of SUVA in bromine incorporation in THM and HAAs.

## References

1. Minear, R.A., Amy, G.L., Eds. *Disinfection By-Products in Water Treatment: The Chemistry of Their Formation and Control*; American Chemical Society, Washington, DC, 1996.
2. *Drinking Water and Health, Vol 7. Disinfectants and Disinfection By-Products*, National Academy of Sciences, National Academy Press, Washington, DC, 1987.

205

3. Glaze, W.H.; Andelman, J.B.; Bull, R.J.; Conolly, R.B.; Hertz, C.D.; Hood, R.D.; Pegram, R.A. *J. AWWA* **1993**, *85* (3), 53.

4. Putnam, S.W.; Graham, J.D. *J. AWWA* **1993**, *85* (3), 57.

5. Pontius, F.W. *J. AWWA* **1996**, *88* (8), 16.

6. Krasner, S.W.; Westrick, J.J.; Regli, S. *J. AWWA* **1995**, *87* (8), 60.

7. Karanfil, T.; Kilduff, J.E.; Schlautman, M.A.; Weber, J.W. Jr. *Environ. Sci. Technol.* **1996**, *30*, 2187.

8. Karanfil, T.; Kilduff, J.E.; Kitis, M.; Wigton, A. *Environ. Sci. Technol.* **1999**, *33*, 3225.

9. Lee, M.; Snoeyink, V.L.; Crittenden, J. *J. AWWA* **1981**, *73*, 440.

10. Weber, W.J., Jr.; Voice, T.C.; Jodellah, A. *J. AWWA* **1983**, *75*, 612.

11. Summers, R.S.; Roberts, P.V. *Journal of Colloid and Interface Science* **1988**, *122*, 367.

12. Summers, R.S.; Roberts, P.V. *Journal of Colloid and Interface Science* **1988**, *122*, 382.

13. Karanfil, T.; Kilduff, J.E. *Environ. Sci. Technol.* **1999**, *33*, 3217.

14. Sontheimer, H.; Crittenden, J.C.; Summers, R.S. *Activated Carbon for Water Treatment*, Second Edition. DVGW-Forschungsstelle, FRG, 1988.

15. Boehm, H.P. *Advances in Catalysis* **1966**, *16*, 179-271.

16. Karanfil, T.; Kilduff, J.E., Kitis, M. In *American Water Works Association Annual Conference Proceedings* **1998**, *3*, 423.

17. Coughlin, R.W.; Ezra, F.S. *Environ. Sci. Technol.* **1968**, *4*, 291.

18. Radovic, L.R.; Silva, I.F.; Ume, J.I.; Menendez, J.A.; Leon y Leon, C.A.; Scaroni, A.W. *Carbon* **1997**, *35*, 1339.

19. Kaneko, Y.; Abe, M.; Ogino, K. *Colloids and Surfaces* **1989**, 211.

20. Pendleton, P.; Wong, S.H.; Shumann, R.; Levay, G.; Denoyel, R.; Rouquerol, J. *Carbon* **1997**, *35*, 1141.

21. Summers, R.S.; Hooper, S.M.; Shukairy, H.M.; Solarik, G.; Owen, D. *J. AWWA* **1996**, *88* (6), 80.

22. Buffle, J. In *Complexation Reactions in Aquatic Systems: An Analytical Approach,* Chalmers, R. A., Masson, M. R., Eds.; Ellis Horwood Series in Analytical Chemistry, Horwood, Chicester, U. K., 1982.

23. Edzwald, J. K.; Van Benschoten, J. E. In *Chemical Water and Wastewater Treatment*, Hahn, H. H., Klute, R., Eds.; Springer-Verlag, Berlin, 1990.

24. Traina, S.J.; Novak, J.; Smeck, N.E. *J. Environ. Qual.* **1990**, *19*, 151.

25. Chin Y.; Aiken, G.; O'Loughlin, E. *Environ. Sci. Technol.* **1994**, *28*, 1853.

26. Li, C.W.; Korshin, G.V.; Benjamin, M.M. *J. AWWA* **1998**, *90* (8), 88.

27. Symons, J.M.; Krasner, S.W.; Sclimenti, M.J.; Simms, L.A.; Sorensen, H.W. Jr.; Spietel, G.E. Jr.; Diehl, A.C. In *Disinfection By-Products in Water Treatment: The Chemistry of Their Formation and Control*, Minear, R.A., Amy, G.L., Eds.; American Chemical Society, Washington, DC, 1996.

28. Amy, G.L.; Tan, L.; Davis, M.K. *Water Res.* **1991**, *25* (2), 191.

29. Krasner, S.W.; Sclimenti, M.J.; Chinn, R.; Chowdhury, Z.K.; Owen, D.M. In *Disinfection By-Products in Water Treatment: The Chemistry of Their Formation and Control,* Minear, R.A.; Amy, G.L., Eds.; American Chemical Society; Washington, DC, 1996.

30. Sinha, S.; Amy, G.L.; Sohn, J. In *American Water Works Association Annual Conference Proceedings* **1997**.

Chapter 14

# Trihalomethanes Formed from Natural Organic Matter Isolates: Using Isotopic and Compositional Data To Help Understand Sources

Brian A. Bergamaschi[1], Miranda S. Fram[1], Roger Fujii[1],
George R. Aiken[2], Carol Kendall[3], and Steven R. Silva[3]

[1]U.S. Geological Survey, California District, Sacramento, CA 95819–6129
[2]U.S. Geological Survey, Water Resources, Boulder, CO 80303
[3]U.S. Geological Survey, Water Resources, Menlo Park, CA 94025

Over 20 million people drink water from the Sacramento-San Joaquin Delta despite problematic levels of natural organic matter (NOM) and bromide in Delta water, which can form trihalomethanes (THMs) during the treatment process. It is widely believed that NOM released from Delta peat islands is a substantial contributor to the pool of THM precursors present in Delta waters. Dissolved NOM was isolated from samples collected at five channel sites within the Sacramento-San Joaquin Rivers and Delta, California, USA, and from a peat island agricultural drain. To help understand the sources of THM precursors, samples were analyzed to determine their chemical and isotopic composition, their propensity to form THMs, and the isotopic composition of the THMs.

The chemical composition of the isolates was quite variable, as indicated by significant differences in carbon-13 nuclear magnetic resonance spectra and carbon-to-nitrogen concentration ratios. The lowest propensity to form THMs per unit of dissolved organic carbon was observed in the peat island agricultural drain isolate, even though it possessed the highest fraction of aromatic material and the highest specific ultraviolet absorbance. Changes in the chemical and isotopic composition of the isolates and the isotopic composition of the THMs suggest that the source of the THMs precursors was different between samples and between isolates. The pattern of variability in compositional and isotopic data for these samples was not consistent with simple mixing of river- and peat-derived organic material.

# INTRODUCTION

Natural organic matter (NOM) present in source water for drinking water forms a variety of disinfection by-products (DBPs) when chlorinated during treatment (*1*). Some of these by-products are believed to be carcinogenic (*2*), and the final concentrations are regulated in finished drinking water (*3*). Typically, the most abundant of these by-products comprises the class of compounds known as trihalomethanes (THMs); the sum of chloroform, and bromodichloromethane, dibromochloromethane, and bromoform (*4*). Environmental factors that determine the extent to which a particular source water forms THMs upon chlorination are the concentration of bromide (*5*), the amount of NOM present in the source water, and the molar reactivity of the NOM with respect to THM formation (*6*). This work focuses on the latter factor—the reactivity of NOM with respect to formation of THMs. It is commonly held that the predominant pathway for THM formation is by reaction of chlorine with aromatic structures present in dissolved humic material (*6*). However, available data show no simple molar relation between the common surrogate for NOM aromaticity, carbon specific ultraviolet absorbance (SUVA) at 254 nanometers, and the amount of THMs formed upon chlorination of waters from the Sacramento-San Joaquin Delta (*7*, *8*). It should be noted that a significant relationship between SUVA and THM formation have been observed elsewhere in specific locations or watersheds (e.g., *8*, *9*), but the relationship does not appear to be universal (*10*).

THM formation is of particular concern for managers of the water-treatment facilities that supply drinking water, originating in the tidal reaches of the Delta, to 22 million people (*11*). The Delta is at the confluence of the Sacramento and San Joaquin rivers, a region largely composed of below-sea-level, peat-rich islands maintained in agricultural production by a network of levees and pumps (*12*). The NOM content and composition of river water changes as it flows through the Delta before being diverted into the California Aqueduct system for use as drinking water in both northern and southern California (*8*, *7*, *11*, *13-15*) or flows into San Francisco Bay. The changes are thought to result, at least in part, from the addition of agricultural return waters from peat-rich Delta islands, waters that contain high concentrations of NOM and aromatic humic substances (*8*, *14*, *15*). High NOM concentrations in Delta water combined with high bromide levels from tidal mixing of brackish water frequently results in source water that potentially forms THMs in excess of regulated levels upon chlorination (*7*, *11*, *13*, *15*).

The purpose of this study is to investigate the relationship between biogeochemical processes in the Delta and the molar capacity of NOM to form THMs. We examined the variability in chemical and isotopic composition in NOM isolates from water samples collected in the Sacramento and San Joaquin rivers, from three locations within the Delta, and from a peat-island agricultural drain.

Although these samples represent a single snapshot in time from each station, if the chemical and isotopic variability of these samples is representative of general Delta processes, we conclude that the NOM in Delta channel water does not represent a simple mixture of river- and peat-derived NOM, and the THM formation potential (THMFP) per unit carbon of Delta water NOM is not necessarily increased by addition of peat-island drainage waters. It should be noted that the molar capacity of

NOM to form THMs is only one factor influencing the amount of THMs produced upon chlorination. NOM concentration is another dominant factor, but not included in this study.

Based on the variability in the carbon isotopic composition of THMs formed upon chlorination of the NOM isolates, we conclude that THM precursors in various fractions of the NOM are different, and suggest that in-channel processes may contribute to the observed variability in the composition of the NOM.

## EXPERIMENTAL METHODS

Samples of about 140 liters (L) were collected by pump from below the surface, near mid-channel at five sites in the Sacramento and San Joaquin rivers and main Delta flow channels (Figure 1) during the summer, when river flows are low. Additionally, an agricultural drainage water sample was collected from a drainage ditch on Twitchell Island. The experimental design uses the Sacramento and San Joaquin rivers and the Twitchell Island drainage samples, respectively, as representatives of upland river and Delta island sources of NOM transported to Delta channel water. The Old River and Middle River sites are located within the Delta and represent a mixture of riverine NOM and NOM added or modified within the Delta (Figure 1). The export site is located just upstream of the facility that pumps water from the Delta into the California Aqueduct, and other drinking-water supply aqueducts (Figure 1). River discharge and NOM content is known to vary considerably in this region, as is the NOM content of agricultural drains (*13*). Also, this is a region dominated by tidal flows (*12*). Thus, since these samples represent a single point in time in a dynamic system, they should be considered as indicative of trends and differences between sources rather than encompassing the breadth of possible variability within the Delta system.

Analytical methods are described in detail elsewhere (*7, 16*). Briefly, NOM was isolated and concentrated from the water samples by sequential extraction on nonionic macroporous resins (XAD-8 and XAD-4; *17*). Materials eluted with base from the XAD-8 and XAD-4 resins are termed hydrophobic acids (HPoA) and hydrophilic acids (HPiA), respectively. DOC measurements were made with a Shimadzu TOC-5000A total organic carbon analyzer. (The use of trade names in this paper is for identification purposes only and does not constitute endorsement by the U.S. Geological Survey.) Elemental composition was determined using a Perkin-Elmer 2400 Series II CHNS/O analyzer. UV measurements were made on filtered samples in a PE Lambda 3B UV/VIS spectrophotometer using a 1 centimeter (cm) cell.

Nuclear magnetic resonance (NMR) spectra for the NOM isolates were obtained by carbon-13 solid-state, cross polarization, magic angle spinning spectroscopy ($^{13}C$ CPMAS) using a 200 megahertz Chemagnetics CMX spectrometer with a 7.5 millimeter diameter probe. The spinning rate was 5000 hertz, the pulse delay was 1 second, the pulse width was 4.5 microseconds for the 90° pulse, and contact times were 1 milliseconds in duration. Data were collected over approximately 2000 transients, and a line broadening of 100 hertz was applied in Fourier transformation

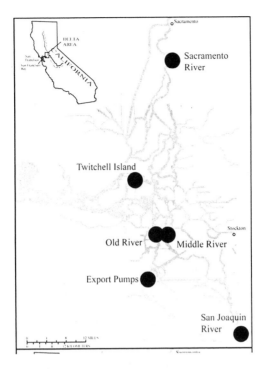

*Figure 1. Location of water sampling sites in the Sacramento-San Joaquin Delta, California. Sacramento and San Joaquin River sites represent the river end-members and Twitchell Island drainage site represents the peat island end-member. Export site is adjacent to the intake pumps for the California Aqueduct.*

of the free induction decay data. We assign resonance in the regions 0-60 parts per million (ppm) as aliphatic, 60-90 ppm as heteroaliphatic, 90-110 ppm as anomeric, 110-160 ppm as aromatic, and 160-190 as carboxylic. Integrated areas in these regions were normalized to the aggregate area and the proportion of each functional assignment expressed as a percentage of the whole.

Aliquots of the reconstituted XAD isolates were reacted with chlorine to form THMs using a modified version of USEPA Method 510.1, following the reactivity-based dosing method (*8, 18, 19*). Briefly, samples were buffered to a pH of 8.1 and dosed with sufficient chlorine to result in a measured chlorine residual of 0.1 to 1.0 mg $Cl_2/L$ at the end of a seven-day incubation at 25°C. Chlorine dosing was calculated using an empirical relationship determined for this type of sample. THM concentrations were measured using a Tekmar ALS2016 and LSC200 purge and trap, and a Hewlett-Packard 5890 II gas chromatograph with an electron capture detector, following a modified version of U.S. EPA Method 502.2. Results are reported as

Table I. DOC concentration and SUVA
for whole-water samples.

| Sample | DOC (mg/L) | SUVA (L/mg-cm) |
|---|---|---|
| Sacramento River | 3.7 | 0.015 |
| San Joaquin River | 4.2 | 0.032 |
| Twitchell Is. Drainage | 17.1 | 0.049 |
| Middle River | 6.1 | 0.028 |
| Old River | 4.4 | 0.042 |
| Export | 4.6 | 0.038 |

NOTE: Dissolved organic carbon (DOC) content and SUVA of whole water samples from locations in the Sacramento-San Joaquin Delta, California, USA. DOC concentrations reported in milligrams of carbon per liter of water (mg/L C); SUVA reported in units of liters per milligram-centimeter and represents UV absorbance normalized to DOC concentration.

millimoles of THM formed per mole carbon (mmol/mol). Reporting THM data on a molar basis rather than a weight basis minimizes the impact of bromine on interpretation of the data, as the brominated species possess a higher mass than their chlorinated analogs. Carbon isotopic ratios of THMs were measured using purge and trap introduction into a gas chromatograph coupled with a combustion furnace and a Micromass Optima continuous flow stable-isotope mass spectrometer ($16$). Results are reported in standard $\delta$ notation, where $\delta^{13}C = 1000*[(^{13}C/^{12}C)_{samp} - (^{13}C/^{12}C)_{std}]/(^{13}C/^{12}C)_{std}$ and V-PDB is the C standard, and atmospheric nitrogen is the N standard.

Mean deviation between replicate THM isotopic measurements was 0.30 per mil (‰) (standard deviation=0.2‰). Replicate analyses for isolate samples were within 0.05‰ for $\delta^{13}C$ and 0.2‰ for $\delta^{15}N$. All replicate elemental analyses were within 1% for concentrations of C and N.

# RESULTS

**Whole Water**

The NOM content of Delta water is a concern for municipalities using the Delta as a source for drinking water. In this study, the NOM concentration was 4.4 milligrams per liter of carbon (mg/L C) in the Old River sample and 6.1 mg/L C for the Middle River sample (Figure 1, Table I). The NOM concentration of the Export sample was 4.6 mg/L C. The San Joaquin River sample and the Sacramento River sample had a NOM content of 4.2 and 3.7 mg/L C, respectively. The Twitchell Island drainage sample possessed a higher NOM concentration (17.1 mg/L C).

UV absorbance at 254 nm is a common parameter used to infer the amount of humic material in river waters, and estimate the amount of THM likely formed during water treatment. The carbon-normalized UV absorbance (SUVA) is used to compare waters with different NOM contents to infer the proportion of aromatic humic material within the NOM pool. SUVA varied by more than a factor of three within the study (Table I). The drainage water sample, as expected, had the highest SUVA (0.049 L/mg-cm), indicating a higher aromatic content. The San Joaquin River sample SUVA value, 0.032 L/mg-cm, was between those of the two Delta channel samples and the SUVA of the Sacramento River was much lower (0.015 L/mg-cm). The Old and Middle River samples SUVA values were 0.028 and 0.042 L/mg-cm, respectively, and the export sample had a SUVA of 0.038 L/mg-cm. This suggests that NOM present in the export sample may have increased in UV absorbance and aromatic humic content during passage through the Delta, in agreement with previous studies (8, 13, 14). Changes in SUVA and THMFP have been the primary evidence previously used to infer that the contribution of peat material altered the humic content of Delta waters.

## NOM Isolates

Isolation of organic material from the aqueous matrix permits comparison of data from a variety of measurements not possible in whole water samples (Table II). Carbon-to-nitrogen ratio (C:N) measurements of HPoA and HPiA isolates in this study revealed that the NOM isolates were universally low in N. C:N can be indicative of the source and diagenetic state of NOM because terrestrial plants generally have C:N greater than 15 while planktonic algae have C:N about 6.5 (20), and diagenesis of organic matter generally results in increasing C:N (21). HPiA C:N values ranged from 23 in the Middle River sample to 35 in the export sample (Table II). HPoA C:N values were more depleted in N than their HPiA counterparts, ranging from 30 in the Old River sample to 62 in the export sample. The highest values were observed in samples from San Joaquin River and the export site.

The carbon isotopic composition of the bulk isolates ( $\delta^{13}C_{BULK}$) varied only slightly among the six samples for the same isolate fraction, but the HPoA and HPiA isolates were distinctly different from each other ($t$-test, $\alpha = 0.05$, P < 0.001 [22]; Figure 2; Table II). The $\delta^{13}C_{BULK}$ of the HPiA isolates averaged -25.9‰, while HPoA isolates averaged 1‰ lighter. The lighter $\delta^{13}C_{BULK}$ value for HPoA isolates may be a result of a higher lignin aromatic carbon content. Lignin, the putative primary precursor for aromatic moieties in humic materials, is isotopically lighter than bulk humic material by about 2‰ (23). This possibility is supported by the higher aromatic content of the HPoA isolates (Table III).

The variability in $\delta^{13}C_{BULK}$ values of the isolates between sites may reflect small variations in the sources of the NOM. The lightest isotopic ratios were observed in the Sacramento River sample and the Twitchell Island drainage sample, consistent with a high C3 terrestrial content. C3 plants are those that use the C3 carbon cycle

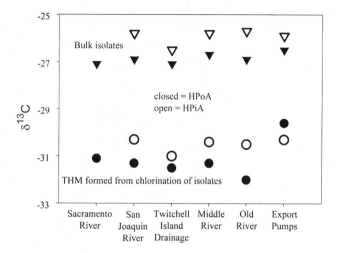

*Figure 2. Carbon isotopic composition of bulk isolates and THMs formed from chlorination of bulk isolates for HPiA and HPoA isolates from water samples from the Sacramento-San Joaquin Delta, California.*

for photosynthetic fixation, resulting in approximately a -27‰ fractionation from the $\delta^{13}C$ of atmospheric $CO_2$. The peat soils were formed largely from the decayed remains of *Scirpus*, a C3 marsh plant endemic to the Delta prior to cultivation of this area by man in the mid 19th century (*24, 25*). The export sample HPoA isolate $\delta^{13}C_{BULK}$ value was heavier than average, perhaps indicative of the contribution of small amounts of heavier, algal- or C4 plant-derived carbon. C4 plants use the C4 carbon cycle for photosynthetic fixation, resulting in a different amount of fractionation, approximately -13‰ from atmospheric $CO_2$.

$\delta^{15}N_{BULK}$ values of HPoA and HPiA isolates varied considerably, ranging over 10‰ between samples, in contrast to the $\delta^{13}C_{BULK}$ results (Table II). The export sample HPiA isolate exhibited the highest $\delta^{15}N$ value (9.7‰), while the HPoA isolate from the Twitchell Island drainage sample exhibited the lowest value (-1.7‰). $\delta^{15}N_{BULK}$ values did not vary systematically among the samples from the six sites nor between the HPiA and HPoA isolates (Table II). The export sample, however, exhibited a nearly 9‰ difference between HPoA and HPiA isolates, indicating distinct sources for the nitrogen found in each isolate.

$^{13}C$ CPMAS NMR analysis of the isolate samples yielded information about the distribution of functional groups within the bulk material and provided a basis for comparison of the compositional differences among samples and between sample isolates. As expected, the aromatic carbon content of the HPoA isolates, which averaged 21%, was distinctly higher than the aromatic carbon content of the HPiA isolates, which averaged only 14% (*t*-test, $\alpha = 0.05$, $P < 0.001$ [*22*]; Table III). The HPiA isolates were relatively enriched in heteroaliphatic and carboxylic carbon

Table II. Chemical and isotopic properties of isolates.

| Sample | $\delta^{13}C_{BULK}$ (‰) | $\delta^{13}C_{THM}$ (‰) | $\Delta\delta^{13}C_{BULK-THM}$ (‰) | $\delta^{15}N_{BULK}$ (‰) | C:N (molar) | STHMFP (mmol THM/ mol DOC) | SUVA (L/mg-cm) |
|---|---|---|---|---|---|---|---|
| | | | HPoA Isolate | | | | |
| Sacramento River | -27.1 | -31.1 | 4.0 | 1.7 | 31 | 9.7 | 0.039 |
| San Joaquin River | -26.9 | -31.3 | 4.4 | 0.4 | 52 | 9.0 | 0.049 |
| Twitchell Is. Drainage | -27.1 | -31.5 | 4.5 | -1.7 | 31 | 8.5 | 0.062 |
| Middle River | -26.7 | -31.3 | 4.6 | 0.9 | 33 | 9.2 | 0.048 |
| Old River | -26.9 | -32.0 | 5.1 | 2.7 | 30 | 9.2 | 0.053 |
| Export | -26.5 | -29.6 | 3.1 | 0.8 | 62 | 8.6 | 0.041 |
| | | | HPiA Isolate | | | | |
| Sacramento River | na | na | na | na | na | na | na |
| San Joaquin River | -25.8 | -30.3 | 4.5 | 3.0 | 33 | 8.0 | 0.032 |
| Twitchell Is. Drainage | -26.5 | -31.0 | 4.5 | 0.2 | 25 | 6.9 | 0.042 |
| Middle River | -25.8 | -30.4 | 4.6 | 0.8 | 23 | 6.9 | 0.034 |
| Old River | -25.7 | -30.5 | 4.8 | 1.3 | 24 | 8.2 | 0.033 |
| Export | -25.9 | -30.3 | 4.4 | 9.7 | 35 | 7.4 | 0.032 |

NOTE: Carbon and nitrogen isotopic compositions, C:N, STHMFPs, and SUVA of the HPoA and HPiA fractions of isolated NOM, and carbon isotopic compositions of THMs formed upon chlorination of these NOM isolates from water samples from locations in the Sacramento-San Joaquin Delta; carbon (C) and nitrogen (N) isotopic ratios reported as per mil (‰) relative to standards; subscript BULK refers to the solid isolate material; subscript THM refers to the THMs; C:N reported in molar units; specific THMFP (STHMFP) reported in millimoles THMs per mole DOC (mmol THM/mol DOC); SUVA reported in units of liters per milligram per centimeter and represents UV absorbance normalized to DOC concentration; na, not analyzed.

compared to the HPoA isolates (Table III). Enrichment in heteroaliphatic and carboxylic carbon and depletion in aromatic carbon in the HPiA compared with the HPoA isolate may indicate that the HPiA material is more degraded.

The highest proportions of aromatic and carboxylic carbon for both the HPoA and HPiA isolates were found in isolates from the Twitchell Island drainage sample (Table III). This result is consistent with a peat origin for the material, as peats in other environments have been shown to be enriched in aromatic and carboxylic carbon (e.g., 26). It is interesting to note that the Delta channel samples from Old River and Middle River did not contain higher proportions of aromatic carbon than the San Joaquin River sample, as may be expected if NOM derived from the peat islands substantially contributed to the bulk NOM values of the Old and Middle River isolates.

Table III.  Functional group distribution for isolates.

| Sample | Carboxylic | Aromatic | Anomeric | Hetero-aliphatic | Aliphatic |
|---|---|---|---|---|---|
| | HPoA Isolate | | | | |
| Sacramento River | 12 | 20 | 8 | 19 | 41 |
| San Joaquin River | 14 | 21 | 9 | 19 | 38 |
| Twitchell Is. Drainage | 15 | 24 | 8 | 17 | 36 |
| Middle River | 14 | 20 | 10 | 17 | 39 |
| Old River | 14 | 21 | 10 | 18 | 36 |
| Export | 12 | 16 | 8 | 18 | 44 |
| | HPiA Isolate | | | | |
| Sacramento River | 15 | 13 | 7 | 23 | 42 |
| San Joaquin River | 13 | 14 | 8 | 22 | 43 |
| Twitchell Is. Drainage | 18 | 18 | 8 | 20 | 36 |
| Middle River | 17 | 13 | 9 | 24 | 37 |
| Old River | 16 | 16 | 8 | 23 | 37 |
| Export | 16 | 13 | 10 | 23 | 37 |

NOTE:  Percentages of carboxylic, aromatic, anomeric, heteroaliphatic, and aliphatic carbon in the HPoA and HPiA fractions of isolated NOM from water samples from locations in the Sacramento-San Joaquin Delta as determined by $^{13}C$ NMR spectroscopy.  Values reported as percentages normalized to 100% for the total peak area from 0 to 190 ppm.  See methods section for details of spectral assignments.

The export sample HPoA isolate was distinctive; it contained more aliphatic and less aromatic carbon than all of the other isolates.  This chemical difference may result from the addition of algal-, agricultural-, or sewage-derived organic material to the water, or from in-channel degradation of organic material.

**THMs**

The carbon specific THMFP (STHMFP) is a measure of the potential of the organic material in the NOM isolates to form THMs when treated with chlorine.  The reactivity of the HPoA isolates averaged 1.5 millimoles THM per mole DOC (mmol/mol) higher than that of the HPiA isolates % ($t$-test, $\alpha = 0.05$, P = 0.003 [22], Table II; Figure 3).  STHMFP values of the HPoA isolates ranged from 8.5 to 9.7 mmol/mol (average = 9.0 mmol/mol) while the STHMFP values of the HPiA isolates were more variable, ranging from 6.9 to 8.2 mmol/mol (average = 7.5 mmol/mol).  The HPoA and HPiA isolates from the Twitchell Island drainage sample exhibited the lowest reactivity, forming 8.5 and 6.9 mmol THM/mol DOC, respectively. The fact that the Twitchell Island drainage isolates were the least reactive is curious in view of the fact that they were the most aromatic of the isolates (Table III) and aromatic structures are the putative precursors for THM compounds.

$\delta^{13}C_{THM}$ values of the THM formed from the isolates averaged 4.4‰ lighter than that of the bulk material (Table II; Figure 2), which is similar to the average difference of 3.5‰ found for 60 other samples of NOM isolates (*16*). The average $\delta^{13}C_{THM}$ formed from the HPoA isolates was 0.6‰ lighter than that from the HPiA isolates, but this difference was not statistically significant (*t*-test, $\alpha = 0.05$ [*22*]).

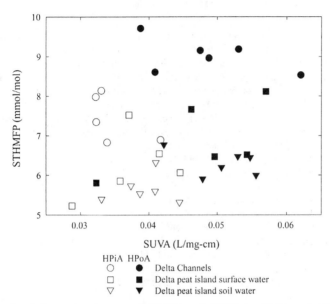

*Figure 3. Relationship between STHMFP and SUVA for HPiA and HPoA NOM isolates from the Sacramento-San Joaquin Delta (Table 2) and a Delta peat island (7). Both STHMFP and SUVA are carbon-normalized values.*

The carbon isotopic fractionations observed between HPoA and HPiA isolates, between THM derived from HPoA and HPiA isolates, and between THM and bulk materials suggest strong intermolecular differences of the THM precursor compounds within the NOM. Different compounds within the NOM carry distinct $\delta^{13}C$ values, and the chlorination reactions are selective in that compounds are attacked and broken down at different rates to form THMs (*10, 16*). The THM formation reactions occur preferentially on isotopically lighter compounds within the NOM.

The difference between the $\delta^{13}C_{THM}$ and $\delta^{13}C_{BULK}$ ($\Delta\delta^{13}C_{BULK-THM}$) does vary among the six samples, which may reflect both distinct sources and different diagenetic histories of the THM precursors in them. HPiA isolate $\Delta\delta^{13}C_{BULK-THM}$ values ranged from 4.4 to 4.8‰, and HPoA $\Delta\delta^{13}C_{BULK-THM}$ values ranged from 3.1 to 5.1‰ (Table II). The highest values for both isolates were in the Old River sample, while the lowest values for both isolates were in the export sample. $\Delta\delta^{13}C_{BULK-THM}$ values for Twitchell Island drainage and San Joaquin River samples were similar and averaged 4.5‰ for both the HpoA and HPiA isolates, while the Sacramento River HpoA sample $\Delta\delta^{13}C_{BULK-THM}$ was 4.0‰. The export sample had the greatest difference between values (1.3‰) for the HpoA and HPiA isolates, supporting the possibility these two isolates contained material from distinctly different sources.

## DISCUSSION

The results presented here provide the opportunity to test the relationship between a variety of chemical parameters and the formation of THMs from selected Delta channel waters. The use of several chemical parameters provides greater power to test assumptions inherent in using any single parameter as a tracer of the source of dissolved organic material or as an indicator of THM formation.

For example, SUVA is the commonly accepted surrogate measurement for humic content and, by extension, STHMFP because numerous studies demonstrated correlation between the two parameters (e.g., 8). SUVA also has been previously applied as a tracer for source waters in the Delta, with high SUVA values indicative of contributions from peat-derived NOM (8, 13). In this study, SUVA values generally were lower in the Sacramento and San Joaquin River samples than the Delta channel and Twitchell Island drainage samples (Tables I and II). This is consistent with previous observations and with the notion that peat-derived, UV-absorbant NOM is present in higher concentration in Delta waters than in upstream river waters. However, contrary to expectations, no relationship between SUVA and STHMFP was observed in the NOM isolates from this study and isolates from an in-depth study of a Delta peat-island agricultural field (Figure 3, 7).

Two assumptions underlie the expected relation between SUVA and THMFP: that the aromatic carbon content is a strong determinant of STHMFP, and that SUVA values largely reflect aromatic carbon content. Using the $^{13}C$ NMR data of the isolates, it is possible to directly explore if the assumptions are valid in these Delta samples. In this study, the Twitchell Island drainage sample, the HpoA isolate with the highest aromatic carbon content, had the lowest STHMFP, while the export sample had the lowest aromatic content of the HpoA isolates, but comparable STHMFP (Table II). Another indication that aromatic carbon content by itself does not determine STHMFP in these samples is the difference between HpoA and HPiA isolates. HPiA isolates contained 33 percent less aromatic carbon than the HpoA isolates, and HPiA isolates also were less likely to form THMs on chlorination. However, STHMFP of HPiA was only 16 percent less than STHMFP of HpoA isolates or only about half the decrease expected if the aromatic carbon content by itself determined STHMFP. In general, we find no relation between aromatic content and STHMFP in these Delta samples and the peat island samples of Fujii and others (7; Figure 4).

This decoupling between aromatic carbon content and STHMFP may be the cause of the generally poor relationship between SUVA and STHMFP in these samples and in other Delta peat island drainage water isolates (Figure 3; 7). A similarly poor relationship between SUVA and STHMFP was observed in surface waters collected from a variety of locations around the United States (*10*). Other researchers have found SUVA to be related to STHMFP (e.g., *8*) in specific locations. We have also previously observed an apparent relationship between SUVA and STHMFP in subsets of Delta isolates, for example, in non-flood San Joaquin River samples (*10*). However, no general relationship exists for a broad array of Delta samples or surface-water samples from throughout the U.S. studied by Fram and colleagues (*10*).

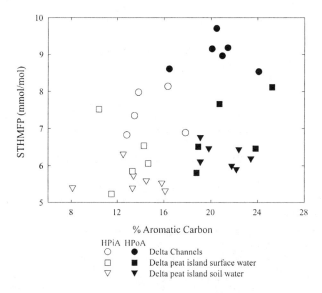

*Figure 4. Relationship between STHMFP and percentage aromatic carbon for HPiA and HPoA NOM isolates from the Sacramento-San Joaquin Delta (Tables II and III) and a Delta peat island (7).*

There is an apparent relationship between SUVA and aromatic carbon content for these samples and the other Delta peat island samples (Figure 5). However, the coefficient of correlation is only 0.65, indicating that aromatic carbon is not the sole variable determining SUVA values. It is likely that there are other organic structures in the NOM that are UV absorbant. Some structures containing carbonyl groups or conjugated double bonds absorb UV radiation near 254 nm, and have molar absorptivity values similar to the aromatic structure band near 254 nm (e.g., *27*). Thus, even if the concentration of these structures in the NOM is only a few percent of the concentration of aromatic structures, the measured SUVA values may be affected. In the case of whole water samples, the presence of inorganic constituents may also affect SUVA values. Inorganic constituents that absorb at a wavelength of 254 nm will cause spuriously high absorption values for the corresponding STHMFP. Also, inorganic constituents can interact with organic constituents so as to lower absorbance (28). Either scenario will tend to degrade any relationship between SUVA and aromatic content (Figure 5) by increasing the number of compositional parameters that determine the SUVA value.

The presence of a fraction of chromophoric aromatic organic material in whole waters or isolates that does not react to form THM will also degrade any relationship between SUVA, STHMFP, and aromatic content (Figures 3 and 4). Presence of an unreactive fraction of organic material would explain, for example, the observation that the Twitchell Island drainage HPoA isolate had the highest SUVA value and aromatic content but the lowest STHMFP among the HPoA isolates (Table II). Variable amounts of a non THM-forming organic fraction might be expected in NOM from different sources and diagenetic histories. This organic material may be reacting with chlorine to form other DBPs, such as haloacetic acids. Reckhow and others (6, 29) found that organic material may preferentially form either THMs or other chlorinated by-products, depending on chemical composition and reaction conditions.

The carbon isotopic data provides evidence for a variable fraction of organic material that does not react to form THMs over the course of the THMFP experiment. We have previously found chemically distinct pools of organic material within isolates that react to form THMs at different rates—some not appreciably reacting over the course of a seven-day formation potential test (10). These pools also have distinct carbon isotopic ratios, the combination of which determines the final $\delta^{13}C_{THM}$ (Table II). The difference between the isotopic value of the bulk material and the isotopic value of the THM ($\Delta\delta^{13}C_{BULK-THM}$) therefore carries information about the amount of unreactive material.

Although the source of the organic material determines its carbon isotopic ratio, experiments with fresh plant decoctions resulted in $\Delta\delta^{13}C_{BULK-THM}$ values that suggest the THM precursors removed during diagenesis are isotopically light, such that diagenesis yields residual samples with lower $\Delta\delta^{13}C_{BULK-THM}$ values (16). Since the reactivity of aromatic organic compounds toward diagenesis and toward chlorination are not likely directly related, in samples of various diagenetic maturities (as indicated by $\Delta\delta^{13}C_{BULK-THM}$) a correlation between aromatic carbon content and STHMFP may not be expected.

The highest values for $\Delta\delta^{13}C_{BULK-THM}$ were observed within the Delta, in the Old and Middle River samples (Table II). Both the River and Export samples had lower values, suggesting these samples contained organic material from isotopically distinct sources or material that was more degraded. Perhaps more importantly, the pattern of $\Delta\delta^{13}C_{BULK-THM}$ values indicates the NOM in the export sample has been modified from its Delta origins by either further degradation or addition of isotopically different material into the THM precursor pool.

More broadly, the isotopic data together with the other chemical measurements indicate that the NOM in Delta channel waters may not be interpreted as a conservative mixture of river- and peat island-derived NOM; this simple scenario does not fully account for the chemical variability among samples in this study. No systematic differences were observed between the Sacramento and San Joaquin River samples and the Old River, Middle River, and export site Delta channel samples in NOM content, proportion of aromatic carbon, C:N, $\delta^{15}N_{BULK}$, STHMFP, $\delta^{13}C_{BULK}$ or $\delta^{13}C_{THM}$. The compositions of the Old River, Middle River, and export samples cannot be accounted for by combinations of the Sacramento River, San Joaquin River, and Twitchell Island drainage samples (Tables I-III; Figure 1). C:N, $\delta^{13}C_{BULK}$, $\delta^{15}N_{BULK}$, and $^{13}C$ NMR measurements of the isolates indicate substantial differences

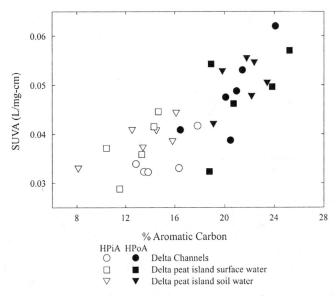

*Figure 5. Relationship between SUVA and aromatic carbon percentage for HPiA and HPoA NOM isolates from the Sacramento-San Joaquin Delta (Tables II and III) and a Delta peat island (7).*

in NOM source and composition, and differences among HPoA and HPiA fractions of the NOM isolates. In order to determine if this indication is robust, sampling and analysis of additional Delta channel water samples and agricultural drainage waters must be conducted.

The export sample was chemically quite distinct from other samples in the study, suggesting that in-channel processes may be more influential than previously thought in determining the NOM content, composition, and THMFP of Delta channel waters. Although the whole-water export sample was relatively high in UV absorbance

(Table I), the isolates had relatively low proportions of aromatic and high proportions of aliphatic carbon (Table III). The $\delta^{13}C_{BULK}$ of the export HPoA sample isolate was the heaviest HPoA carbon isotopic value measured, and the $\delta^{13}C_{THM}$ of the THM formed from this sample was 1.5‰ heavier than the average (Table II). These results suggest that material with a heavier carbon isotopic composition had influenced the export sample. Possible sources include material from algal, wastewater or agricultural sources. The high C:N values for the isolates (Table II), however, suggest fresh algal or waste-water material was not the source of the heavier carbon. Although evocative, these results, must be confirmed using samples collected over different seasons, river flow conditions, and points in the tidal cycle.

## CONCLUSIONS

The following conclusions are supported by the data obtained from the samples analyzed. The limited number of samples in this study cannot constrain the considerable temporal and spatial variability within the Delta and thus the conclusions should be regarded as provisional until a broader sampling and analysis is accomplished.

1. The variability in chemical composition of the Delta channel water isolates, as indicated by significant differences in $^{13}C$ CPMAS NMR spectra and C:N measurements, was not consistent with simple mixing of river- and peat-derived NOM.

2. For these samples, there was no significant increase in the aromatic carbon content of NOM isolates between the Sacramento and San Joaquin River samples and the Delta channel samples (Old River, Middle River, and export), even though SUVA did increase. These results indicate that changes in UV absorbance may be decoupled from THMFP and aromatic carbon contents in this environment.

3. The lowest STHMFP observed was in the peat island agricultural drainage sample, even though this sample had the highest SUVA and aromatic carbon content, suggesting that the addition of this peat island water into Delta channels may not necessarily increase the channel water STHMFP.

4. The higher aromatic carbon content and STHMFP and lower $\delta^{13}C$ value of THMs formed from the HPoA isolates suggest that THMs form from different chemical sources in HPoA than is the case in HPiA isolates.

## ACKNOWLEDGMENTS

The authors are grateful for the support of the U.S. Geological Survey Toxics Substances Hydrology Program and the U.S. Geological Survey Drinking Water Initiative. Earlier versions of this manuscript benefited from careful reviews by K. Kuivila, C. Kratzer, C. Sanchez, and two anonymous reviewers.

# LITERATURE CITED

1. Rook, J. J. *Water Treatment Exam.* **1974,** *23* (2), 234-280.
2. *Drinking water regulations and health advisories;* 822-R-94-001; U.S. Environmental Protection Agency: Washington, DC, 1994; 10 p.
3. U.S. Environmental Protection Agency. *Fed. Reg.* **1998,** *63,* 241, part IV.
4. Krasner, S. W.; McGuire, M. J.; Jacangelo, J. G.; Patania, N. L.; Reagan, K. M.; Aieta, E. M. *J. Am. Water Works Assoc.* **1989,** *81* (8), 41-53.
5. Symons, J. M.; Krasner, S. W.; Simms, L. A.; Sclimenti, M. J. *J. Am. Water Works Assoc.* **1993,** *85* (1), 51-62.
6. Reckhow, D. A.; Singer, P. C.; Malcolm, R. L. *Environ. Sci. Technol.* **1990,** *24,* 1655-1664.
7. Fujii, R.; Ranalli, A. J.; Aiken, G. R.; Bergamaschi, B. A. *Dissolved organic carbon concentrations and trihalomethane formation potentials in waters from agricultural peat soils, Sacramento-San Joaquin Delta, California - Implications for drinking-water quality;* Water-Resources Investigations Report 98-4147, U.S. Geological Survey: Reston, VA, 1998; 75 p.
8. Krasner, S. W.; Croué, J.-P.; Buffle, J., Perdue, E. M. *J. Am. Water Works Assoc.* **1996,** *88* (6), 66-79.
9. Rathbun, R. E. *Arch. Environ. Contam. Tox.* **1996,** *30,* 156-162.
10. Fram, M. S.; Fujii, R.; Weishaar, J. L.; Bergamaschi, B. A.; Aiken, G. R. In *Con-tamination of Hydrologic Systems and Related Ecosystems*; Morganwalp, D. W., Buxton, H. T., Eds.; U.S. Geological Survey Water Resources Investigations Report 99-4018B; U.S. Geological Survey, Reston, VA, 1999; pp. 423-430.
11. Krasner, S. W.; Sclimenti, M. J.; Means, E. G. *J. Am. Water Works Assoc.* **1994,** *86* (6), 34-47.
12. *Sacramento-San Joaquin Delta Atlas;* California Department of Water Resources: Sacramento, CA, 1993; 121 p.
13. *Five-year report of the municipal water quality investigations program: Summary and findings during five dry years, January 1987—December 1991*; California Department of Water Resources: Sacramento, CA, 1994; 119 p.
14. Amy, G. L.; Thompson, J. M.; Tan, L.; Davis, M. K.; Krasner, S. W. *J. Am. Water Works Assoc.* **1990,** *82* (1), 57-64.
15. *Delta island drainage investigation report, water years 1987 and 1988;* California Department of Water Resources; Interagency Delta Health Aspects Monitoring Program, California Department of Water Resources: Sacramento, CA, 1990; 110 p.
16. Bergamaschi, B. A.; Fram, M. S.; Kendall, C.; Silva, S. R.; Aiken, G. R.; Fujii, R. *Org. Geochem.* **1999,** *30,* 835-842.
17. Aiken, G. R.; McKnight, D. M.; Thorn, K. A.; Thurman, E. M. *Org. Geochem.* **1992,** *18,* 567-573.
18. Krasner, S. W.; Sclimenti, M. J. *Natural organic matter in drinking water: origin, characterization, and removal;* Natural Organic Matter Workshop, Chamonix, France, American Water Works Association Research Foundation: Denver, CO, 1994; 109 p.

222

19. California Department of Water Resources Bryte Laboratory's quality assurance and quality control manual; California Department of Water Resources: Sacramento, CA, 1994; 63 p.
20. Hedges, J. I.; Clark, W. A.; Cowie, G. L. Limnol. Ocean. 1988, 33, 1116-1136.
21. Hedges, J. I.; Clark, W. A.; Cowie, G. L. Limnol. Ocean. 1988, 33, 1137-1152.
22. Zar, J. H. Biostatistical analysis; Prentice-Hall: New York, NY, 1984; 718 p.
23. Benner, R.; Fogel, M. L.; Sprague, E. K.; Hodson, R.E. Nature 1987, 329, 708-710.
24. Mason, H. L. The flora of the marshes of California; University of California, Berkeley Press: Berkeley, CA, 1957; 878 p.
25. Atwater, B. F. Ph.D. thesis, University of Delaware, Dover, DE, 1980.
26. Hatcher, P. G.; Simoneit, B. R. T.; Mackenzie, F. T.; Neumann, A. C.; Thorstenson, D. C.; Gerchakov, S. M. Org. Geochem. 1982, 4, 93-110.
27. Skoog, D. A.; Holler, F. J.; Nieman, T. A. Principals of instrumental analysis; Harcourt Brace: New York, NY, 1998; 849 p.
28. Thomas, M. J. K. Ultraviolet and visible spectroscopy; John Wiley & Sons: New York, NY, 1996; 229 p.
29. Reckhow, D. A.; Singer, P. C. In Water chlorination: chemistry, environmental impacts and health effects; Jolley, R. L.; et al., Eds.; Lewis Publishers: Chelsea, MI, 1985; Vol. 5, 1575 p.

Chapter 15

# A Comprehensive Kinetic Model for Chlorine Decay and Chlorination By-Product Formation

John N. McClellan[1], David A. Reckhow[1], John E. Tobiason[1], James K. Edzwald[1], and Darrell B. Smith[2]

[1]Department of Civil and Environmental Engineering, University of Massachusetts, 18 Marston Hall, Amherst, MA 01003
[2]South Central Connecticut Regional Water Authority, 90 Sargent Drive, New Haven, CT 06511

A model based on a simplified conceptual reaction mechanism that predicts chlorine decay and the formation of THMs and HAAs was developed and calibrated. The form of the model is a system of differential equations (solved numerically). The model was calibrated using a low SUVA, low alkalinity, low bromide water. Excellent results were obtained when the model was tested using a water similar to the calibration water.

## Introduction

Most U.S. water utilities use chlorine to comply with CT requirements, and it is general practice to maintain a low-concentration chlorine residual in water distribution systems. This provides essential public health protection by preventing communication of waterborne infectious disease. Chlorine is a strong oxidant that reacts with the natural organic matter (NOM) present in all natural waters. Although chlorine is invaluable in the protection of public health against waterborne disease, some by-products of the reactions between chlorine and NOM may represent long-term health risks to water consumers. The trihalomethanes (THMs) and haloacetic acids (HAAs) are two important classes of potentially harmful chlorination by-products that are currently regulated by the USEPA, and therefore of particular interest to water utilities.

Water suppliers that practice chlorine disinfection face conflicting objectives in providing adequate microbial protection while minimizing the formation of harmful chlorinated organic by-products. Chlorine/NOM reactions may proceed over a period of days under conditions typical in water distribution systems and variations in the source water quality and in environmental conditions such as temperature and pH can affect the rates and extents of by-product forming reactions. As a result of chemical and hydraulic dynamics, the chemical composition of water can vary substantially in time and space in distribution systems. Models that can capture these dynamics would be powerful tools for optimizing water quality in distribution systems.

The typical form of a mechanistic model is a system of differential equations where reactant concentrations are represented explicitly. Although mechanistic models are more cumbersome from a computational standpoint, they have some potential advantages compared to the empirical models that have been used in most previous chlorine/by-product modeling efforts. The first advantage is flexibility. Mechanistic models are well suited for application to complex situations that arise in practice, such as modeling reactive by-product species, "booster" chlorination, or distribution systems supplied by multiple sources of different quality. Mechanistic models are expected to be more robust than empirical models (*1*). A model that is based on the physics of a system will be as robust as the underlying physical mechanism, but there is no basis for assuming that an empirical model can be extrapolated beyond the range of values in the calibration data. Finally, development and testing of a mechanistic model serves to confirm or refute the validity of the proposed mechanism and thus contributes to our scientific understanding of the process being modeled. It is therefore desirable for kinetic models to reflect the underlying chemistry to the extent possible. Mechanistic modeling has been considered impractical due to the complexity of NOM molecules and NOM/chlorine reaction pathways, requiring complex computations. However, advances in computer technology have now brought this level of computing power to the general public. The goal of this research is to develop a kinetic model that captures the important mechanistic features of chlorine/NOM reactions.

# Background

## Existing Models

Efforts to model chlorine decay in natural waters began before 1950 (*2,3*) and a number of studies have been conducted where chlorine decay was modeled in the contexts of drinking water, waste water, and power plant cooling water. The most widely used form for modeling chlorine decay has been exponential decay or first-order (the reaction rate has first-order dependence on chlorine concentration). Several investigators have presented empirical models for THM formation as a function of other parameters such as UV absorbance, TOC, pH, temperature and bromide concentration (4-8). These are linear models (in log-transformed variables) that have been fit to large experimental data sets by multivariate linear regression procedures. This approach has also been used to model HAA formation (8). Models for chlorine decay where the reaction rates have first-order dependence on the concentrations of both NOM sites and chlorine (second-order dependence on reactant concentrations overall) have been proposed by MacNeill (9), Jedas-Hecart et al. (10), and Clark (11, 12). The models proposed by MacNeill (9) and Clark (11, 12) also predict THM formation.

Vasconcelos et al. (13) compared several chlorine decay models by fitting them to data sets from bottle tests of eleven different treated waters. Goodness of fit (as measured by the coefficient of determination $R^2$) for the widely used first-order model was generally poor. For other models compared by Vasconcelos et al. (14), reasonably good fits were achieved with wide variations in the parameter values for the various data sets.

## Reaction Mechanisms

The aqueous chlorination of organic compounds has been studied by a number of researchers (14-23). The organic substrates investigated included various pure ketones, organic acids, and aromatics as well as organic fractions extracted from natural waters. Some key findings that bear on the development of a mechanistic model are summarized in this section.

Under conditions typical in water treatment, HOCl and OCl⁻ predominate and $Cl_2$ is not present in significant concentrations. HOCl is expected to be more reactive as an oxidant and halogenating agent than OCl⁻ (24). The structural features within NOM molecules that have been implicated as important chlorine consumers and by-product precursors include activated aromatics (particularly those with *meta* hydroxyl substituents) and some oxygen-rich ketones and keto-acids (17-22). For several ketones and keto-acids, the rate of reaction with halogens is first-order in ketone and not sensitive to halogen concentration except at very low levels (14-16). The reason is a rate-limiting initial dissociation of a proton to form enolate ion, the reactive species. In the classic haloform reaction of halogens with methyl ketones, enolate formation is followed by a faster halogen addition. After the initial halogenation the remaining protons are more acidic so that the second and third ionizations and halogenations proceed rapidly followed by base-catalyzed hydrolysis to form haloform (24).

For orcinol and resorcinol, the O⁻ substituted species react much faster with chlorine than the corresponding OH substituted species. Nevertheless, the completely protonated species are predominant and are expected to be the most significant participants in chlorine reactions over the pH range of interest in water treatment (23). Acid-base chemistry would play a more important role for aromatics with more acidic substituents, and a first-order rate-limiting step similar to the rate-limiting step of the haloform reaction could be envisioned.

Most of the reaction mechanisms that have been proposed for chloroform formation pass through a trichloromethyl ketone intermediate (*17, 20, 22*). Chlorine concentration and pH can affect the speciation of products even where reactions of the intermediate are not rate-limiting steps in the overall reaction (*22, 25*).

The objective is to capture the most important elements of the reaction, i.e. the elements that control the overall reaction rate, and those that affect the proportions of products that are formed. Based on the reaction mechanisms that have been proposed, it seems appropriate to include two conceptual reaction pathways in the model: one with second-order kinetics (i.e. the reaction rate has first-order dependence on both chlorine and NOM site concentrations), and one with a rate-limiting first-order (in NOM site concentration) step similar to the haloform reaction. Chlorine consumption and by-product formation rates will be determined as functions of the reaction rates through these conceptual pathways.

## Experimental

A series of bench-scale experiments was conducted to characterize the effects of temperature, pH, and reactant concentrations on chlorine decay and by-product formation in a treated water. These experiments were conducted using three samples collected at different times from the same source, filter effluent of the Lake Gaillard Water Treatment Plant, No. Branford, Conn. This plant treats an impounded surface water source. The process includes alum coagulation, flocculation, and anthracite/sand filtration. Table I gives UV absorbance and TOC values for each of the samples. Measured alkalinity values for these samples are not available, but a typical value for the source water is 10-15 mg/L as $CaCO_3$.

**Table I. Lake Gaillard WTP Sample NOM Characteristics**

| Date Sample Collected | UV Abs. at 254 nm (cm$^{-1}$) | TOC (mg/L) |
|---|---|---|
| 7/1/97 | 0.030 | 1.91 |
| 11/14/97 | 0.028 | 2.11 |
| 7/21/98 | 0.029 | 1.97 |

A series of bottle tests was conducted with each sample. For these tests, four-liter batches were prepared by adjusting pH and temperature to the desired values. Samples were diluted with deionized water to vary the NOM concentration for certain sets. Batches were then chlorinated and dispensed into 300 mL bottles. Table II shows the ranges of pH, temperatures, TOC concentrations and chlorine doses used for the bench scale experiments. Bottles were stored headspace-free in the dark at the desired temperature and removed at intervals for measurement of free chlorine, THMs, and HAAs. Chlorine measurements were made using the DPD Titrimetric Method (*26*). Haloacetic acids samples were prepared using the acidic methanol derivitization method of Xie et al. (*27*). Trihalomethane samples were prepared by micro-extraction into pentane. Haloacetic acid and trihalomethane samples were analyzed by gas chromatography with electron capture detection.

**Table II. Conditions for Bench Scale Experiments**

| Parameter | Range of values |
|---|---|
| pH | 6.3-9.0 |
| Temperature (°C) | 2-25 |
| Chlorine dose (mg/L) | 1.3-4.0 |
| Reaction time (hours) | 0.1 –172 |
| TOC (mg/L)[1] | 1.1-2.1 |

[1]TOC was varied from the values given in Table I by dilution with deionized water.

## Model Description

Three classes of reactive functional groups within the NOM are hypothesized. These are: (1) Sites that react with chlorine instantly relative to the time scale of interest (minutes to days). These reactions are treated as constants (e.g. "instantaneous chlorine demand") in the model. (2) Sites that react with chlorine (HOCl or OCl⁻) where the first (rate-limiting) step is second-order (first-order in NOM and first-order in chlorine). These are called $S_1$ sites. (3) $S_2$ sites, where there is a rate-limiting initial step that is first-order in NOM, leading to an active form ($S_2^-$) that participates in a faster second-order reaction with HOCl. The prototype for the $S_2$ reaction pathway is the classical haloform reaction of ketones with halogens, where the rate-limiting step (first-order in ketone) is a proton dissociation to form enolate ion. For the $S_1$ and $S_2$ pathways, a series of faster chlorine-consuming steps follow, producing halogenated and oxidized organic compounds, $CO_2$, and Cl⁻. The amount of formation of each by-product includes an initial amount (representing "instantaneous" formation) plus the sum of specified fractions of the total site consumption through the $S_1$ and $S_2$ pathways. The rate of chlorine consumption is equal to the total rate of site consumption through each pathway multiplied by stoichiometric coefficients representing consumption in intermediate fast steps. The conceptual reaction mechanism is depicted in schematic form in Figure 1.

Equation 1.1 represents the $S_1$ pathway reaction rate, and equations 1.2, 1.3, and 1.4 represent rates related to the $S_2$ pathway.

$$R_1 = k_1'[S_1][Cl_2] \tag{1.1}$$

$$R_{21f} = k'_{21f} [S_2H]$$
(1.2)

$$R_{21r} = k'_{21r} [S_2^-]$$
(1.3)

$$R_{22} = k'_{22} [S_2^-][HOCl]$$
(1.4)

The terms $R_1$, $R_{21f}$, $R_{21r}$ and $R_{22}$ are rates, $k'_1$, $k'_{21f}$, $k'_{21r}$, and $k'_{22}$ are apparent rate constants, and $[S_1]$, $[S_2H]$, and $[S_2^-]$ represent NOM site concentrations. The subscript "1" is used for terms associated with the $S_1$ pathway. The subscripts "21f" and "21r" are associated with the forward and reverse components respectively of the $S_2H$ proton dissociation, and the subscript "22" is associated with the HOCl-$S_2^-$ reaction.

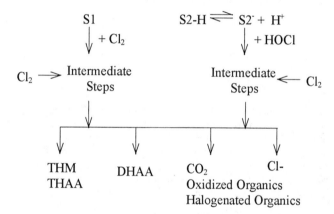

*Figure 1. Conceptual Reaction Mechanism*

Equations 2 represent the rates of consumption of the reactants (chlorine and NOM sites),

$$\frac{-d[Cl_2]}{dt} = n_1 R_1 + n_2 R_{22}$$
(2.1)

$$-\frac{d[S_1]}{dt} = R_1$$
(2.2)

$$-\frac{d[S_2H]}{dt} = R_{21f} - R_{21r}$$
(2.3)

$$-\frac{d[S_2^-]}{dt} = R_{2f} - R_{2lf} + R_{22} \tag{2.4}$$

where $n_1$ and $n_2$ are coefficients that account for chlorine consumption in intermediate steps and the other terms are as described above for equations 1. The formation kinetics of di- and tri- halogenated HAA species were found to be quite different, so they have been treated as separate groups, with "DiHAA" representing the sum of di-halogenated HAA species, and "TriHAA" representing the sum of tri-halogenated HAA species (in the low-bromide Lake Gaillard water, DCAA and BCAA are the only di-halogenated species that form in measurable concentrations, and TriHAA is approximately equal to TCAA). Total trihalomethane is denoted as "TTHM." Equations 3 describe the rates of formation of TTHM, TriHAA, and DiHAA. The $\alpha$ factors represent the fractions of site consumption through each pathway that result in formation of the respective products.

$$\frac{d[TTHM]}{dt} = \alpha_{1,THM} R_1 + \alpha_{2,THM} R_{22} \tag{3.1}$$

$$\frac{d[DiHAA]}{dt} = \alpha_{1,DiHAA} R_1 + \alpha_{2,DiHAA} R_{22} \tag{3.2}$$

$$\frac{d[TriHAA]}{dt} = \alpha_{1,TriHAA} R_1 + \alpha_{2,TriHAA} R_{22} \tag{3.3}$$

Equations 1, 2 and 3 comprise the basic framework of the model. The system of equations must be integrated using numerical methods. For a particular water and set of environmental conditions, a set of apparent rate constants $k'$, factors $n$ and $\alpha$, and initial site concentrations $[S_1]$, $[S_2H]$, and $[S_2^-]$ can be determined by fitting to experimental data. However, for a particular water, the apparent rate constants and factors may vary with environmental conditions. An important objective is to capture these effects in the model. To achieve this objective, a mechanistic treatment of the effects of temperature and pH was incorporated. Equation 4.1 relates the apparent rate constant $k'_1$ to fundamental rate constants, including temperature and pH effects. Here, $k_{11}$ and $k_{12}$ are the rate constants for the reaction of $S_1$ sites with HOCl and OCl$^-$ respectively, and $\alpha_{HOCl}$ and $\alpha_{OCl^-}$ are pH-dependent distribution factors for the chlorine species. The term $\theta_1$ with reaction temperature $T$ and reference temperature $T_{ref}$ is the standard engineering representation of the Arrhenius expression.

$$k'_1 = (k_{11}\alpha_{HOCl} + k_{12}\alpha_{OCl^-})\theta_1^{T-T_{ref}} \tag{4.1}$$

Equations 4.2, 4.3, and 4.4 describe the apparent rate constants $k'_{21f}$, $k'_{21r}$, and $k'_{22}$.

$$k'_{21f} = k_{21f}\theta_{2f}^{T-T_{ref}} \tag{4.2}$$

$$k'_{21r} = k_{21r}[H^+]\theta_{2r}^{T-T_{ref}} \tag{4.3}$$

$$k'_{22} = k_{22}\theta_{22}^{T-T_{ref}} \tag{4.4}$$

Equations 5 and 6 describe the distribution factors that relate the relative proportions of products to temperature and pH. Our experimental work suggests that substantial THM and TriHAA formation occurs through competitive pathways. A common pathway leading to a penultimate intermediate that may form either THM by hydrolysis or TriHAA by oxidation (*22*) is hypothesized. The following distribution factors account for the effects of temperature and pH on this hydrolysis/oxidation competition (equations 5):

$$\alpha_{1,P} = k_{P,HOCl}[HOCl]/(k_P + k_{P,OH}[OH^-] + k_{P,HOCl}[HOCl]) \tag{5.1}$$

$$\alpha_{2,P} = (k_P + k_{P,OH}[OH^-])/(k_P + k_{P,OH}[OH^-] + k_{P,HOCl}[HOCl])\theta_P^{T-T_{ref}} \tag{5.2}$$

where $k_P$ represents uncatalyzed hydrolysis, $k_{P,OH}$ represents base-catalyzed hydrolysis, $k_{P,HOCl}$ represents oxidation, $\alpha_{1,P}$ is the fraction that follows the oxidation pathway, and $\alpha_{2,P}$ is the fraction that follows the hydrolysis pathway. The distribution factors for the individual by-product species are linear combinations of the factors given in equa-tions 5. The general form of the distribution factors for individual species is given in equations 6, where the generic subscript DBP denotes TTHM, DiHAA, or TriHAA.

$$\alpha_{1,DBP} = \beta_{1,DBP} + \beta_{11,DBP}\alpha_{1,P} + \beta_{12,DBP}\alpha_{2,P} \tag{6.1}$$

$$\alpha_{2,DBP} = \beta_{2,DBP} + \beta_{21,DBP}\alpha_{1,P} + \beta_{22,DBP}\alpha_{2,P} \tag{6.2}$$

In equations 6, the $\beta$ factors are species-specific constants. The factors $\alpha_{1,DBP}$ and $\alpha_{2,DBP}$ describe the fractions of the $S_1$ and $S_2$ pathways respectively that result in the formation of a particular DBP species. In practice, not all of the right hand side terms are included for an individual species: $\beta_1$ and $\beta_2$ are included in the DiHAA terms, $\beta_1$, $\beta_{11}$, and $\beta_{21}$ are included in the TriHAA terms, and $\beta_{12}$, and $\beta_{22}$ are included in the TTHM terms.

## Calibration and Computation

As stated above, the form of the proposed model is a set of differential equations with known initial conditions that must be solved numerically. Calibration and computational issues are discussed in this section. Two software packages were used for computations and data analysis in this work: the *Scientist* data fitting package (*28*) and "DBP Viewer," a program written specifically for this work.

### Calibration

In order for numerical optimization routines to converge on a solution with the proposed model, initial parameter estimates that are very close to their optimum values are required. We utilized Windows-based computer graphics to make initial parameter estimates. The DBP Viewer program was developed to allow adjustment of the model parameters with slide bars while displaying experimental data with the model predictions superimposed. Initial estimates made with the DBP Viewer program were utilized with the *Scientist* (*28*) software package to find values that resulted in minimum sums of squared residuals. For purposes of model fitting, all response variables (chlorine, TTHM, DiHAA and TriHAA concentrations) were normalized with respect to their average values in the calibration data set. The parameters were optimized with respect to all response variables simultaneously with no weighting. Model parameters and estimated values are presented in Table III.

There are a total of 20 parameters listed in Table III. Of these, 16 are adjustable, with values estimated by fitting to bench scale data. Rate $R_{21r}$ is assumed to be diffusion limited, and constant $k_{21r}$ is therefore estimated based on a literature value (*29*). The intermediate step chlorine consumption coefficients $n_1$ and $n_2$ cannot be determined independently from the other rate constants and have been set arbitrarily. The values for $n_1$ and $n_2$ are rough estimates of the chlorine consumed in forming chloroform from dihydroxybenzene and methyl ketone respectively. The by-product formation parameter $k_P$ is not independent of the parameters $k_{P,OH}$ and $k_{P,HOCl}$ and is therefore set arbitrarily to unity.

The parameters listed in Table III are intended to apply generally, or at least to similar (i.e. low SUVA, low bromide, low alkalinity) waters. The nature and concentration of the NOM in a particular water is captured in the model by the initial concentrations of $S_1$ and $S_2$ sites, initial concentrations of by-product species (representing "instantaneous" formation), and by an estimate of instantaneous chlorine demand. These values serve as initial conditions for solving the system of differential equations presented in equations 2 through 6.

**Table III. Model Parameters**

| Parameter | Units | Fitted Value |
|---|---|---|
| $S_1$ Path | | |
| $k_{11}$ | $\mu M^{-1}\ hr^{-1}$ | 0.0347 |
| $k_{12}$ | $\mu M^{-1}\ hr^{-1}$ | $9.70 \times 10^{-4}$ |
| $\theta_1^{T-20°\ C}$ | T in °C | 1.0996 |
| $S_2$ Path | | |
| $k_{21f}$ | $hr^{-1}$ | $7.18 \times 10^{-3}$ |
| $k_{21r}^{1}$ | $\mu M^{-1}\ hr^{-1}$ | $1.44 \times 10^{7}$ |
| $k_{22}$ | $\mu M^{-1}\ hr^{-1}$ | 0.854 |
| $\theta_{21}$ | $\theta^{T-20}$: T in °C | 1.089 |
| $Cl_2$ Demand | | |
| $n_1^{2}$ | Dimensionless | 8 |
| $n_2^{2}$ | Dimensionless | 4 |
| Byproduct Formation | | |
| $k_P^{2}$ | Dimensionless | 1 |
| $k_{P,HOCl}$ | $\mu M^{-1}$ | 0.0620 |
| $k_{P,OH}$ | $M^{-1}$ | $9.82 \times 10^{6}$ |
| $\theta_P$ | $\theta^{T-20}$: T in °C | 1.007 |
| $\beta_{12,THM}$ | Dimensionless | 0.201 |
| $\beta_{22,THM}$ | Dimensionless | 0.0956 |
| $\beta_{1,TCAA}$ | Dimensionless | 0.0387 |
| $\beta_{11,TCAA}$ | Dimensionless | 0.0272 |
| $\beta_{21,TCAA}$ | Dimensionless | 0.0777 |
| $\beta_{1,DCAA}$ | Dimensionless | 0.0346 |
| $\beta_{2,DCAA}$ | Dimensionless | 0.0345 |

[1]Rate assumed to be diffusion-limited. Not treated as adjustable parameter, approximate value from reference (*29*)

[2]Fixed arbitrarily, not treated as adjustable parameter.

No method is available for determining these values directly, so they must be estimated experimentally. At a minimum, a scaled-down version of the bottle tests described above is required, with measurements of instantaneous by-product and chlorine concentrations, and several measurements of chlorine concentration at intervals, so that initial NOM site concentration parameter values can be determined by least-squares fitting.

## Numerical Schemes

The *Scientist* package *(28)* includes several numerical routines including an explicit (Euler) scheme, a fourth-order Runge-Kutta scheme, and a scheme designed for stiff equations called the EPISODE scheme. We found that the EPISODE scheme produced the best combination of accuracy and computational speed. The scheme used in the DBP Viewer program is a second-order mid-point scheme with a time step of 0.1 hours. Under this scheme, explicit response variable estimates at the time-step mid points are used to advance the solution.

## Model Comparison

### Calibration Data Set Comparison

The proposed ("mechanistic") model was compared with two other model forms using the calibration (Lake Gaillard) data set. These were: the "power function" form and the "second-order" form. The power function models are linear functions of TOC, UV absorbance, pH, temperature, chlorine dose and reaction time in log-transformed variables. Bromide concentration was omitted as an independent variable in this comparison because of the low bromide concentration in the Lake Gaillard water. Power function models have been widely used for predicting THM and HAA formation *(7, 8)*. In this comparison, the power function form was also used for predicting chlorine decay.

The second-order form for chlorine decay presented by Clark *(11, 12)* is given in equation 7, where $C$ represents chlorine concentration, $T$ is THM concentration, and $D$, $\mu$ and $K$ are constants.

$$C = C_o \frac{1-K}{1-Ke^{-\mu t}} \qquad 7.1$$

$$T = D(C_o - C) \qquad 7.2$$

Clark *(12)* presented models for determining the constants $D$, $\mu$ and $K$ in equations 7 as empirical functions of pH, temperature, TOC and initial chlorine concentration. For the Lake Gaillard data set comparison, HAA formation was modeled using this form in addition to chlorine decay and THM formation.

This comparison was intended to evaluate the capacity of the model forms to capture the effects of changes in pH, temperature, initial chlorine concentration, and NOM concentration for a specific water. This comparison does not reflect the generality of the models. Note that for this comparison, published parameter values for the power function and second-order models were not utilized. Each model was calibrated using the Lake Gaillard data set. The power function model was calibrated by multiple linear regression using log-transformed variables. The mechanistic model and the second-order model were calibrated using a numerical least-squares minimization routine (28). Results of the comparison are presented in Table IV. Table IV shows that the mechanistic model resulted in the highest coefficients of determination, although all three models fit the data well.

**Table IV. Comparison of Models: Calibration Data Set[1]**

|  | Model Form | | |
| --- | --- | --- | --- |
|  | Mechanistic | Power Function | Second-Order |
| Number of adjustable parameters | 21 | 28 | 25 |
| Coefficient of determination $R^2$: | | | |
| Chlorine predictions | 0.95 | 0.92 | 0.89 |
| TTHM predictions | 0.95 | 0.90 | 0.85 |
| DiHAA predictions | 0.91 | 0.87 | 0.83 |
| TriHAA predictions | 0.88 | 0.83 | 0.81 |
| All predictions | 0.92 | 0.88 | 0.85 |

[1]Size of data set: 770 response variable observations (approximately 200 each of chlorine, TTHM, DiHAA, and TriHAA).

Figure 2 shows typical chlorine decay and by-product formation curves, with predictions of the mechanistic model superimposed on experimental observations. Figures 3, 4, and 5 show residual plots for the mechanistic model, the power function model, and the second-order model respectively. These plots show residuals (defined as predicted value minus observed value) versus predicted values for the calibration data set.

## External Data Sets

The mechanistic, power function, and second-order models were tested and compared using two external data sets. These data sets were developed from bench scale experiments similar to the experiments used to develop the calibration data set. The waters used were filter effluent from the Robert E. McQuade Water Treatment Plant in Andover, Mass. (30), and Colorado River water collected at the Central

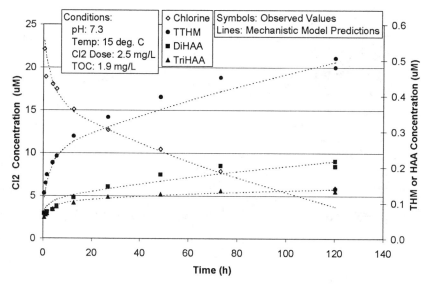

*Figure 2. Typical chlorine decay and by-product formation curves.*
*Calibration data set.*

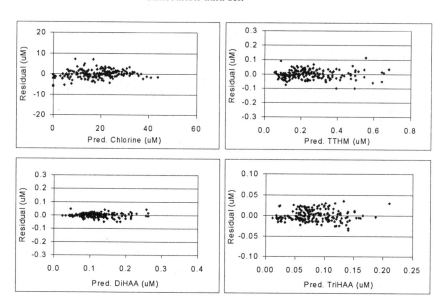

*Figure 3. Calibration data set residual plots: mechanistic model.*

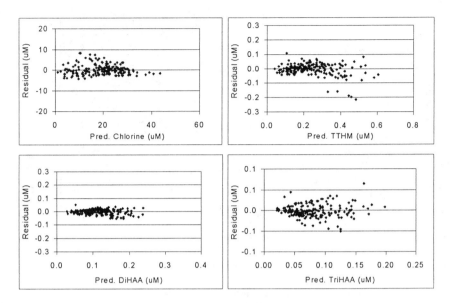

*Figure 4. Calibration data set residual plots: power function model.*

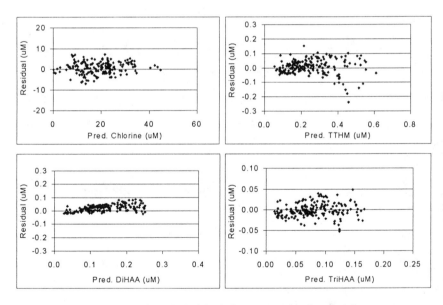

*Figure 5. Calibration data set residual plots: second order model.*

Arizona Project (CAP) intake at Scottsdale, Arizona (*31*). The Robert E. McQuade plant treats a blend of pond water and Merrimack River water. The process includes ozonation, alum coagulation, flocculation, settling, and GAC/sand filtration. Table V shows source water characteristics. Table VI shows the range of pH, temperature, chlorine dose, TOC, and reaction time values in each of the data sets.

**Table V. Andover and CAP Water Characteristics**

| Parameter | Andover water | CAP water |
|---|---|---|
| TOC (mg/L) | 1.7[1] | 4.2 |
| UV Abs. At 254 nm | 0.03[1] | 0.053 |
| Alkalinity (mg/L as CaCO₃) | <25[1] | 120 |

[1]Typical value. Measured value for sample not available.

**Table VI. Andover and CAP Data Sets: Range of Conditions**

| | Range of values | |
|---|---|---|
| Parameter | Andover data | CAP data |
| pH | 8.0-9.0 | 5.5-9.5 |
| Temperature (°C) | 8-20 | 2-15 |
| Chlorine dose (mg/L) | 1.3-3.0 | 1.7-2.7 |
| Reaction time (hours) | 0.5–168 | 0.03-119 |
| TOC (mg/L) | 0.9-1.7[1] | 4.2 |
| Size of data set[2] | 416 | 140 |

[1]Typical value. Measured value for sample not available.

[2]Total number of response variable observations.

As stated previously, a chlorination experiment is required to apply the mechanistic model. The results from one bottle test experiment for each of the three water samples (two Andover samples and one CAP sample) were set aside and used for this purpose. This data was used to set initial concentration values in the mechanistic model for each of the water samples. The power function and second-order forms were fine-tuned for each of the three external samples by scaling to fit the set-aside bottle test results (after complete calibration using the Lake Gaillard data). This sample-specific fine-tuning is not normally performed when applying the power function and second-order forms, but it is included here in order to compare the models on an equal basis. In addition, the second-order model for chlorine decay presented by Clark (*12*) and the power function models for THM and HAA species formation presented by Montgomery Watson (*8*) ("literature models") were compared using published parameter values and no sample-specific fine tuning. The character and concentration of the NOM is captured in these models by TOC and UV absorbance (for the Montgomery Watson models) terms.

Table VII shows the results of these tests in terms of percents of "successful" predictions. Successful predictions are arbitrarily defined as those where the observed values are within the following ranges of model predictions: ±2.8 µM (0.2 mg/L) for chlorine, ±0.034 µM (4 µg/L) for TTHM, ±0.023 µM (3 µg/L) for DiHAA, and ±0.012 (2 µg/L) for TriHAA. Figures 6, 7, 8, and 9 show residual plots for the mechanistic, power function, second-order, and literature models respectively, for the Andover data set. Figures 10, 11, 12, and 13 show residual plots for the mechanistic, power function, second-order, and literature models respectively, for the CAP data set. Table VII shows that the results from the mechanistic, power function, and second-order models were very good and comparable for the Andover data, where the water is similar to the Lake Gaillard water used for calibration (i.e., low SUVA, low bromide, low alkalinity). The results were not as good for the CAP data set. The literature models, without the benefit of sample-specific fine tuning, did not perform as well for either data set.

## Conclusions

A model based on a simplified conceptual reaction mechanism that predicts chlorine decay and the formation of THMs and HAAs was developed and calibrated. The form of the model is a system of differential equations and initial concentrations, solved numerically. Previous empirical models relied on measurable parameters such as UV absorbance or TOC to capture the character and concentration of the reactive NOM. The mechanistic model described here includes terms that represent the concentrations of reactive NOM sites, but no relationships have been developed to determine these concentrations as functions of measurable surrogates. The initial concentrations of NOM sites must therefore be established indirectly. The "fine-tuning" procedure involves chlorinating the water to be modeled and making a limited number of response variable (chlorine and by-product) measurements. Initial NOM site concentrations are established by least-squares fitting using the chlorination test results. Therefore the mechanistic model cannot be compared on an equal basis with empirical models as they are typically used (i.e., without the benefit of sample-specific fine-tuning). In this work, the mechanistic model and empirical model forms were calibrated and fine-tuned using the same data for purposes of comparison. The mechanistic model, fine-tuned empirical models, and empirical models used with published parameter values and no fine-tuning were compared using two external data sets.

This work shows that mechanism-based modeling of chlorine reactions in natural waters is feasible. Excellent results were achieved when the model was tested using a water of similar quality to the water used for calibration. The performance of the mechanistic model and the fine-tuned empirical models was

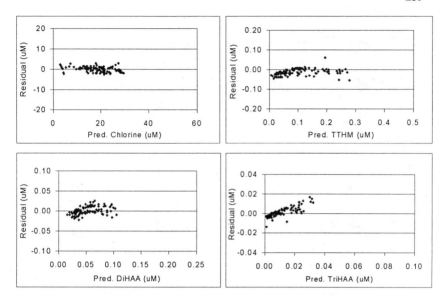

**Figure 6. Residual plots: mechanistic model, Andover data set.**

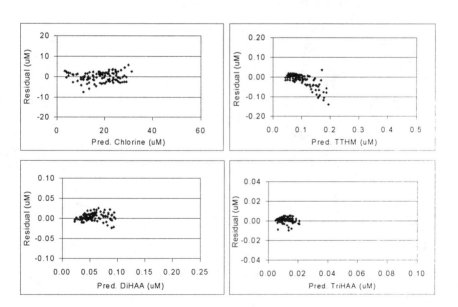

**Figure 7. Residual plots: power function model, Andover data set.**

240

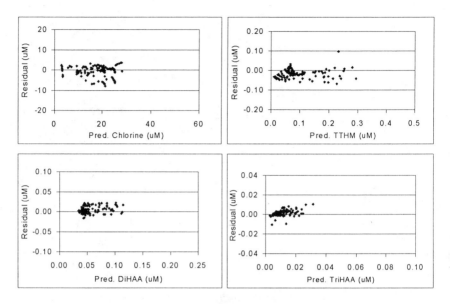

*Figure 8. Residual plots: second-order model, Andover data set.*

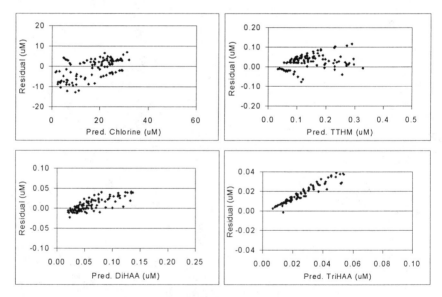

*Figure 9. Residual plots: Clark (12) (chlorine predictions) and Montgomery Watson (8) (THM and HAA predictions) models, Andover data set.*

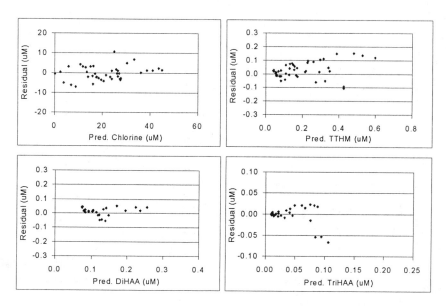

*Figure 10. Residual plots: mechanistic model, CAP data set.*

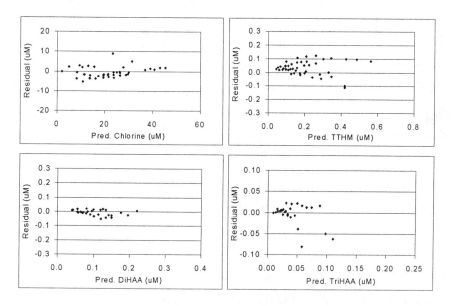

*Figure 11. Residual plots, power function model, CAP data set.*

242

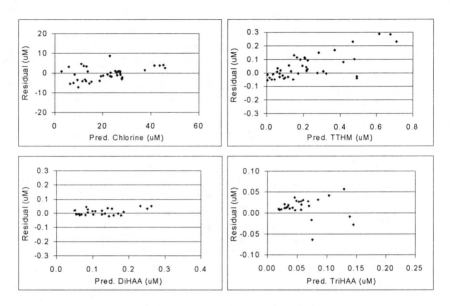

Figure 12. Residual plots, second-order model, CAP data set.

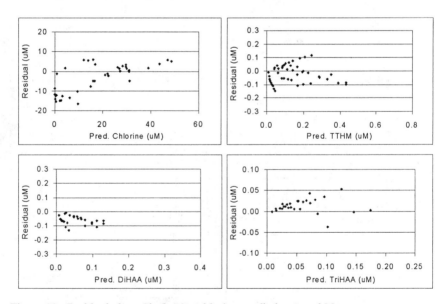

Figure 13. Residual plots, Clark (12) (chlorine predictions) and Montgomery
Watson (8) (THM and HAA predictions) models, CAP data set.

Table VII.  Model Comparison: External Data Sets[1]

| | Model | | | |
|---|---|---|---|---|
| | Mechanistic | Power function | Second-order | Literature[2] |
| Percent "successful"[3] predictions | | | | |
| Andover Data Set | | | | |
| Chlorine | 97 | 79 | 75 | 34 |
| TTHM | 87 | 75 | 74 | 51 |
| DiHAA | 98 | 98 | 100 | 80 |
| TriHAA | 96 | 100 | 100 | 31 |
| Overall | 94 | 88 | 87 | 49 |
| CAP Data Set | | | | |
| Chlorine | 59 | 73 | 53 | 38 |
| TTHM | 49 | 44 | 38 | 31 |
| DiHAA | 56 | 74 | 74 | 7 |
| TriHAA | 59 | 59 | 30 | 52 |
| Overall | 55 | 61 | 47 | 32 |

[1]Andover data set: Filter effluent from the Robert E. McQuade WTP, Andover, Mass. CAP data set: Laboratory filtered Colorado River water.

[2]Clark second-order model (*11*) used for chlorine predictions, Montgomery Watson power function models (*7*) used for THM and HAA predictions.

[3]"Successful" prediction: Observed values within the following ranges of predicted values: chlorine predictions, ±2.8 µM (0.2 mg/L); TTHM predictions, ±0.034 µM (4 µg/L); DiHAA predictions, ±0.023 µM (3 µg/L); TriHAA predictions, ±0.012 µM (2 µg/L).

comparable. The fine-tuned models (mechanistic and empirical) performed better than the empirical models used with published parameter values and no fine tuning.

The primary advantage of the mechanistic model described here compared to traditional empirical models is flexibility to handle more complex situations such as "booster" chlorination or modeling distribution systems with multiple sources of different quality. The mechanistic model could be easily modified (by addition of decomposition terms) to predict reactive by-product species.

## Future Work

The work presented here represents a first step in developing a generally applicable mechanism-based model for chlorine/by-product reaction kinetics in drinking water treatment. Future goals include the following: (1) Conducting experiments to characterize the effects of bromide, and incorporation of these effects in the model. (2) Modeling other by-products including individual THM and HAA species. (3) Refining the treatment of fast reactions that are treated as constants in the current model. (4) Developing relationships between directly measurable parameters (e.g., UV absorbance or TOC) and the concentrations of reactive NOM sites.

## Acknowledgments

The funding for this research provided by the South Central Connecticut Regional Water Authority is gratefully acknowledged. The authors would like to thank John Affinito, Tom Whitbread and Jeff Johnson at the Lake Gaillard Water Treatment Plant and John Pollano at the Robert E. McQuade Water Treatment Plant for their cooperation in providing water samples used in this work. The authors would also like to thank Paul Westerhoff at Arizona State University for providing the Central Arizona Project water data set.

## Acronyms

| | |
|---|---|
| BCAA | bromochloroacetic acid |
| CAP | Central Arizona Project |
| DCAA | dichloroacetic acid |
| DBP | disinfection by-product |
| DHAA | sum of di-halogenated HAA species |
| HAA | haloacetic acid |
| NOM | natural organic matter |
| SUVA | specific UV absorbance |
| TCAA | trichloroacetic acid |
| THAA | sum of tri-halogenated HAA species |
| THM | trihalomethane |
| TOC | total organic carbon |
| TTHM | total trihalomethane (sum of four species) |
| USEPA | United States Environmental Protection Agency |
| UV | ultra-violet |

## List of Symbols

| | |
|---|---|
| $\alpha_{1,THM}\ \alpha_{2,THM}\ \alpha_{1,TriHAA}\ \alpha_{2,TriHAA}$ $\alpha_{1,DiHAA}\ \alpha_{2,DiHAA}$ | Factors relating by-product formation and NOM site consumption |
| $\alpha_{HOCl},\ \alpha_{OCl^-}$ | PH-dependent distribution factors for the chlorine species |

| | |
|---|---|
| $\alpha_{1P}, \alpha_{2P}$ | Distribution factors accounting for pH and oxidant concentration effects on by-product speciation |
| $\beta_{12,THM}, \beta_{22,THM},$ $\beta_{1,TriHAA}, \beta_{11,TriHAA}, \beta_{21,TriHAA},$ $\beta_{1,DiHAA}, \beta_{2,DiHAA}$ | Factors representing the fractional contributions of NOM site classes, and uncatalyzed, oxidation, and hydrolysis pathways to the formation of individual by-product species |
| $\theta_1, \theta_{2f}, \theta_P$ | Rate adjusting coefficients for temperature effects |
| $[Cl_2]$ | Free chlorine concentration |
| $[HOCl], [OCl^-]$ | Hypochlorous acid, hypochlorite concentrations |
| $[S_1], [S_2H], [S2^-]$ | NOM site concentrations |
| $k_1, k_{21f}, k_{21r}, k_{22}$ | Rate constants |
| $k'_1, k'_{21f}, k'_{21r}, k'_{22}$ | Apparent rate constants |
| $k_P, k_{P,OH}, k_{P,HOCl}$ | Coefficients representing the relative importance of uncatalyzed, hydrolysis, and oxidation pathways in by-product formation |
| $n_1, n_2$ | Factors relating chlorine demand and NOM site consumption |
| $R_1, R_{21f}, R_{21r} R_{22}$ | Reaction rates |
| $T, T_{ref}$ | Temperatures |
| $R^2$ | Coefficient of determination |

# References

1. Box, G. E. P.; Hunter, W. G.; Hunter, J. S. *Statistics for Experimenters*, John Wiley and Sons: New York, 1978.
2. Taras, M. J. *J. Am. Water Works Assoc.* **1950,** *42,* 462.
3. Feben, D.; Taras, M. J. *J. Am. Water Works Assoc.* **1951,** *43,* 922.
4. Engerholm, B. A.; Amy, G. L. *J. Am. Water Works Assoc.* **1983,** *75,* 418.
5. Urano, K.; Wada, H.; Takemasa, T. *Water Res.* **1983,** *17,* 1797.
6. Edzwald, J. K. In *Organic Carcinogens in Drinking Water,* N. Ram, E. J. Calabrese, R. F. Christman, Eds.; John Wiley and Sons: New York, 1986; pp 15-28.
7. Amy, G. L.; Chadik, P. A.; Chowdhury, Z. K. *J. Am. Water Works Assoc.* **1987,** *79* (7), 89.
8. Montgomery Watson. *Final Report: Mathematical Modeling of the Formation of THMs and HAAs in Chlorinated Natural Waters,* American Water Works Association, Denver, CO, 1983.
9. MacNeill, A. L. Master's Degree Project, Department of Civil and Environmental Engineering, University of Massachusetts, Amherst, MA, 1994.

246

10. Jadas-Hécart, A.; El Morer, A.; Stitou, M.; Bouillot, P.; Legube, B. *Water Res.* **1992**, *26*, 1073.

11. Clark, R. M. *Journal of Environmental Engineering*, ASCE **1998**, *124*, 16.

12. Clark, R. M. *Journal of Environmental Engineering*, ASCE **1998**, *124*, 1203.

13. Vasconcelos, J. J.; Grayman, W. M.; Kiene, L.; Wable, O.; Biswas, P.; Bhari, A.; Rossman, L. A.; Clark, R. M., Goodrich, J. A. *Characterization and Modeling of Chlorine Decay in Distribution Systems*, AWWA Research Foundation, Denver, CO, 1996.

14. Bell, R. P.; Gelles, E.; Moller, E. *Proceedings of the Royal Society of London* **1949**, *198*, 308.

15. Bell, R. P.; Gelles, E. *Proceedings of the Royal Society of London* **1952**, *210*, 310.

16. Bell, R. P.; Fluendy, M.A.D. *Transactions of the Faraday Society* **1963**, *59*, 1623.

17. Rook, Johannes J. *Environ. Sci. Technol.* **1977**, *11*, 479.

18. Christman, R. F.; Johnson, J. D.; Haas, J. R.; Pfaender, F. K.; Liao, W. T.; Norwood, D. L.; Alexander, H. J. In *Water Chlorination: Environmental Impacts and Health Effects*, R. L. Jolley et al., Eds., Ann Arbor Science Publishers: Ann Arbor, MI, 1978; Vol. 2, pp 15-28.

19. Norwood, D. L.; Johnson, J. D.; Christman, R. F.; Haas, J. R.; Bobenrieth, M. J. *Environ. Sci. Technol.* **1980**, *14*, 187.

20. Boyce, S. D.; Hornig, J. F. *Environ. Sci. Technol.* **1983**, *17*, 202.

21. Larson, R. A.; Rockwell, A. L. *Environ. Sci. Technol.* **1979**, *13*, 325.

22. Reckhow, D. A.; Singer, P. C. In *Water Chlorination: Environmental Impacts and Health Effects*, R. L. Jolley et al., Eds., Ann Arbor Science Publishers: Ann Arbor, MI, 1985; Vol. 5, pp 1229-1257.

23. Rebenne, L. M.; Gonzalez, A. C.; Olson, T. M. *Environ. Sci. Technol.* **1996**, *30*, 2235.

24. Morris, J. C. In *Water Chlorination: Environmental Impacts and Health Effects*, R. L. Jolley et al., Eds., Ann Arbor Science Publishers: Ann Arbor, MI, 1985; Vol. 5, pp 701-711.

25. Gurol, M. T.; Wowk, A.; Myers, S.; Suffet, I. H. In *Water Chlorination: Environmental Impacts and Health Effects*, R. L. Jolley et al., Eds., Ann Arbor Science Publishers: Ann Arbor, MI, 1983; Vol. 4, pp 269-284.

26. *Standard Methods for the Examination of Water and Wastewater*, 18th ed. American Public Health Association: Washington, D.C., 1992.

27. Xie, Y.; Reckhow, D. A.; Springborg, D. *J. Am. Water Works Assoc.* **1998**, *90* (4), 131.

28. *Scientist,* Version 2.0, software package and user's manual, MicroMath: Salt Lake City, UT, 1995.

29. Brezonik, P. L. *Chemical Kinetics and Process Dynamics in Aquatic Systems*, Lewis Publishers: Ann Arbor, MI, 1994.

30. Bjorn, A. Department of Civil and Environmental Engineering, University of Massachusetts at Amherst, unpublished results, 1997.

31. Westerhoff, P., et al. Arizona State University, unpublished results, 1999.

# CHEMISTRY OF ALTERNATIVE DISINFECTANTS

Chapter 16

# The Influence of Dissolved Organic Matter Character on Ozone Decomposition Rates and $R_{ct}$

Michael S. Elovitz[1,2], Urs von Gunten[1], and Hans-Peter Kaiser[3]

[1]Swiss Federal Institute for Environmental Science and Technology (EAWAG), Ueberlandstrasse 133, CH–8600 Dübendorf, Switzerland
[2]Water Supply and Water Resources Division/NRMRL/ORD, U.S. Environmental Protection Agency, 26 West M. L. King Drive, Cincinnati, OH 45268
[3]Zürich Water Works (Wasserversorgung), Postfach CH–8023 Zürich, Switzerland

The effects of DOM character on ozonation of natural waters and solutions of DOM isolates were investigated. Batch kinetic investigations measured $O_3$ decomposition rate constants and $R_{ct}$ values. $R_{ct}$ describes the ratio of •OH concentration to $O_3$ concentration, and thus provides an indirect measurement of the transient concentration of •OH during ozonation. Experiments were conducted on a variety of natural waters, for two natural waters over the course of a year, and in a series of well-characterized DOM isolate solutions. $O_3$ decomposition rate constants and $R_{ct}$ varied as much as two orders of magnitude in both the series of natural waters and DOM isolate solutions. Correlation analyses between the kinetic parameters and water quality and DOM descriptors were performed to identify possible surrogates for rate constants and $R_{ct}$. Depending on the experimental series (i.e. natural waters, seasonal variations, DOM isolates), different potential surrogates were identified.

## Introduction

Operation of chemical-based disinfection processes requires a balance between high disinfectant exposure to insure pathogen inactivation, and paradoxically minimal disinfectant exposure to control disinfection byproduct (DBP) formation. With the recent enactment of the Stage 1 Disinfectants/Disinfection By-Product (D/DBP) and Interim Enhanced Surface Water Treatment (IESWT) rules, and the scheduled follow-up to these rules *(1)*, the balance between these control objectives will be even more difficult to maintain. Ozonation has been considered an alternative to conventional chlorination practices because of its typically greater disinfection potential and lower propensity to form halogenated DBPs. In addition, ozone-based technologies have the added capability to oxidize micropollutants by the hydroxyl radical (•OH), which is formed during decomposition of $O_3$. With this in mind, possible future regulation for the drinking water pollutants—i.e., Drinking Water Contaminant Candidate List (DWCCL) *(1)*—could promote the role of $O_3$ and •OH during water treatment. Ozonation, however, also leads to DBPs, the most deleterious being bromate, formed during ozonation of bromide-containing waters. The mechanism for bromate formation is complex and involves both molecular $O_3$ and •OH. With a recent addendum to the bromate formation topic, the synergistic roles of •OH and $O_3$ in bromate formation has become even more apparent *(2)*. Hence, to optimize an ozone-based treatment process, it is necessary to insure disinfection (predominantly through $O_3$) and pollutant oxidation (both $O_3$ and •OH), while also minimizing bromate formation (via both $O_3$ and •OH reactions). This in turn requires explicit knowledge of the concentration of both oxidants, •OH and $O_3$.

The need to ascertain both •OH and $O_3$ concentration during ozonation has motivated the use of mechanistic based kinetic models to predict $O_3$ decomposition, •OH formation, and bromate formation. Though these models have shown limited success in predicting ozonation performance in "pure" water systems, they have not proven accurate for modeling ozonation in real waters containing naturally occurring dissolved organic matter (DOM) *(3)*. The principal crux for applying the mechanistic models to DOM-containing waters is accounting for both $O_3$-DOM reactions and •OH-DOM reactions within the context of the complex radical-type chain mechanism for $O_3$ decomposition. DOM can act as a direct consumer of molecular $O_3$, as an initiator of the $O_3$ decomposition cycle, as a •OH scavenger that leads to dead-end products (*inhibition* of $O_3$ decomposition, i.e., chain termination), or as a •OH scavenger that leads to enhanced $O_3$ decomposition (*promotion* of $O_3$ decomposition, i.e., chain propagation); for a review, see *(4,5)*. Because DOM is a complex mixture of unresolved carbonaceous materials, it is virtually impossible to write discrete rate expressions for these DOM reactions. Therefore, it is difficult to adapt the mechanistic models to make *a priori* predictions when DOM is involved.

## *The $R_{ct}$ Concept*

In a departure from these *a priori* mechanistic models, we have developed an experimental approach, called the $R_{ct}$ concept, for indirectly measuring •OH concentrations during ozonation of real waters. The transient •OH concentration is indirectly measured by use of a •OH-probe compound (Figure 1, for a review on this topic see *(6)*). The •OH-probe reacts only with •OH, and not significantly with $O_3$ or other reaction species. In addition, the concentration of the probe compound is kept very low such that the amount of •OH scavenged by the probe is insignificant compared to the scavenging occurring through reaction with DOM and carbonate. Thus, the decrease in concentration of a probe compound is proportional to the overall •OH action.

In our studies, we used *p*-chlorobenzoic acid (pCBA) for the •OH probe compound ($k_{•OH/pCBA} = 5 \times 10^9 \ M^{-1}s^{-1}$ *(7,8)*; ($k_{O3/pCBA} \leq 0.15 \ M^{-1}s^{-1}$ *(9)*). Following the conceptual chemical model in Figure 1 where $k_{pCBA}[pCBA] \ll \Sigma k_{Si}[S_i]$, then,

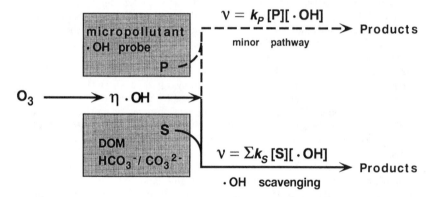

Figure 1. Reaction scheme for the formation of •OH from $O_3$ decomposition and the subsequent quenching of •OH by scavengers (S, major pathway) and the probe compound or micropollutant (P, minor pathway).

$\ln([pCBA]/[pCBA]_o) = -k_{\cdot OH/pCBA}I[\cdot OH]dt$, where the term $I[\cdot OH]dt$ represents the time-integrated concentration of $\cdot OH$, which is equal to the $\cdot OH$-exposure or the $\cdot OH$-ct. At each time during the reaction that we measure pCBA concentration (and subsequently calculate $I[\cdot OH]dt$), we can also measure $O_3$ concentration. The area under the $[O_3]$-vs-time curve, $I[O_3]dt$, is the time-integrated ozone concentration, which is equal to the true $O_3$-exposure or $O_3$-ct. These two determinations are described in the literature (5). Our spin on these two measurements was to define a term, $R_{ct}$, which describes the ratio of $\cdot OH$-exposure and $O_3$-exposure (or $\cdot OH$-ct and $O_3$-ct), i.e., $R_{ct} = I[\cdot OH]dt/I[O_3]dt$. Substitution of these two terms gives $R_{ct} = -\ln([pCBA]/[pCBA]_o) / (k_{\cdot OH/pCBA}I[O_3]dt)$. Hence, $R_{ct}$ can be backed out from the slope of the plot of "$\ln([pCBA]/[pCBA]_o)$" versus "$I[O_3]dt$." The $R_{ct}$ concept has its roots in the $\Omega$-concept developed by Hoigné et al. (10), but because it is a kinetic approach, the $\cdot OH$-exposure, $O_3$-exposure, and $R_{ct}$ values can be calculated as a function of ozonation time. Like the $\Omega$ parameter, for a given set of reaction conditions, $R_{ct}$ is a measurable parameter describing ozonation performance. This enables different waters, or reaction conditions, to be compared with respect to the observed $\cdot OH/O_3$ dynamics.

In our earlier publication (11), we showed that for ozonation of both model and natural waters the rate of ozone depletion was first-order in $[O_3]$, and the concentration of pCBA decreased in parallel with decreasing ozone concentration. More importantly, it was shown that for a given water and set of reaction conditions, $R_{ct}$ is a constant value during most of the reaction. When $R_{ct}$ is constant, it follows that $R_{ct}$ is also a measure of the ratio of the concentrations of $\cdot OH$ and $O_3$ (i.e. $R_{ct} = [\cdot OH]/[O_3]$). Hence, by calibrating a water with respect to its $R_{ct}$ and its ozone decay kinetics, $\cdot OH$-exposure and $\cdot OH$ concentrations can be calculated from $O_3$-exposure and $O_3$ concentrations.

One important aspect of the $O_3$ depletion kinetics and the $R_{ct}$ plots is that there is often two or more reaction phases during ozonation. In many cases, there is a rapid initial consumption of $O_3$—the "instantaneous" $O_3$ consumption (12)—lasting seconds to a few minutes, followed by a secondary reaction phase. The first-order kinetic plots and $R_{ct}$ plots both show a distinct break where the initial phase kinetically wanes, and the secondary phase begins to dominate. The rapid initial phase is difficult to study due to fast kinetics and short reaction time-scales. The secondary reaction phase is comparatively easy to study. Therefore, the $O_3$ decomposition kinetics (first-order in $O_3$ concentration, see above), $O_3$ and $\cdot OH$ exposures discussed in this paper pertain to the secondary reaction phase.

## Experimental

All chemicals were reagent grade or analytical grade when available. Stock solutions of chemicals were made up in nanopure water ($\geq 18$ M$\Omega$) or double-distilled water. Glassware was rinsed with pre-ozonated double-distilled water prior to use. Concentrated ozone stock solutions were produced by passing $O_3$-containing oxygen through double-distilled water that was cooled in an ice bath. Concentration of dissolved $O_3$ in the $O_3$-stock solutions was determined by UV absorbance at

258 nm ($\varepsilon = 3000$ $M^{-1}cm^{-1}$). Solutions of approximately 60 mg/L (1.25 mM) $O_3$ were routinely produced. The concentration of dissolved $O_3$ in reaction solutions was determined by the Indigo method *(13)*. The concentration of the *in situ* •OH-probe compound, *p*-chlorobenzoic acid (pCBA), was monitored by reverse-phase HPLC with UV detection at 230nm (detection limit < 0.025 $\mu$M). Measurements of solution pH were performed with a combination pH electrode (Metrohm) calibrated with Titrisol buffers (Merck). Details of all analytical methods are given in reference *(11)*.

## Natural Water Systems

Natural waters were collected from different water treatment plants (WTPs) around Switzerland, returned to the lab on ice, filtered through a 0.45 $\mu$m filter (cellulose nitrate, Sartorius), and stored at 4°C until use. The original water quality parameters (DOC, pH, UV absorbance, and alkalinity/hardness) were measured by the staff at the Waterworks of Zürich, or the participating WTP, on replicate samples using routine procedures (Table I). Samples from Lake Zürich (Lengg) and the Sihl River (in Zürich) were taken monthly to examine seasonal changes in water quality and ozonation. Lake Zürich water was taken from the WTP intake pipe at a depth of 32 m in Lake Zürich. Lake Zürich is a deep, oligotrophic, seasonally stratified lake and the water quality at this depth of the lake does not change appreciably. The measured physical and chemical properties of the raw water are relatively constant (13-month average±s.d. – DOC: 1.3±0.08 mg/L; carbonate alkalinity: 2.53±0.04 mM $HCO_3^-/CO_3^{2-}$ (253 mg/L as $CaCO_3$); $UV_{254}$: 2.4±0.1 AU/m; pH 7.82±0.11). The Sihl River water was collected at the surface of a shallow river draining a large watershed of mixed residential and agricultural use and a wastewater treatment plant. Sihl River water demonstrated greater variability in water quality and reactivity in the ozonation system (Table I). In all cases except for Sihl River water, the natural water was not altered except for the standard procedures as described below. Sihl River water was consistently more reactive than most waters, and therefore it was diluted by 50% for all experiments to enable more accurate kinetic measurements.

Reaction solutions were prepared by addition of an inorganic buffer (10 mM borate) to the filtered water, and adjustment to the desired pH with NaOH/HCl. The natural carbonate alkalinity of the waters was not altered except for minor changes caused by buffering the water to pH 8, and 50% dilution in the case of Sihl River water. Reaction bottles were fitted with a "bottletop" dispenser, and were submerged in a thermostated bath to control temperature to within ±0.2°C. All reactions were performed at pH 8.0 and 15°C. Prior to each experiment, a small amount of pCBA was added to the vessel to yield a concentration of 0.25-0.5 $\mu$M. Ozonation reactions were performed in batch mode and were initiated by addition of $O_3$ stock solution while stirring (Teflon coated stirbar). At selected reaction times, the dispenser was used to withdraw a sample, which was dispensed directly into Indigo solution thus quenching the ozone reaction. The sample was measured for $O_3$ and pCBA concentrations as described above. For all experiments reported here, a single $O_3$ dose corresponding to 1mg/L applied $O_3$ was used.

Table I. Water Quality Data and Experimental Results for Ozonation of Natural Waters.

| City | Source | $pH^a$ | $T^a$ (C) | DOC (mg/L) | Alkalinity (mM $HCO_3^-$) | $UV_{254}$ ($m^{-1}$) | $k_{sec}$ ($s^{-1}$)$^b$ | $R_{ct}^{sec\ b}$ | $[O_3]dt^c$ ($M\cdot min$)/ ($mg/L\cdot min$) | $[\cdot OH]dt^c$ ($M\cdot min$) |
|---|---|---|---|---|---|---|---|---|---|---|
| Zürich-Lengg[h] | Lake Zürich[d] | 7.8 | 6.1 | 1.3 | 2.5 | 2.4 | $2.0\times10^{-3}$ | $1.5\times10^{-8}$ | $9.8\times10^{-5}$ / 4.7 | $1.5\times10^{-12}$ |
| Horgen | Lake Zürich[e] | 7.67 | 5.7 | 1 | 2.8 | 3.1 | $9.5\times10^{-4}$ | $9.5\times10^{-9}$ | $2.3\times10^{-4}$ / 11 | $1.9\times10^{-12}$ |
| Küsnacht | Lake Zürich[f] | 7.41 | 6.3 | 1.1 | 2.8 | 3.3 | $1.0\times10^{-3}$ | $1.0\times10^{-8}$ | $2.1\times10^{-4}$ / 10 | $2.2\times10^{-12}$ |
| Zürich-Limmat[j] | Limmat River[g] | 8.34 | 14.5 | 1.5 | 2.1 | 2.4 | $2.1\times10^{-3}$ | $1.8\times10^{-8}$ | $4.5\times10^{-5}$ / 2.2 | $8.3\times10^{-13}$ |
| Kreuzlingen | Lake Constance | 7.85 | 9.4 | 1.2 | 2.5 | | $1.4\times10^{-3}$ | $1.2\times10^{-8}$ | $1.5\times10^{-4}$ / 7.0 | $1.7\times10^{-12}$ |
| Lugano | Lake Lugano | 7.8 | 6.9 | 1 | 2.4 | 2.6 | $6.3\times10^{-4}$ | $1.2\times10^{-9}$ | $4.0\times10^{-4}$ / 19 | $8.3\times10^{-13}$ |
| Bern | groundwater | 7.4 | 7.0 | 0.7 | 6.7 | 1.2 | $1.4\times10^{-4}$ | $7.0\times10^{-10}$ | $1.8\times10^{-3}$ / 85 | $1.5\times10^{-12}$ |
| Biel | | 7.67 | 7.9 | 1.6 | 3.6 | 5.8 | $2.4\times10^{-3}$ | $1.0\times10^{-8}$ | $9.4\times10^{-5}$ / 4.5 | $7.9\times10^{-13}$ |
| Lausanne | Lake de Brêt | 7.48 | 14.8 | 3.2 | 3.4 | 10.6 | $1.5\times10^{-2}$ | $4.0\times10^{-8}$ | $1.3\times10^{-5}$ / 0.6 | $2.4\times10^{-13}$ |
| Porrentruy | groundwater | 7.2 | 10.8 | 0.9 | 5.4 | | $2.7\times10^{-4}$ | $1.3\times10^{-9}$ | $7.9\times10^{-4}$ / 38 | $1.2\times10^{-12}$ |
| Zürich-Sihl[j] (50% diluted)[k] | Sihl River | 8.3 [8.2-8.5] | 11 [0.2-21] | 1.5 [1.0-2.3] | 1.6 [1.3-1.8] | 3.5 [2.2-5.5] | $6.1\times10^{-3}$ [2.5-17$\times10^{-3}$] | $5.8\times10^{-8}$ [2.4-12$\times10^{-8}$] | $2.5\times10^{-5}$ / 1.2 | $1.4\times10^{-12}$ |
| Vevey | Lake Geneva | 8.11 | 6.8 | 0.8 | 2.64 | 2.3 | $6.2\times10^{-4}$ | $4.4\times10^{-9}$ | $5.0\times10^{-4}$ / 24 | $2.9\times10^{-12}$ |

Notes: All experiments performed at pH 8, 15C, and natural carbonate alkalinity. [a] Natural temperature and pH of water at time of collection. [b] $O_3$ decay rate and $R_{ct}$ for secondary reaction phase. [c] $O_3$-exposure and $\cdot$OH-exposure for secondary reaction phase. [d] North side of Lake Zürich. [e] South side of Lake Zürich, water taken from 32-m depth. [f] West side of Lake Zürich. [g] River draining the surface water from Lake Zürich. [h] 13-month average (6.96-6.97). [j] Data from 10.96 and [min-max of 13-month range] when applicable. [k] Sihl River water was diluted 50% with double distilled water.

## DOM Isolates

DOM isolates, predominately hydrophobic acids extracted by an XAD-8 resin, were collected from various water sources (Table II). International Humic Substance Society (IHSS) Suwannee River fulvic acid (SRF) and humic acid (SRH) (hydrophobic acids), and a single XAD-4 isolate (hydrophilic acid) were also investigated. The DOM isolates had been analyzed for specific UV absorbance at 254 nm ($SUVA_{254}$) and 280 nm ($SUVA_{280}$), specific fluorescence (SFluor), weight-averaged molecular weight (Mw), and elemental composition (C,H,N,O), and characterized by $^{13}C$-NMR spectrometry for aromatic, aliphatic, and carboxyl carbon percentages. The isolates were obtained from George Aiken and Paul Westerhoff (USGS, Boulder, CO and Arizona State University, Tempe, AZ, respectively) and details of the procedures for isolation and characterization are given in references *(14,15)* and references therein.

DOM isolates, received by our lab as freeze-dried solids, were redissolved in double-distilled water to produce concentrated stock solutions (*ca.* 40 mg/L). Ozonation experiments were also conducted in batch-mode. Reaction conditions were approximately the same for all samples: 2 mg/L $O_3$, 2 mg/L DOC, 2.5 mM ($HCO_3^-/CO_3^{2-}$) carbonate alkalinity, pH 7.0 (2.5 mM phosphate buffer), and 20°C. These reactions were performed at pH 7, as opposed to pH 8 for the natural waters, to slow the rate of $O_3$ decomposition so that several reaction half-lives could be measured. Pre-ozonated double-distilled water was acidified to *ca.* pH 2 with 2.5 mM $H_3PO_4$ and degassed to remove all carbonate (as $CO_2$). Aliquots of the concentrated DOM stock solutions were added to achieve a DOC concentration of approximately 2 mg/L (as carbon), and $NaHCO_3$ and HCl/NaOH added to achieve a total carbonate concentration of 2.5 mM ($HCO_3^-/CO_3^{2-}$) and pH 7.0. The •OH probe compound, pCBA, was also added to a concentration of *ca.* 0.5 µM. The DOC concentration of the final reaction solution was then measured. Final DOC concentrations differed slightly from 2 mg/L, but this variability was accounted for by normalizing all kinetic parameters to the actual DOC concentration (DOC-normalized). Reactions were initiated and sampled as described above.

## Results and Discussion

### $R_{ct}$ as a Function of Water Source

For the 12 waters studied, $O_3$ decomposition rate constants ($k_{sec}$) and $R_{ct}$ values ranged over two orders of magnitude (Table I). Overall, $O_3$ decomposition rate constants varied from $1.4 \times 10^{-4}$ to $1.5 \times 10^{-2}$ $s^{-1}$, and $R_{ct}$ ([•OH]/[$O_3$]) varied from $7.0 \times 10^{-10}$ to $5.5 \times 10^{-8}$. The two waters that showed the highest and lowest reactivity are relative extremes with respect to DOC concentration and alkalinity. The Bern water, which is a groundwater, has a low DOC concentration and very high alkalinity, which leads to enhanced $O_3$ stabilization ($t_{1/2} = 83$ min). The water from

TABLE II. Geographic Locations from where DOM Isolates were Obtained, Abbreviations for DOM Isolates, and the Operationally Defined DOM Fraction.

| Geographic Location | abbr. | Fraction |
|---|---|---|
| Groundwater sample (near) Lake Shingobee, MN | GW8 | XAD-8 |
| Lake Fryxell, Antarctica | FXL | XAD-8 |
| Williams Lake, MN | WLL | XAD-8 |
| Lake Shingobee, MN | SHL | XAD-8 |
| Yakima River at Kiona, WA | KNR | XAD-8 |
| Ohio River, OH | OHR | XAD-8 |
| Calif. State Project Water, CA | SP4 | XAD-4 |
| Calif. State Project Water, CA | SP8 | XAD-8 |
| Silver Lake, CO | SLW | XAD-8 |
| Ohio River, OH | OHR | XAD-8 |
| Suwannee River, GA[a] | SRF | Fulvic acid[b] |
| Suwannee River, GA[a] | SRH | Humic acid[b] |

Notes: [a] Obtained from International Humic Substances Society (Golden, CO);
[b] Humic and fulvic acid separation acid/base precipitation.

Lausanne-Lake de Brêt, a highly eutrophic lake, has the highest DOC concentration in the survey, but only a moderate alkalinity, and thus $O_3$ decomposition is very rapid ($t_{1/2}$ = 46 sec). While the case for widely ranging decomposition rates is well known, the equally broad variation in $R_{ct}$ is not well documented. Bablon et al. *(4)* mentioned ratios of [•OH]/[$O_3$] that vary about $1 \times 10^{-10}$ - $1 \times 10^{-8}$; however, the authors do not provide references for these values. Also, Haag and Yao *(8)* used pCBA as a •OH probe compound to measure steady-state [•OH]/[$O_3$] ratios during *continuous* ozonation experiments. In contrast to our findings, they reported [•OH]/[$O_3$] ratios that varied only a factor of four, from $0.8 \times 10^{-7}$ - $2.7 \times 10^{-7}$, for ten different waters ranging from surface waters to waste waters. Differences in reaction conditions and experimental procedure between their study and ours could account for the differing conclusions *(16)*.

Six of the waters (Zürich-Lengg, Horgen, Küsnacht, Limmat River, Kreuzlingen, and Lugano) have moderate and closely matched DOC concentrations ($1.2 \pm 0.2$ mg/L) and alkalinities ($2.5 \pm 0.3$ mM $HCO_3^-/CO_3^{2-}$) (Table I). These waters allow a suitable comparison of DOM reactivity, with lesser influence from differences in DOC concentration and alkalinity. Comparing these six waters only, there is a three-fold difference in $O_3$ depletion rates and an order of magnitude variation in $R_{ct}$ values. There is also an order of magnitude spread in $O_3$-exposure and •OH-exposure. This subset of six waters indicates that DOM differences can cause considerable differences in reaction rates and more so in •OH-exposure and $O_3$-exposure.

Within this subset of six waters, four waters (Zürich-Lengg, Horgen, Küsnacht, and Limmat River) all come from Lake Zürich, though from different locations around the 88 $km^2$ lake (6-25 km distances between locations). Lake Zürich is large and relatively unpolluted, but there are conceivable differences in water quality at these locations due to localized environmental factors such as run-off and point-source pollution. Nevertheless, these four waters have closely matched water qualities and ozonation reactivity. It appears, therefore, that their common large-scale biogeochemistry (limestone geology, temperate climate) account for their similar reactivity. This deduction is further supported by results for the Lake Constance (Kreuzlingen) water. Lake Constance (539 $km^2$) lies *ca.* 60 km northeast of Lake Zürich, is also oligotrophic and relatively unpolluted, and has a similar large-scale biogeochemistry as Lake Zürich. Once again, despite potential local environmental factors, the major water quality parameters and ozonation parameters are very similar to the four Lake Zürich waters, thus supporting the concept that the large-scale biogeochemical factors dictate the behavior of these waters towards ozonation. As a contrasting example, the water from Lake Lugano, which has a general water quality similar to the Lake Zürich and Constance waters, had a significantly smaller $k_{sec}$ and $R_{ct}$ compared to the aforementioned waters. Lake Lugano (49 $km^2$) lies *ca.* 200 km south of Lake Zürich, on the south side of the Alps, and the Lugano area is dominated by a granitic geology and a more southern Mediterranean climate. Thus, considering its similar pH, alkalinity, and DOC concentration to the Lake Zürich and Constance waters, the marked difference in its $O_3$ reactivity suggests that the DOM character in Lake Lugano water is quite different as a consequence of different biogeochemical influences.

Finally, it was of interest to explore property-activity relationships to identify potential surrogate parameters for $O_3$ decomposition rate constants and $R_{ct}$. Figure 2 is

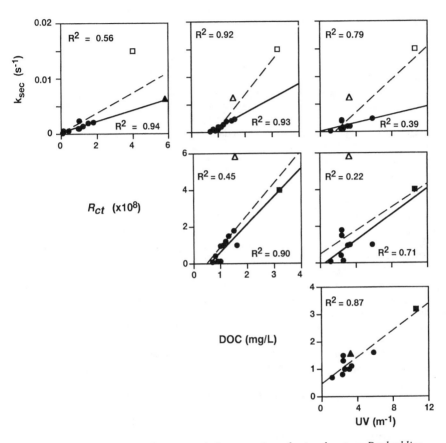

*Figure 2. Scatter plot correlation matrix for ozonation of natural waters. Dashed line and upper left hand corner R² value includes all waters. Solid line and lower right hand corner R² value (where applicable) excludes one or both of Sihl River water (▲) or Lake de Brêt water (■) – filled symbol denotes inclusion, open symbol denotes exclusion.*

a scatter plot correlation matrix for $k_{sec}$ and $R_{ct}$ with common water quality parameters. Correlations that include all waters (dashed lines) were generally poor. The exception was the correlation between $k_{sec}$ and DOC concentration ($R^2 = 0.92$). However, despite the good correlation coefficient, visual inspection of the plot indicates that even this correlation is poor. Two waters, Lausanne-Lake de Brêt (■) and Sihl River (▲), demonstrated much greater reactivity compared to the others. It is not clear if the DOM character or some other specific property of the water is responsible for these waters anomalously high $O_3$ reactivity. Excluding one or both of these waters (open symbols when excluded, filled symbols when included), some correlations clearly improved (solid lines). For example, excluding Lake de Brêt water, there was a good correlation between $k_{sec}$ and $R_{ct}$. This is understandable if variation in $R_{ct}$ is due mainly to changes in $O_3$-exposure (as opposed to •OH-exposure, Table I), which is inversely proportional to the rate of $O_3$ decay. For correlations of $k_{sec}$ and $R_{ct}$ with the various possible surrogates, DOC concentration demonstrated the best fit. Although $UV_{254}$ correlated well with DOC concentration, $UV_{254}$ provided a relatively weak correlation. SUVA showed no correlation with these waters ($R^2 < 0.10$, data not shown in Figure 2).

## $R_{ct}$ as a Function of Seasonal Variation in Water Quality

A separate survey was conducted testing two waters, the deep-water from Lake Zürich (Lake Zürich-Lengg) and Sihl River water (50% diluted), once a month for a period of a year. As mentioned above, the Lake Zürich water was taken at a 32-m depth, below the thermocline of the seasonally stratified lake, and water quality was relatively constant. In contrast, the Sihl River has major fluctuations in flow, which in addition to other factors, such as agricultural practices in the watershed, non-point runoff, and wastewater influx from a wastewater treatment plant upstream, could affect both concentration and character of the DOM. As above, for ozonation experiments with these waters, the temperature and pH were controlled to 15°C and pH 8.0. Therefore, reaction variability was primarily limited to seasonal differences in DOM reactivity, DOC concentration, and alkalinity (though alkalinity varied only slightly – $2.53\pm0.04$ mM and $1.6\pm0.2$ mM as $HCO_3^-$ for Lake Zürich and Sihl River (50%), respectively).

For the Lake Zürich water, results for the period June 1996 – June 1997 show only a minor variation in reaction rate constants ($1.97\pm0.5\times10^{-3}$ s$^{-1}$) and $R_{ct}$ values ($1.48\pm0.30\times10^{-8}$) (Figure 3 a,b). Considering the relatively constant water quality (see the *Experimental* section), this suggests that the Lake Zürich deep-water has little seasonal variation in the character of the DOM. Since the water quality parameters, $k_{sec}$, and $R_{ct}$ values were relatively constant for the 13-month period, it is not surprising that the •OH-exposure and $O_3$-exposures were also relatively constant (data not shown). As a consequence, minimal calibration experiments are needed to provide the ozone kinetics and $R_{ct}$ values necessary to model and control the ozonation process throughout the year.

For the same time period, the Sihl River water (50%) demonstrated a significant variation in $O_3$ decay rate constants ($2.5\times10^{-3}$ s$^{-1}$ - $1.7\times10^{-2}$ s$^{-1}$) and $R_{ct}$

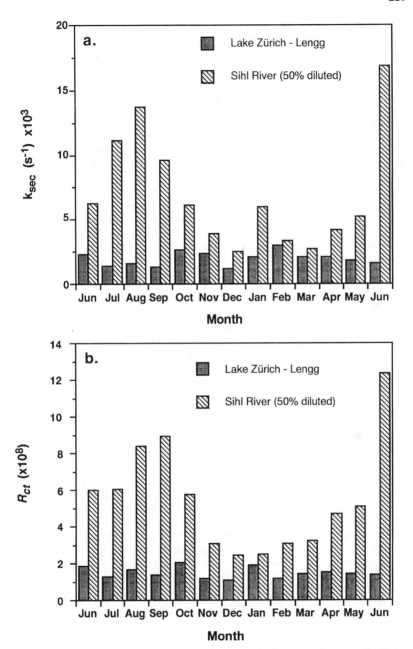

*Figure 3a,b. (a) Ozone decomposition rate constants ($k_{sec}$) and (b) $R_{ct}$ values for Lake Zürich-Lengg and Sihl River waters for the period June 1996 – June 1997; pH 8.0, 15°C, 1 mg/L $O_3$ dose.*

$(2.4 \times 10^{-8} - 12.4 \times 10^{-8})$ (Figures 3a,b). In contrast to the Lake Zürich deep-water, the Sihl River is subject to greater temporal variation in DOM character and concentration, as indicated by the variation in DOC concentration and $UV_{254}$ (Table I). WTPs operating with river waters will require more frequent measurement for $O_3$ kinetics and $R_{ct}$ values.

Figure 4 shows a scatter plot correlation matrix analysis for the Sihl River water for $k_{sec}$, $R_{ct}$, and $UV_{254}$, DOC concentration, and SUVA. There is a reasonable correlation between $k_{sec}$ and $R_{ct}$, for the same reasons as explained above. Correlations between $k_{sec}$ and water quality descriptors demonstrate the best correlation with $UV_{254}$, slightly weaker correlation with DOC concentration, and poor correlation with SUVA. Correlation of $R_{ct}$ with these descriptors show the same trend. In accord with the standard interpretation of a $UV_{254}$ correlation, this suggests that the $\cdot OH/O_3$ dynamics are governed by reactions with aromatic and conjugated moieties in the DOM. The reasonable correlation between DOC and $UV_{254}$ suggests that despite temporal variation in DOC concentration, the fraction of $UV_{254}$ absorbing DOM does not change appreciably for Sihl River water. In other words, the character of the DOM appears to change less than simply the concentration of the DOM. This behavior has been observed by others for different water types *(17)*.

## $R_{ct}$ as a Function of DOM Character/Properties

DOM of whole waters is a mixture of numerous operationally defined DOM fractions (e.g., hydrophobic acids, hydrophobic bases, hydrophilic acids, etc.,). Analogous to chlorination chemistry, it is believed that $O_3$ is most reactive with the aromatic and conjugated moieties present predominantly in the hydrophobic (humic) fractions of DOM. Therefore, it can be inferred that the humic fraction present in natural waters accounts for much of the variability in ozonation observed in whole waters. It was of interest then to investigate if the humic fraction, defined by XAD-8 resin isolation, displayed the same degree of variability in $\cdot OH/O_3$ dynamics as observed in the whole waters. Our study was an extension, for the purpose of measuring $R_{ct}$, of a similar study of ozonation of DOM isolates by Westerhoff et al. *(15)*.

Eleven DOM isolates were investigated: eight XAD-8 isolates, one XAD-4 isolate (SP4) corresponding to the non-humic fraction of California State Project water XAD-8 isolate (SP8), and IHSS Suwannee River fulvic (SRF) and humic (SRH) acids isolates (Table II). Our results for the $O_3$ decomposition and correlation analyses mirror those of Westerhoff et al. *(15)* who studied ozonation of some of the same isolates under slightly different conditions. As observed for whole waters, $O_3$ decomposition kinetics were first-order in $O_3$ concentration for all samples. Also, two-phase kinetics (a fast initial phase followed by a slower secondary phase) was observed in all cases (e.g. Figure 5). Thus, even specific DOM fractions display significant heterogeneity in $O_3$ and $\cdot OH$ reactivity as exhibited in whole waters. DOC-normalized rate constants ($k_{doc}$) for the main secondary reaction phase ranged more than two orders of magnitude – from $2.1 \times 10^{-4}$ to $4.1 \times 10^{-2}$ L/mg/s (Figure 6a). $R_{ct}$ plots for all isolates showed a linear region for the secondary reaction phase (e.g. Figure 5, inset). As with rate constants of $O_3$ decomposition, $R_{ct}$ also ranged over more than two orders of magnitude – from $6.8 \times 10^{-9}$ to $5.3 \times 10^{-7}$ (Figure 6b). The SRF and SRH results were significantly higher relative to the other isolates. Excluding these two samples from the dataset, rate constants varied from $2.1 \times 10^{-4}$ to $1.6 \times 10^{-3}$ L/mg/s (a factor of eight) and $R_{ct}$ varied $6.8 \times 10^{-9}$ to $3.6 \times 10^{-8}$ (a factor of five).

$O_3$ decomposition rate constants were compared with the DOM surrogates $SUVA_{254}$; $SUVA_{280}$; specific fluorescence (SFluor); average molecular weight (Mw); aromatic-, carboxyl- and aliphatic-carbon content; and elemental ratios (Figure 7). For the entire set of DOM isolates, no correlation was found for either $k_{doc}$ or $R_{ct}$ because SRF and SRH isolates were so much more reactive (Figure 6). Excluding only SRH values, $SUVA_{254}$, $SUVA_{280}$, and Mw gave correlation coefficients $R^2 > 0.70$ for $k_{doc}$ (Figure 7, $R^2$ values in upper corner of plots). However, visual inspection of the plots shows that SRF still appears to be an outlier. For $k_{doc}$, excluding both SRH and SRF, $SUVA_{254}$ and $SUVA_{280}$ gave correlations with $R^2 > 0.70$ (Figure 7, solid lines and $R^2$ values in lower corner of plots). Correlation analyses between $R_{ct}$ and the DOM parameters resulted reasonable correlations with $SUVA_{254}$, $SUVA_{280}$, and SFluor when SRH and SRF isolates were excluded from the analyses.

For both $k_{doc}$ and $R_{ct}$ correlations, the SRF and SRH isolates are visible outliers showing much greater reactivity relative to the other isolates. The Suwannee River DOM is characteristic of a lignin derived material, higher in aromatic and phenol moieties than the other DOM isolates studied. However, if this is the principal characteristic of the SRH and SRF material that differentiates them from the other isolates, none of the surrogates that offer some degree of measurement of aromaticity was able to provide a correlation that included SRH and SRF. This implies that Suwannee River aquatic humic material possesses other ozonation reaction sites that are not accounted for by the DOM surrogates evaluated here.

## Conclusions and Implications

The focus of this paper was to examine the effects of DOM character on $O_3$ decomposition rates and $R_{ct}$ in natural waters as well as model systems containing DOM isolates. The effects of DOM character were examined for 12 different natural waters, in two natural waters over a 13 month period, and in a series of solutions of 11 well-characterized DOM isolates collected from different water sources.

The survey of natural waters, covering a moderate variety of water types (rivers, ground waters, oligotrophic lakes, eutrophic lakes), exhibited a 100-fold difference in $k_{sec}$ and $R_{ct}$. These results express the realistically large variation in $k_{sec}$ and $R_{ct}$ that could be encountered for ozonation of real waters. Correlation analyses of $k_{sec}$ and $R_{ct}$ with various water quality parameters revealed that DOC concentration was the only reasonable surrogate. Despite the fact that DOC correlated well with $UV_{254}$, neither $UV_{254}$ nor SUVA afforded a good correlation.

The two waters monitored for seasonal variation in water quality, $k_{sec}$, and $R_{ct}$ were in marked contrast to one another. The Lake Zürich deep water remained relatively constant with respect to water quality, and consequently ozonation behavior. Therefore, although this is a surface water, it is apparently well-buffered with respect to small-scale influences and seasonal changes. For the Sihl River water, $k_{sec}$ and $R_{ct}$ changed significantly in response to changing water quality over the 13

262

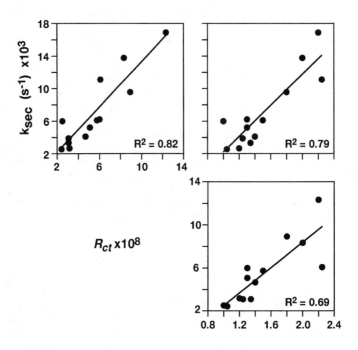

*Figure 4. Scatter plot correlation matrix for ozonation of Sihl River water*

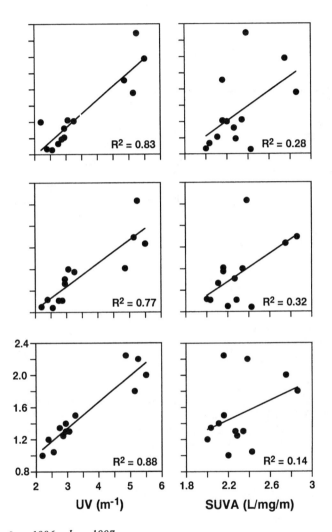

*June 1996 – June 1997.*

264

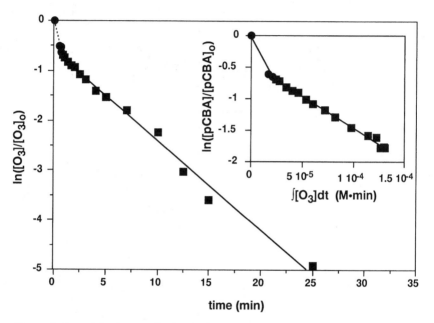

*Figure 5. First-order kinetic plot for $O_3$ decomposition and corresponding $R_{ct}$ plot (inset) for ozonation of Ohio River (OHR) DOM isolate, pH 7.0, 20°C, 2.5 mM $HCO_3^-$, 2 mg/L $O_3$ dose, ≈2 mg/L DOC. Circles (•) indicate initial reaction phase; squares (■) indicate secondary reaction phase.*

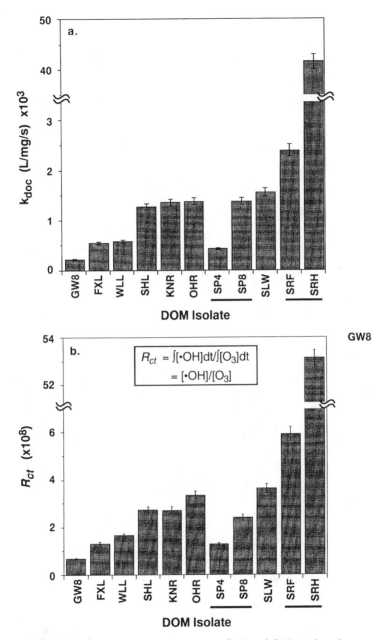

*Figure 6a,b. (a) O₃ decomposition rate constants (k_doc) and (b) R_ct values for ozonation of DOM isolates; pH 7.0, 20°C, 2.5 mM HCO₃⁻, 2 mg/L O₃ dose, ≈2 mg/L DOC.*

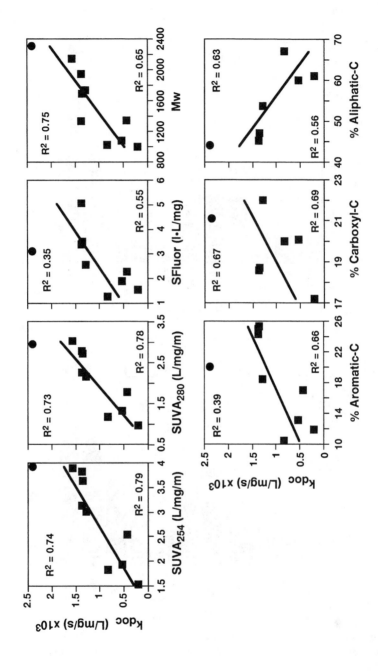

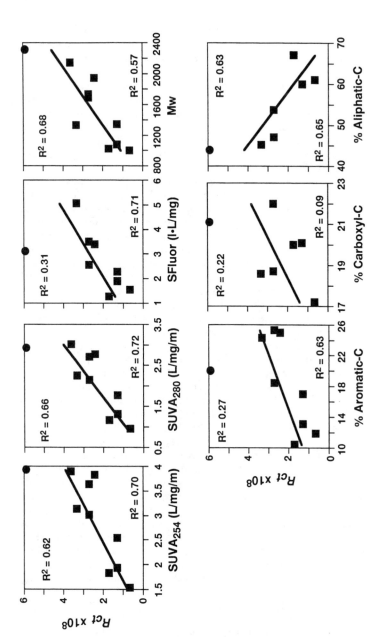

Figure 7. Scatter plot correlation matrix for ozonation of DOM isolates. SRH isolate excluded for all correlations. Upper $R^2$ value includes SRF isolate (●). Solid line and lower $R^2$ value excludes SRF isolate. Correlations with elemental ratios were always poor and are not included in the figure.

month period. Correlation analyses showed that $k_{sec}$ and $R_{ct}$ both correlated best with $UV_{254}$, though also reasonably well with DOC concentration. A strong correlation between UV and DOC suggests that seasonal changes in the Sihl River water are expressed as changes in DOC concentration rather than major changes in DOM character.

Evaluation of $k_{doc}$ and $R_{ct}$ for the series of aquatic humic substance isolates perhaps best demonstrated that the nature of the DOM can affect ozonation. The Suwannee River isolates (SRH and SRF) possessed one to two orders of magnitude greater $k_{doc}$ and $R_{ct}$ values than the other isolates. Comparing $k_{doc}$ and $R_{ct}$ with various DOM property parameters again indicated that SRH and SRF isolates are anomalously reactive with $O_3$. Excluding these isolates, $k_{doc}$ and $R_{ct}$ generally correlated best with SUVA (at either 254nm or 280nm).

Overall, the investigations demonstrate a wide range of •OH and $O_3$ dynamics as a function of different DOC concentration and DOM character. Depending on the pertinent variable, e.g., temporal effects, geographical effects, natural water versus fractionated water, different surrogates provided a better property-activity relationship. However, in the experiments investigating both the series of natural waters and the series of DOM isolates, there were distinct outliers demonstrating higher reactivity with $O_3$ relative to the other samples. Considering these outliers and the current lack of a consistent property-activity relationship for $O_3$-DOM interactions, it points to the need to approach ozonation-DOM water treatment research from more than one angle to provide tools for optimizing water treatment.

First, there is the large-scale, comprehensive approach similar to the natural waters and DOM isolates experiments. These studies involve a (large) variety of waters/DOM types from different locations, employ various DOM characterization methods, and try to identify pertinent models for, or correlations between, $O_3$ reactivity and DOM character. Despite the considerable advancements in studying the structure of DOM, and in understanding $O_3$ chemistry in natural systems, the current models still do not provide the level of precision that individual utilities might need to meet the more stringent regulations. Nevertheless, these studies continue to provide insights into the nature of these reactions and identify new areas for further research.

An alternative approach is aimed at understanding the specific ozonation process in a specific water of interest. Different utilities will have different concerns over temporal variations in DOM and general water quality (e.g. Lake Zürich-Lengg water versus Sihl River water). A thorough characterization of a utility's water quality and the corresponding ozonation of the water under realistic conditions can provide more precise tools for process optimization. In practice, the water-specific models may take on the same empirical quality as the global models; however, the degree of confidence in model prediction would be much greater. This is nicely illustrated by the $k_{sec}$ and $R_{ct}$ results for the Sihl River water where the Sihl River water is poorly correlated with the other waters investigated (Figure 2), whereas the suitable property-activity relationship in Figure 4 can provide a reasonable prediction of $k_{sec}$ and $R_{ct}$.

This paper has shown that both $O_3$ decomposition rates and $R_{ct}$ are strongly dependent on DOM character, and previously we have shown equally strong dependence on water quality parameters such as temperature and pH *(16)*. The results add to our general understanding of $O_3$ and •OH interactions for different DOM

269

materials, but they also emphasize the limitations of applying global models to modeling ozonation processes in real waters. The $R_{ct}$ parameter discussed in this paper is one of several tools that characterizes the ozonation process. The $R_{ct}$ concept enables measurement of •OH concentration and •OH-exposure in addition to $O_3$ concentration and $O_3$-exposure during ozonation. It has great application for predicting oxidation of micropollutants *(18,19)* (particularly $O_3$ resistant micropollutants), bromate formation *(2)*, and possibly in the future, control of other ozonation byproducts (e.g. aldehydes, bromo-organics) in the water of interest. While not necessarily providing vital information for predicting microbial disinfection, $R_{ct}$ could factor into an expanded version of the Integrated Disinfection Design Framework *(20)* that included micropollutant oxidation and ozonation byproduct formation.

# Literature Cited

1.  Pontius, F.W. *Jour. AWWA* 1999, *91*, 46-58.
2.  von Gunten, U.; Oliveras, Y. *Environ. Sci. Technol.* 1998, *32*, 63-70.
3.  Song, R. Ph.D. dissertation, University of Illinois at Urbana-Champaign, Urbana-Champaign, IL, 1996.
4.  Bablon, G.; Bellamy, W.D.; Bourbigot, M.-M.; Daniel, F.B.; Doré, M.; Erb, F.; Gordon, G.; Langlais, B.; Laplanche, A.; Legube, B.; Martin, G.; Masschelin, W.J.; Pacey, G.; Reckhow, D.A.; Ventresque, C. In *Ozone in Water Treatment: Applications and Engineering*; Langlais, B., Reckhow, D.A., Brink, D.R., Eds.; Lewis Publishers: Chelsea, MI, 1991; pp 11-132.
5.  Hoigné, J. In *Quality and Treatment of Drinking Water II*; Hrubec, J., Ed.; Springer-Verlag: Belin-Heidelberg, 1998; Vol. 5, Part C; pp 83-141.
6.  Hoigné, J. *Wat. Sci. Tech.* 1997, *35*, 1-8.
7.  Neta, P.; Dorfman, L.M. *Adv. Chem. Ser.* 1968, *81*, 222-230.
8.  Haag, W.R.; Yao, C.C.D. Proceedings of the Eleventh Ozone World Congress; San Francisco, CA, USA, 1993.
9.  Yao, C.C.D.; Haag, W.R. *Wat. Res.* 1991, *25*, 761-773.
10. Hoigné, J.; Bader, H. *Ozone Sci. Eng.* 1979, *1*, 357-372.
11. Elovitz, M.S.; von Gunten, U. *Ozone Sci. Eng.* 1999, *21*, 239-260.
12. Hoigné, J.; Bader, H. *Ozone Sci. Eng.* 1994, *16*, 121-134.
13. Hoigné, J.; Bader, H. *Wat. Res.* 1981, *15*, 449-456.
14. Westerhoff, P.; Song, R.; Amy, G.; Minear, R. *Ozone Sci. Eng.* 1997, *19*, 33-49.
15. Westerhoff, P.; Aiken, G.; Amy, G.; Debroux, J. *Wat. Res.* (in press).
16. Elovitz, M.S.; von Gunten, U. *Ozone Sci. Eng.* (submitted).
17. Owen, D.M.; Amy, G.L.; Chowdhury, Z.K.; Paode, R.; McCoy, G.; Viscosil, K. *Jour. AWWA* 1995, *87* (1), 46-63.
18. Acero, J.L.; Stemmler, K.; von Gunten, U. *Environ. Sci. Technol.* (submitted).
19. von Gunten, U.; Elovitz, M.; Kaiser, H.-P. *J. WSRT—Aqua* (in press).
20. Bellamy, W.D.; Finch, G.R.; Haas, C.N. *Integrated Disinfection Design Framework*; AWWA Research Foundation: Denver, CO, 1998.

# Chapter 17

# Removal Simulation for the Radiation-Induced Degradation of the Disinfection By-Product Chloroform

Thomas Tobien[1], William J. Cooper[1], and Klaus-Dieter Asmus[2]

[1]Department of Chemistry, University of North Carolina at Wilmington, 601 South College Road, Wilmington, NC 28403
[2]Radiation Laboratory, University of Notre Dame, Notre Dame, IN 46556

Reaction mechanistic assumptions have been successfully utilized to simulate chloroform removal experiments via a large-scale electron beam treatment. The success is mainly based on a careful re-determination of rate constants regarding reactions of chloroform with hydrated electrons ($[1.38\pm0.10] \times 10^{10}$ [M s]$^{-1}$) and hydrogen atoms ($[7.85\pm0.80] \times 10^7$ [M s]$^{-1}$). In addition, electron attachment rate constants for several other halogenated methanes are reported to help assessment of possible methodical errors in our present evaluation, i.e., bromodichloromethane ($[2.1\pm0.1] \times 10^{10}$ [M s]$^{-1}$), chlorodibromomethane ($[2.0\pm0.1] \times 10^{10}$ [M s]$^{-1}$), bromoform ($[2.6\pm0.1] \times 10^{10}$ [M s]$^{-1}$), and methylene chloride ($[5.45\pm0.25] \times 10^9$ [M s]$^{-1}$).

Chlorination of drinking water is widely used to control microbial growth and, in short, disinfect water to make it potable. Unfortunately, chlorine also reacts with naturally occurring humic substances in the water to form chloroform ($CHCl_3$) among other compounds. Chloroform is a B2 carcinogen and, therefore, regulated by the US EPA in drinking water to a maximum contamination level (MCL) of 0.1 mg L$^{-1}$. Strategies to economically suppress disinfection by-products are in high demand. Economically viable removal of humic substances as a precursor has not yet being accomplished. Activities today are directed toward a removal of chloroform after chlorination and a reduction of its formation utilizing, e.g., membranes, ozonation, etc. Some examples are aeration (*1*), or air stripping, which does not remove but relocate the problem, adsorption onto activated carbon(*2*), again a non-destructive method, ozone treatment, or a combination of ozone, hydrogen peroxide and UV light (*3, 4*), or photolysis using $TiO_2$ (*5, 6*). The latter destructive methods utilize the oxidative power of hydroxyl radicals ($^{\cdot}$OH). In their reaction with $CHCl_3$ a hydrogen atom is abstracted leaving the C–Cl bond intact at this initial stage (7-10).

$$CHCl_3 \quad + \quad {}^{\cdot}OH \quad \longrightarrow \quad {}^{\cdot}CCl_3 \quad + \quad H_2O \qquad (1)$$

Hydrated electrons ($e_{aq}^{-}$) generated by, e.g., an electron beam process, on the other hand, initiate the breakdown of chloroform via a reductive C–Cl bond cleavage:

$$CHCl_3 \quad + \quad e_{aq}^{-} \quad \longrightarrow \quad {}^{\cdot}CHCl_2 \quad + \quad Cl^{-} \qquad (2)$$

In oxygenated systems both radicals will be subject to peroxidation in which oxygen reacts with the carbon-centered initial radicals resulting in further degradation to free chloride ions and fully dechlorinated end products ($CO_2$, $CO$, $HCO_2^{-}$, etc.) (11).

The work presented below describes simulations of large scale chloroform removal experiments by means of an electron beam. The original experimental data published in two earlier papers still lack a successful simulation (12, 13). In particular, these electron beam studies indicated a possibly lower rate constant (factor of 2–3) than the published value (14) for the reaction of $CHCl_3$ with hydrated electrons. Similarly, we were concerned about the accuracy of the rate constant for the hydrogen atom induced process. Those crucial rate constants were, therefore, re-examined. To assess possible methodical errors rate constants have also been determined/re-investigated for the reduction of some other halogenated alkanes by hydrated electrons. The overall outcome of our present study is a very accurate simulation-based description of the experimental removal of chloroform.

## Experimental Section

*Materials*

Dichloromethane (methylene chloride, Aldrich, 99.9%, hplc grade), trichloro-methane (chloroform, sample 1: Fisher Scientific, certified, stabilized with 0.6% ethanol; sample 2: Sigma-Aldrich, 99.9%, hplc grade, ACS, stabilized with 1% ethanol), tetrachloromethane (carbon tetrachloride, Fisher Scientific, certified ACS), tribromomethane (bromoform, Aldrich, 99+%), chlorodibromomethane (Aldrich, 98%), bromodichloromethane (ACROS, 98+%, stabilized with $K_2CO_3$), thioanisole (TA, Aldrich, 99%), 2-propanol (*iso*-propanol, Aldrich, 99.5%, hplc grade) and 2-methyl-2-propanol (*tert*-butanol, Fisher Scientific, certified) were obtained at the highest purity available and used as received for the pulse radiolysis experiments. The water in all pulse radiolysis experiments was deionized in a multi-stage water purification system (18.2 M$\Omega$ resistance, 120 $\mu$g L$^{-1}$ TOC).

*Pulse Radiolysis Method*

A Model TB-8/16-1S linear electron accelerator, providing 5-50 nanosecond pulses of 8 MeV electrons and generating radical concentrations of 1-3 $\mu$M per pulse in all investigated systems, was used for the pulse radiolysis experiments. A detailed description of the experimental setup at the Radiation Laboratory, University of Notre Dame ($15$), as well as the basic details of the equipment and data analysis ($16$, $17$), have been given elsewhere. The dosimetry was based on the oxidation of 0.01M thiocyanate anions ($SCN^-$) to $(SCN)_2^{\cdot-}$ in aqueous, $N_2O$-saturated solutions at pH 7. Absorptions are given in $G\varepsilon$ units, referring to a $G\varepsilon$-value of 46,400 [molecules $(100 \text{ eV})^{-1} M^{-1} cm^{-1}$] for $(SCN)_2^{\cdot-}$ ($18$).

All experiments were carried out in aqueous solutions. In such systems, the electron pulse produces $^\cdot OH$ and $^\cdot H$ radicals, hydrated electrons $e_{aq}^-$, the molecular products $H_2$ and $H_2O_2$ and protons $H_{aq}^+$. Yields are given in terms of G-values which represents the number of species per 100 eV or approximate µmoles per 10 J deposited energy ($19$). The actual figures listed in equation (3) refer to measured yields in dilute substrate solutions.

$$H_2O \xrightarrow{\text{radiolysis}} 2.7 \; ^\cdot OH, 0.6 \; ^\cdot H, 2.6 \; e_{aq}^-, 0.45 \; H_2, 0.7 \; H_2O_2, 2.6 \; H_{aq}^+ \quad (3)$$

For selective monitoring the reaction of hydrated electrons with the substrate, 0.5 M 2-methyl-2-propanol (*tert*-butanol, *t*-BuOH) was added to the solution in order to scavenge $^\cdot OH$-radicals ($k_4 = 6 \times 10^8 \text{ [M s]}^{-1}$) according to equation (4) ($20$):

$$(CH_3)_3COH + ^\cdot OH \longrightarrow ^\cdot CH_2(CH_3)_2COH + H_2O \quad (4)$$

To study the reaction between the substrate and hydrogen atoms the solution was acidified to pH 1 by addition of $HClO_4$. Under these conditions the hydrated electrons are rapidly converted to H-atoms according to equation (5) ($k_5 = 2.3 \times 10^{10}$ $\text{[M s]}^{-1}$) ($21$):

$$e_{aq}^- + H^+ \longrightarrow ^\cdot H \quad (5)$$

To avoid further interfering reactions all solutions were purged with nitrogen in order to remove oxygen since the latter is readily reduced by hydrated electrons and hydrogen atoms.

*Kinetic Model and Simulation Procedure*

The kinetic model and all subsequent simulations refer to experimentally obtained results of large scale electron beam experiments ($12$, $13$). Typically, 3,000 gallons (11,370 L) of substrate solutions in drinking water were irradiated and the disappearance of the substrate monitored. A detailed description of the Electron Beam Research Facility can be found elsewhere ($22$).

In dilute substrate solutions equation (3) describes the net result at about $10^{-7}$ seconds after highly accelerated electrons have passed through the aqueous phase. Subsequently, the three active species ($^\cdot OH$, $e_{aq}^-$, $H^\cdot$) will react among themselves and/or with suitable substrates. A collection of all relevant reactions with their corresponding bimolecular rate constants is available in the literature ($13$).

The initial concentration of the radicals, at any experimental dose, was obtained from the G-values in equation (3). The concentrations of $OH^-$ and $H^+$ ions were evaluated from the pH, while carbonate and bicarbonate concentrations were calculated from the total alkalinity and pH. For each experimental run, pH, total alkalinity, dissolved oxygen, dissolved organic carbon (DOC) and nitrate ion concentrations were measured and their values used in the simulation, since these species constitute the most important sinks for the initial radicals in natural waters.

The computer program "MAKSIMA-CHEMIST" provided by Atomic Energy of Canada, Ltd., was applied to simulate removal efficiencies. Details of the integration algorithm and validation tests can be found elsewhere (23-25). The input to the kinetic model includes a list of all reacting species, their initial concentrations (via experimental measurements), the reaction rate constants (experimental and literature values) and the applied dose rates. The dose rates were assumed to be constant for the duration of 0.091 seconds, which is the residence time of the substrate solution in the irradiation beam. The predictability of our model greatly depends upon the ability to account for all relevant reactions with their proper rate constants and on the accuracy of the experimentally determined concentrations.

## Results and Discussion

### Pulse Radiolysis of Chloroform

*Reaction with Hydrated Electrons*

Chloroform and its intermediates generated upon reaction with hydrated electrons do not exhibit any UV-visible absorptions in the wavelength range from 250 nm to 750 nm. Therefore, the strong absorption of hydrated electrons, peaking at 720 nm, was utilized to determine the bimolecular rate constant for the reaction:

$$CHCl_3 \quad + \quad e_{aq}^- \quad \longrightarrow \quad Products \qquad (6)$$
$$720 \text{ nm}$$

In oxygen-free, 0.5 M *tert*-butanol solutions, the exponential decay of hydrated electrons was monitored in the presence of different amounts of substrate. The decay curves fit an exponential rate law and a plot of the respective first order rate constants versus different substrate concentrations yields a straight line as shown in Figure 1. From the slope of this relationship we determined the bimolecular rate constant of $(1.38\pm0.10) \times 10^{10}$ $(M \ s)^{-1}$ for reaction (6). Both chloroform samples obtained from Fisher Scientific and Sigma contained 0.6% v/v (0.10 M) or 1% v/v (0.17 M) stabilizer ethanol. While ethanol efficiently reacts with ·OH radicals and ·H atoms, its reaction with hydrated electrons occurs with a rate constant of only $1.2 \times 10^3$ $(M \ s)^{-1}$ (26). Due to this very low value and an ethanol concentration in the aqueous phase of approximately 1/70 or 1/116 of that of chloroform any contribution from ethanol to the observed hydrated electron decay is negligible.

**274**

Our present rate constant value is considerably lower compared to a reported one of $3.0 \times 10^{10}$ (M s)$^{-1}$ by Hart et al. ($14$), but it is in very good agreement with the value of $1.0 \times 10^{10}$ (M s)$^{-1}$ predicted by Mak et al. in their earlier electron beam study.

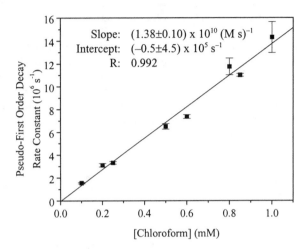

*Figure 1. Plot of pseudo-first-order rate constants for a hydrated electron signal decay at different chloroform concentrations and of chloroform from different chemical suppliers: Fisher (■) and Sigma (●)*

*Reaction with Hydrogen Atoms*

At pH 1 all hydrated electrons are transformed into hydrogen atoms according to reaction 5. Unlike $e_{aq}^{-}$, hydrogen atoms do not exhibit an absorption in the UV-visible range of 250 nm to 750 nm. Therefore, in order to monitor a reaction of H˙ with chloroform a competition method has to be employed. A suitable competitor reacts with hydrogen atoms and its product shows an absorption. In the past chemicals like benzoquinone, or benzoic acid were used for this purpose. Most of them are either oxygen-sensitive or their purification is difficult to achieve. Recently, a reliable method utilizing thioanisole (TA) was introduced by our group to study hydrogen atom competition reactions ($27$). TA is less oxygen sensitive and, as a liquid, can easily be handled under exclusion of surrounding air. In its reaction with hydrogen atoms it generates a transient absorption with a maximum at 365 nm attributable to the H-atom addition product (equation (7)).

$$CH_3S\text{-}C_6H_5 \quad + \quad \text{˙}H \quad \longrightarrow \quad products \qquad (7)$$
$$365 \text{ nm}$$

In the presence of chloroform this absorption is lowered by the fraction of H˙ which reacts with chloroform.

$$CHCl_3 \quad + \quad {}^{\bullet}H \quad \longrightarrow \quad products \qquad (8)$$

This competition can be expressed mathematically (28):

$$\frac{Abs_0(TA(H)^{\bullet})}{Abs(TA(H)^{\bullet})} = 1 + \frac{k(CHCl_3 + H^{\bullet})}{k(TA + H^{\bullet})} \frac{[CHCl_3]}{[TA]} \qquad (9)$$

where $Abs_0(TA(H)^{\bullet})$ is the absorption of the hydrogen atom adduct to thioanisole without chloroform present and $Abs(TA(H)^{\bullet})$ the absorption of the same adduct in the presence of chloroform. A plot of the absorption ratio $Abs_0(TA(H)^{\bullet})/Abs(TA(H)^{\bullet})$ versus the concentration ratio $[CHCl_3]/[TA]$ yields the rate constant ratio $k(CHCl_3+{}^{\bullet}H)/k(TA+{}^{\bullet}H)$ as slope. (See Figure 2.)

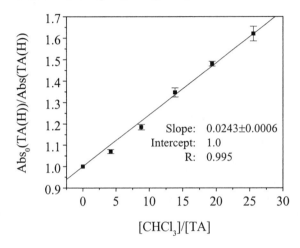

*Figure 2. Plot of an absorption ratio $Abs_0(TA(H)^{\bullet})/Abs(TA(H)^{\bullet})$ at 365 nm for the product of hydrogen atom attachment to thioanisole versus the concentration ratio $[CHCl_3]/[TA]$*

With $k_7=3.24 \times 10^9$ (M s)$^{-1}$ (27), the rate constant for hydrogen atoms reacting with chloroform can be determined to $k_8=(7.85\pm0.20) \times 10^7$ (M s)$^{-1}$. The error limit listed represents that for the weighted curve through all points. It does not take into account the uncertainty concerning the $CHCl_3$ concentration. If the latter is assumed to be in the order of the generally accepted $\pm10\%$ error limit for radiation chemical data, then $k_8 = (7.85\pm0.80) \times 10^7$ (M s)$^{-1}$. A further competing reaction, namely, that of chloroform stabilizer ethanol with hydrogen atoms, can be neglected. For the chloroform sample from Fisher Scientific, the molar ratio $n(CHCl_3):n(EtOH)$ is calculated to be 116:1. Taking into account a rate constant of $1.3 \times 10^7$ (M s)$^{-1}$ (29) for EtOH + H$^{\bullet}$, an ethanol contribution to $k_8$ will be below 1% and, therefore, smaller than the experimental error.

## Chloroform Removal Simulation

The chloroform removal simulations, which are described below, are based on experiments of Mak et al. ($12$, $13$). The initial radicals $^{\cdot}OH$, $^{\cdot}H$ and $e_{aq}^-$, generated by electron beam irradiation, are reacting with all the solutes present in the drinking water, depending on their respective reactivity towards these substrates and their concentrations. A set of relevant reactions with corresponding rate constants has been developed by Mak et al. and can be found in the literature ($13$). In addition, the following chloroform substrate reactions have been considered. See Table I.

**Table I. Chloroform Substrate Reactions with Initial Radicals $^{\cdot}OH$, $^{\cdot}H$ and $e_{aq}^-$ and its Intermediates with Oxygen**

| Reactions | Rate Constant $(M\ s)^{-1}$ | Reference |
|---|---|---|
| $CHCl_3 + e_{aq}^- \rightarrow {}^{\cdot}CHCl_2 + Cl^-$ | $1.25 \times 10^{10}$ | This work |
| $CHCl_3 + {}^{\cdot}OH \rightarrow {}^{\cdot}CCl_3 + H_2O$ | $5.0 \times 10^6$ | ($8$) |
| $CHCl_3 + {}^{\cdot}H \rightarrow {}^{\cdot}CCl_3 + H_2$ | $7.85 \times 10^7$ | This work |
| ${}^{\cdot}CCl_3 + O_2 \rightarrow CCl_3O_2^{\cdot}$ | $3.3 \times 10^9$ | ($30$) |
| ${}^{\cdot}CHCl_2 + O_2 \rightarrow CHCl_2O_2^{\cdot}$ | $4.7 \times 10^9$ | ($31$) |

Reactions of the carbon-centered radicals with oxygen, although not being relevant for the initial scavenging, have been included, since oxygen is one of the compounds competing for hydrated electrons and hydrogen atoms and its concentration will be affected by the peroxidation reactions. Table II lists all solutes in drinking water with their experimental concentrations ($12$), which have been accounted for in the simulations.

**Table II. Solutes in Drinking Water with Their Concentrations ($12$)**

| Solute | Concentration mg $L^{-1}$ | M |
|---|---|---|
| Alkalinity (as $CaCO_3$) | 46.00 | $9.20 \times 10^{-4}$ |
| Chloride $Cl^-$ | 41.55 | $1.19 \times 10^{-3}$ |
| Bromide $Br^-$ | 0.29 | $3.63 \times 10^{-6}$ |
| Sulfate $SO_4^{2-}$ | 20.80 | $2.18 \times 10^{-4}$ |
| Nitrate $NO_3^-$ (as N) | 0.36 | $2.57 \times 10^{-5}$ |
| DOC (as C) | 6.00 | $5.00 \times 10^{-4}$ |
| DO (as $O_2$) | 5.60 | $1.75 \times 10^{-4}$ |

The computer program "MAKSIMA CHEMIST" numerically integrates all given differential rate equations and calculates chloroform concentrations depending on the applied dose. Experimental and simulated removal of chloroform from drinking water at three different pH are plotted in Figure 3a – c.

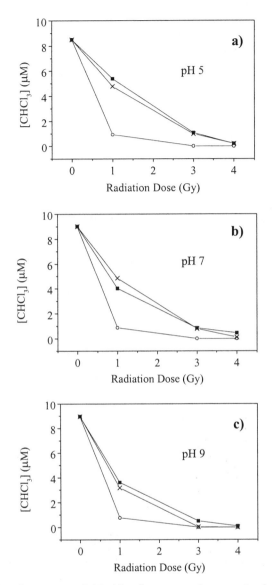

*Figure 3. Plot of experimental (■) chloroform removal versus simulated data based on newly determined (x), and earlier (o) results at a) pH 5, b) pH 7, and c) pH 9*

As shown in Figures 3a – c, experimental chloroform removal results are very close to those simulated curves which are based on our newly determined rate constants. These findings constitute a major improvement when compared to earlier simulations by Mak et al. (*13*). Furthermore, it confirms their assumption of an error in the original rate constant value for a reaction of chloroform which governs its main removal pathway. (See reaction (6).) Our results add confidence to predictions of removal efficiencies for different water qualities and will help to estimate costs for drinking water purification on the basis of an electron beam treatment.

## Pulse Radiolysis Results for Additional Halogenated Methanes

In the preceding section we have shown the importance of correctly determined rate constants. As part of a larger study, which aims for development of a kinetic model for the electron beam process and evaluation of its application to drinking water treatment and hazardous waste site remediation we have also determined rate constants for the reaction of hydrated electrons with several other halogenated methanes. These values were obtained under similar experimental conditions as described above. A plot of pseudo-first-order decay rate constants of hydrated electrons versus different halogenated methanes concentrations is shown in Figure 4.

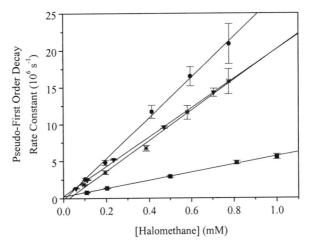

*Figure 4. Plot of pseudo-first-order rate constants for a hydrated electron signal decay at different halomethane concentrations: CHBr$_3$ (●, intercept: –2.1 x 10$^5$ s$^{-1}$; R = 0.999), CHBr$_2$Cl (▼, intercept: 3.0 x 10$^5$ s$^{-1}$; R = 0.9997), CHBrCl$_2$ (*, intercept: – 5.4 x 10$^5$ s$^{-1}$; R = 0.997), CH$_2$Cl$_2$ (■, intercept: 2.4 x 10$^5$ s$^{-1}$; R = 0.999)*

The slopes in Figure 4 represent bimolecular rate constants for a reaction of halogenated methanes with hydrated electrons. A comparison of experimentally determined rate constants with literature values can be found in Table III.

**Table III. Bimolecular Rate Constants for Halogenated Methanes Reacting with Hydrated Electrons**

| Reactions of $e_{aq}^-$ with | Bimolecular Rate Constants $(M\ s)^{-1}$ | | |
|---|---|---|---|
| | Experimental | Literature | |
| Methylene chloride | $(5.45\pm0.25) \times 10^9$ | $6.3 \times 10^9$ | (32) |
| Bromodichloromethane | $(2.1\pm0.1) \times 10^{10}$ | not found | |
| Chlorodibromomethane | $(2.0\pm0.1) \times 10^{10}$ | not found | |
| Bromoform | $(2.6\pm0.1) \times 10^{10}$ | $1.0 \times 10^{10}$ | (33) |

From the experimentally determined rate constants in Table III no general trend can be derived, which could be interpreted in terms of a methodical error in our rate constant evaluations. The values for methylene chloride are in reasonable agreement with the published value, considering the generally accepted 10% error limit for values utilizing pulse radiolytic methods. The value for bromoform is more than double the literature value, but makes sense within the series of the other brominated alkanes investigated here, for which no rate constants were available so far. Assuming bromide to be the leaving halogen in the mixed halogenated compounds (because of a weaker carbon-bromine bond when compared to carbon-chlorine), one should expect similar rate constants as for bromoform. Indeed, all three experimental values are similar, also with respect to a literature value of $2.0 \times 10^{10}$ $(M\ s)^{-1}$ for the $e_{aq}^-$ induced reduction of dibromomethane (34).

## Conclusions

In this work, it has been shown that chloroform removal from drinking water by means of an electron bean treatment can be simulated most accurately. Our re-evaluated rate constants for reactions of chloroform with hydrated electrons and hydrogen atoms play a key role and were the major reason for previously unsuccessful attempts. Methodical errors during our rate constant re-determination can practically be ruled out on the basis of a comparison of additionally re-determined rate constants for other halogenated methanes.

In conclusion to our introductory remarks, one should now be able to quantify capital costs more accurately regarding a commercial implementation of the electron beam process since the energy costs required to destroy chloroform can be predicted. From an environmental point of view the implementation of both processes, namely, a contaminant destruction by a radiation chemical treatment and chlorination to a much lower extent to prevent a microbial regrowth makes good sense.

280

## Acknowledgments

This work was supported by the National Science Foundation grant BES 9796296 and grant DMI-9531275 through a subcontract from Science Research Laboratory, Inc., and, concerning K.D.A., by the Office of Science of the U.S. Department of Energy. The use of the Notre Dame Radiation Laboratory Linear Accelerator facility was essential for the completion of this work. We gratefully acknowledge their cooperation as well as the Miami-Dade Water and Sewer Authority for the original data.

## Literature Cited

1.  Adams, J. Q.; Clark, R. M. *J. Am. Water Works Assoc.* **1991**, *83*, 49.
2.  Njam, I. N.; Snoeyink, V. L.; Lykins, B. W., Jr.; Adams, J. Q. *J. Am. Water Works Assoc.* **1991**, *83*, 65.
3.  Langlais, B.; Reckhow, D. A.; Brink, D. R. *Ozone in Water Treatment: Application and Engineering*; Lewis Publishers: Chelsea, MI, 1991, pp. 27.
4.  Hoigne, H.; Bader, H. *Water Res.* **1983**, *17*, 173.
5.  Pruden, A. L.; Ollis, D. F. *Environ. Sci. Technol.* **1983**, *17*, 626.
6.  Korman, C.; Bahnemann, D. W.; Hoffman, M. R. *Environ. Sci. Technol.* **1991**, *25*, 494.
7.  Haag, W. R.; Yao, C. C. D. *Environ. Sci. Technol.* **1992**, *26*, 1005-13.
8.  Rezansoff, B. J.; McCallum, K. J.; Woods, R. J. *Can. J. Chem.* **1970**, *48*, 271-6.
9.  Anbar, M.; Meyerstein, D.; Neta, P. *J. Chem. Soc., Pt. B* **1966**, 742-7.
10. Chutny, B. *Collect. Czech. Chem. Commun.* **1966**, *31*, 358-61.
11. Asmus, K.-D.; Bahnemann, D.; Krischer, K.; Lal, M.; Moenig, J. *Life Chemistry Reports* **1985**, *3*, 1-15.
12. Mak, F. T.; Cooper, W. J.; Kurucz, C. N.; Nickelsen, M. G.; Waite, T. D. In *Disinfection By-Products in Water Treatment*; Minear, R.M., Amy, G.A., Ed.; CRC Lewis Publishers: Boca Raton, FL, 1996, pp 131-50.
13. Mak, F. T.; Zele, S. R.; Cooper, W. J.; Kurucz, C. N.; Waite, T. D.; Nickelsen, M. G. *Wat. Res.* **1997**, *31*, 219-28.
14. Hart, E. J.; Gordon, S.; Thomas, J. K. *J. Phys. Chem.* **1964**, *68*, 1271-4.
15. Schuler, R. H. *Radiat. Phys. Chem.* **1996**, *47*, 9-17.
16. Asmus, K.-D. *Methods Enzymol.* **1984**, *105*, 167-71.
17. Asmus, K.-D.; Bonifacic, M. In *Excercise and Oxygen Toxicity*; Sen, C.K., Packer, L., Haenninen, O., Ed.; Elsevier: New York, 1994, pp 1-47.
18. Schuler, R. H.; Patterson, L. K.; Janata, E. *J. Phys. Chem.* **1980**, *84*, 2088-89.
19. Buxton, G. V.; Greenstock, C. L.; Helman, W. P.; Ross, A. B. *J. Phys. Chem. Reference Data* **1988**, *17*, 513-886.
20. Wolfenden, B. S.; Willson, R. L. *J. Chem. Soc., Perkin Trans. 2* **1982**, 805-12.
21. Gordon, S.; Hart, E. J.; Matheson, M. S.; Rabani, J.; Thomas, J. K. *J. Am. Chem. Soc.* **1963**, *25*, 1375.

22. Kurucz, C. N.; Waite, T. D.; Cooper, W. J. *Radiat. Phys. Chem.* **1995**, *45*, 299-308.

23. Carver, M. B.; Hanley, D. V.; Chapin, K. R. *MAKSIMA-CHEMIST, A Program for Mass Action Kinetic Simulated Manipulation and Integration Using Stiff Techniques, Chalk River Nuclear Laboratories Report*, Report 6413, Atomic Energy of Canada, Ltd.: Chalk River, Canada, 1979, pp 1-28.

24. Gear, C. W. *Comm. ACM* **1971**, *14*, 176.

25. Schwarz, H. A. *J. Phys. Chem.* **1962**, *66*, 255.

26. Park, H.-R.; Getoff, N. *Z. Naturforsch., A, Phys. Sci.* **1992**, *47A*, 985-91.

27. Tobien, T.; Cooper, W. J.; Nickelsen, M. G.; Pernas, E.; O'Shea, K. E.; Asmus, K.-D. *Environ. Sci. Technol.* **1999** (submitted).

28. Spinks, J. W. T.; Woods, R. J. *An Introduction to Radiation Chemistry*; 2nd ed; John Wiley & Sons: New York, 1976, pp. 1-504.

29. Smaller, B.; Avery, E. C.; Remko, J. R. *J. Chem. Phys.* **1971**, *55*, 2414-8.

30. Moenig, J.; Bahnemann, D.; Asmus, K.-D. *Chem. Biol. Interact.* **1983**, *47*, 15-27.

31. Emmi, S. S.; Beggatio, G.; Casalbore, G.; Fuochi, P. G. Haloalkyl radicals: Formation and reactivity of mono- and dichloromethylperoxy radicals. In *Proc. Fifth Tihany Symposium on Radiation Chemistry*; Sept 1982; Siofok, Hungary: Akad. Kiado, Budapest.

32. Balkas, T. I. *Int. J. Radiat. Phys. Chem.* **1972**, *4*, 199-208.

33. Lal, M.; Mahal, H. S. *Radiat. Phys. Chem.* **1992**, *40*, 23-6.

34. Hayes, D.; Schmidt, K. H.; Meisel, D. *J. Phys. Chem.* **1989**, *93*, 6100-9.

Chapter 18

# Influence of Natural Organic Matter on Bromate Formation During Ozonation of Low-Bromide Drinking Waters: A Multi-Level Assessment of Bromate

Christopher J. Douville and Gary L. Amy

Department of Civil, Architectural, and Environmental Engineering, University of Colorado at Boulder, Boulder, CO 80309–0421

A multi-level approach is used to assess bromate formation. The size, structure and functionality of natural organic matter (NOM) and its role in bromate formation is being investigated via a nationwide survey of ozonation facilities, bench-scale ozonation and simultaneous NOM characterization of source waters, as well as scale-up comparison testing between bench-, pilot-, and full-scale ozone contactors. Initial results indicate that many utilities will be faced with the challenge of optimizing their ozonation process in order to achieve the desired *Cryptosporidium* inactivation, that may become a consequence of the proposed Stage 2 Disinfectant/Disinfection By-Product (D/DBP) Rule, along with compliance of the existing Stage 1 bromate standard of 10 µg/L. Ongoing work will continue to show that a solid understanding of the character of the NOM will enable utilities to predict how NOM will either inhibit or promote bromate formation.

## Relevance of Research

*Cryptosporidium* may become the target organism for disinfection of drinking water. Because this organism is much more resistant to disinfection than *Giardia*, water treatment plants may soon have to adjust their disinfection strategy to achieve an appropriate log-kill of *Cryptosporidium* (*1*). Many existing ozonation facilities will need to increase their ozone dosage if *Cryptosporidium* disinfection is required. Many treatment plants have made the decision to implement ozone within their process, with many more upgrades and new constructions involving ozone anticipated. As the movement towards ozonation facilities is gaining momentum, a close eye must be kept on the balance between obtaining effective disinfection while minimizing the formation of ozone by-products. Of particular interest is the problematic compound bromate (BrO₃⁻). In Figure 1, a schematic portrays the tradeoff issue between the acute risk of *Cryptosporidium* and the chronic risk associated with bromate. In November of 1998, the USEPA regulated bromate in drinking water with a maximum contaminant level (MCL) of 10 µg/L as part of Stage 1 of the Disinfectant/Disinfection By-Product (D/DBP) Rule (*2,3*). Due to the associated cancer risk of bromate, this standard may be lowered to 5 µg/L (or lower) in Stage 2 of the D/DBP Rule (*2,4*).

## Players in the Bromate Formation Game

As with any formed by-product, identification of the significant precursor(s) is vital to understanding it's formation and subsequently developing a minimization/control strategy. A large amount of research has been conducted on bromate in order to understand which disinfection treatment conditions and water quality constituents are factors in bromate formation. Important treatment conditions include ozone dose (measured as applied or transferred ozone dose), contact time, and water temperature. Increased levels of ozone dose and contact time as well as an increase in water temperature will ultimately result in higher levels of bromate (*5,6*). However, microorganism inactivation is also enhanced at higher temperatures.

An obvious precursor is the inorganic constituent bromide (Br⁻) and many researchers have shown that with everything else being equal, increased levels of bromide contribute to elevated bromate formation (*5-7*). Research has shown that pH is a major factor affecting bromate formation, with higher bromate values resulting as pH increases (*5,6*). The parameter dissolved organic carbon (DOC) has been studied in previous research and is known to participate in reactions with ozone. DOC or natural organic matter (NOM) has been shown to exert an ozone demand (*8*). Some research has shown that high levels of DOC have resulted in increased levels of bromate (*7*). DOC can be dissected further into classes of NOM. Bromate formation work has been conducted using fractions and isolates of NOM to expand the depth of study (*9,10*). Additionally, NOM has been shown to influence the formation of hydroxyl radicals (OH•) and ultimately bromate formation (*9*). Since the OH• pathway has been shown to be the dominant bromate formation pathway (*10*), promotion or inhibition of OH• will influence bromate chemistry. Understanding NOM's composition and character and its associated role in bromate formation is the focus of this research.

## Focus of Research

There is a need to expand the understanding of how and to what degree certain components of NOM will react with ozone that may inhibit or promote bromate formation. By utilizing a suite of NOM characterization techniques, the composition and functionality of the multi-faceted NOM compounds can be linked to bromate levels. With an increased understanding of the influence of NOM, disinfection practices can be more appropriately optimized. The USEPA Surface Water

Treatment Rule (SWTR) established the "CT" system for disinfection credit (*11*), and provides tables that link CT to corresponding levels of *Giardia* inactivation (*4*). Inactivation requirements are currently being developed for *Cryptosporidium* inactivation as part of the Enhanced Surface Water Treatment Rule (ESWTR) (*12*). The challenge for treatment plants will be to realize the CT/*Cryptosporidium* inactivation level required, while at the same time acknowledging the associated CT versus bromate relationships. This is also one of the major challenges for the upcoming Stage 2 D/DBP and ESWTR regulation development.

A nationwide survey of bromide in drinking water yielded an average value of approximately 80 μg/L (*13*). The research discussed here has focused on low- to moderate-bromide waters, designated as 10 to 100 μg/L levels. Up until recently, low-bromide waters were not targeted for study during bromate formation research. However, the likely onset of higher doses of ozone now creates a potential bromate issue with even low-bromide source waters. Low-bromide waters at treatment plants that ozonate are now a legitimate concern in terms of bromate.

## Scope of Work

A multi-level scope of work has been used to investigate bromate formation. Specifically, the components are:

- A nationwide bromate survey of full-scale ozonation facilities
- NOM characterization of source waters
- Simultaneous bench-scale ozonation of the source waters to measure bromate formation
- A scale-up comparison study between full-, pilot- and bench-scale ozone contactors

The details of each component are described below.

### Bromate Survey

The nationwide bromate survey, entitled the "AWWARF Bromate Survey," was devised to understand bromate occurrence (formation) at the full-scale during realistic current treatment levels of ozone application. Twenty-four full-scale ozonation facilities are represented in the survey, including 21 from the US and 3 from France. Two sampling campaigns (June 1998 and November 1998) were conducted to obtain sample pairs of "before" and "after" ozone application. Pre-, intermediate-, and dual-ozonation facilities (both pre- and intermediate-ozonation) are represented within the participants. Samples were analyzed for DOC, ultraviolet light absorbance at 254 nanometers (UVA$_{254}$), alkalinity (ALK), ammonia (NH$_3$-N), bromide ("before" ozone sample) and bromate ("after" ozone sample). Treatment conditions were also obtained from the utility at the time of sampling; information consisted of ozone dose (transferred and/or applied), contact time estimates (hydraulic residence time (HRT) and/or $t_{10}$), pH and temperature.

## NOM Characterization

The goal of this component is to use a suite of characterization techniques to discern information about size, structure and functionality of NOM. Raw water from each of the three participating utilities was subjected to analyses of DOC and $UVA_{254}$ to yield specific UVA (SUVA), UV spectrum (200 to 400 nm), high performance size exclusion chromatography (HPSEC) to give an estimate of apparent molecular weight (MW), and the XAD-8/-4 fractionation protocol which fractionates NOM into operationally defined classes of percent hydrophobic (XAD-8 adsorbable), transphilic (XAD-4 adsorbable), and hydrophilic DOC (neither XAD-8 nor XAD-4 adsorbable) (*14*). Additionally, differential UV spectra ($\Delta$UVA) for the waters were developed. A differential UV spectrum, which provides insight to the relative degree of ozone reactivity, can be obtained by subtracting the UV spectrum of an ozonated water sample (after ozonation) from that of an un-ozonated sample (before ozonation). The analyses were performed on bulk water samples (no isolates or fractions). At the same time of the NOM characterization analyses, the general water quality of the waters was assessed (i.e. alkalinity, $NH_3$-N, pH, bromide).

## Bench-Scale Ozonation

True-batch ozonation (100% transfer efficiency) was used to determine bromate formation potentials ($BrO_3^-$-FPs) for the same three waters subjected to NOM characterization. A bench-scale reactor system (a modified graduated cylinder with a sample port) was used under standard conditions of temperature (20 degrees C), pH (7.0) and ozone to DOC dose ratio (2:1 mass based ratio). A 1.0 mM phosphate buffer was used to stabilize the experimental pH and bring the total volume of liquid in the reactor to 500 mL, thus incurring a dilution of bulk water. This dilution creates a new water quality matrix, and is cited as such. A 500 mL total volume was used because of the high volume necessary to obtain samples for kinetic data. For waters #2 and #3, dilution was also used to create equalization of DOC so that only the NOM properties would be influential in bromate formation. For these waters, experimentation was performed with a 1.0 mg/L DOC value in the reactor for each experiment. The reactor setup is shown in Figure 2. Samples of ozone residual were taken over time to generate an ozone decay curve as the ozone reacted with the NOM. The area under the decay curve can be mathematically integrated to yield an estimate of CT called ozone exposure (OE). After completion of the reaction (i.e. an ozone residual of 0.0 mg/L), the $BrO_3^-$-FP was measured. With this data, the CT versus bromate relationship was evaluated. Each separate true-batch experiment resulted in one data point; multiple data points reflect replication efforts at different ozone doses.

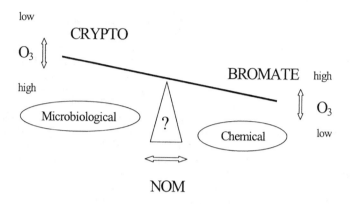

Figure 1. A risk schematic showing the Cryptosporidium versus bromate tradeoff.

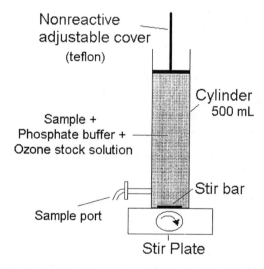

Figure 2. The true-batch ozonation reactor used for bench-scale bromate formation experiments.

## Scale-Up Comparison Study

A scale study was performed to compare full-, pilot-, and bench-scale ozone contactors in terms of calculated CT versus $BrO_3^-$ formation. One of the aforementioned participating utilities also participated in the scale-up comparison study that involved on-site testing of full- and pilot-scale contactors and laboratory testing of a bench-scale ozone contactor. The bench-scale contactor was a novel continuous-flow system. In this reactor a bulk, undiluted water sample was continuously ozonated with dissolved ozone bubbles created by a frit as it flowed through a 380 mL glass column. The contactor has internal sample recirculation to create a completely mixed system as well as a water jacket for experimental temperature control. Full- and pilot-scale tests were performed at the facility to generate estimates of CT (as calculated per SWTR guidelines). Aliquots of ozonated water (quenched) were shipped to the university lab to determine bromate concentrations. Bench-scale experimentation took place in the university lab. To obtain different estimates of CT using the bench-scale system, separate experiments were run at a variety of liquid flow rates which resulted in a range of CT estimates, each with its respective $BrO_3^-$-FP. These CT estimates were calculated by multiplying the average ozone residual of the water times the HRT, and will be referred to as ozone exposure. The different scales were compared once each CT versus bromate relationship was derived and analyzed simultaneously.

# Data Acquisition

All of the analytical work mentioned in the paper was performed at the University of Colorado at Boulder (CU), with the exception of the low-level bromate analyses (< 2.0 µg/L) which were conducted at the University of Illinois at Urbana-Champaign (UI) (15). Final data was generated using calibration curves that were derived from appropriate, lab-grade standards. DOC was measured with a Sievers 800 TOC analyzer which utilizes the UV/persulfate oxidation method. UV analyses were conducted on a Shimadzu UV-VIS 160 spectrophotometer. Ammonia analysis was performed using a HACH DR-2000 spectrophotometer and the Nessler method. Bromide and bromate values were obtained using a Dionex DX-300 ion chromatography system and an AS9-HC analytical column (high capacity). The detection limit at CU for both bromate and bromide was approximated at 2 µg/L by the lowest concentration of standard that was consistently detected. Dissolved ozone concentrations were obtained by the indigo method using HACH Accu-Vac vials and a DR-2000 spectrophotometer. HPSEC work was performed using a Shimadzu high-pressure liquid chromatography system and a Waters Protein Pak 125 column (16). At least 10% triplication was applied to the samples within each analytic tool.

# Results and Discussion

## Bromate Survey

The intent of the survey was to assemble a representative cross-section of ozonation utilities as participants, with the hopes that data from the survey would yield insight to full-scale occurrence and formation of bromate. A geographic distribution of U.S. survey participants can be found in Figure 3. Although participation from California appears to be a geographic bias, the larger number of participants is somewhat justified due to the high amount of ozone usage within the state.

Results from the survey are presented from two separate rounds of sampling. The first round was a late spring sampling session (June), while the second round was during the late fall (November). Minimum, mean, and maximum values of all the measured parameters for both rounds are presented in Table I. The data represents the water quality at the point of ozone application. There are several noteworthy results. First, the average formed bromate decreased from round 1 to round 2 (4.8 μg/L versus 3.6 μg/L). Potential reasons for this observed trend could be the decrease in average water temperature (June versus November sampling dates), or a significantly lower average level of bromide in the source water (66 μg/L versus 49 μg/L). Bromate levels did range significantly, from below the detection limit of

**Table I. Summary Results of the AWWARF Bromate Survey**

| Parameter | Round 1 [a] | | | Round 2 [b] | | |
|---|---|---|---|---|---|---|
| (Units) | Min | Mean | Max | Min | Mean | Max |
| DOC (mg/L) | 1.3 | 3.1 | 8.6 | 1.0 | 3.3 | 8.9 |
| $UVA_{254}$ (cm$^{-1}$) | 0.012 | 0.053 | 0.260 | 0.011 | 0.058 | 0.247 |
| $\Delta UVA_{254}$ (cm$^{-1}$) | -0.010 | 0.044 | 0.181 | 0.004 | 0.041 | 0.117 |
| SUVA (L/mg-m) | 0.9 | 2.1 | 3.4 | 0.9 | 2.2 | 3.2 |
| $NH_3$-N (mg/L) | nd [e] | 0.17 | 0.70 | 0.01 | 0.16 | 1.11 |
| ALK (mg/L) | 8.7 | 111 | 270 | 6.8 | 98 | 280 |
| Br$^-$ (μg/L) | 6.4 | 66 | 180 | 3.7 | 49 | 150 |
| Br$^-$ conversion [c] | 0% | 8% | 39% | 0% | 7% | 57% |
| $BrO_3^-$ (μg/L) [d] | <0.4 (n=4) | 4.8 | 19 | <0.3 (n=2) | 3.6 | 36 |

NOTE: [a] n=22 for statistics during round 1 sampling.
[b] n=21 during round 2 sampling.
[c] Bromide conversion is the percent mass of bromide that was converted to bromate.
[d] Low level samples were sent to University of Illinois for bromate analysis. Detection limit was 0.4 μg/L for round 1 and 0.3 μg/L for round 2, determined by calculation.
[e] nd designation for below the detection limit of 0.01 mg/L for $NH_3$-N.

0.3 µg/L to well over 30 µg/L. Typically, bromate formation for waters low in bromide with current treatment (based on *Giardia* inactivation, at ambient pH) is less than the Stage 1 bromate MCL. Also, under current treatment conditions, typically less than 10 percent of the bromide is converted to bromate (on average).

The average organic character, in terms of DOC, $UVA_{254}$ and SUVA, did not appear to change significantly between the sampling dates. The waters in this study (at the point of ozonation) were not high in humic content, as the SUVA was less than or equal to 2.5 L/mg-m (on average). Typically, the waters in this study had moderate alkalinity (around 100 mg/L on average). With limited data interpretation to this point, there doesn't seem to be any clear trends in the correlation of one particular parameter to bromate levels in finished water. Multiple-variable analyses may prove to be more informative and further explain bromate's dependence on certain combinations of water quality parameters and/or treatment conditions.

### NOM Characterization

Analyses were performed on three different raw, bulk waters from participating utilities. An initial task was to perform a general water quality assessment of the waters in terms of DOC, $UVA_{254}$, pH, alkalinity, ammonia, and bromide ion. Results of these parameters can be found in Part A of Table II. It can be seen that each of the waters is different from the others in terms of the six parameters analyzed. Water #1 is a low bromide, low alkalinity water with high DOC and $UVA_{254}$, moderate SUVA, neutral pH, and average ammonia level. A moderate bromide ion concentration (72 µg/L) is the standout parameter for water #2. This water is has a high alkalinity and pH, average DOC and $UVA_{254}$, and low SUVA and ammonia level. Water #3 has the lowest DOC value of the three waters (1.9 mg/L) and intermediate values of the other measured parameters.

The suite of NOM characterization techniques was performed on the three waters. Table III summarizes the results of the analyses, which reveal much more about the organic component in water than a simple DOC analysis. The data from the table can be interpreted to reveal a relative organic comparison of the three waters. Organic matter in water #1 yielded the highest SUVA, apparent MW, and percent hydrophobic NOM, while at the same time contained the lowest percent of transphilic and hydrophilic NOM. Conversely, water #2 contained organic matter with the highest percentage of transphilic and hydrophilic NOM but the lowest level of SUVA, apparent MW, and percent hydrophobic NOM. Interestingly enough, water #3 had intermediate values of all the above mentioned organic measurements. SUVA is an important parameter because of its relationship to humic content. Typically, a water with high SUVA contains mainly humic material, while low SUVA waters contain largely nonhumic material (*17*). A graphical display of the NOM fractions of the waters can be found in Figure 4. It is important to observe that while the percent NOM fraction distribution does not appear to be drastically different from one water to the next, the mass of hydrophobic NOM for water #1 is by far the dominant fraction on an absolute DOC basis.

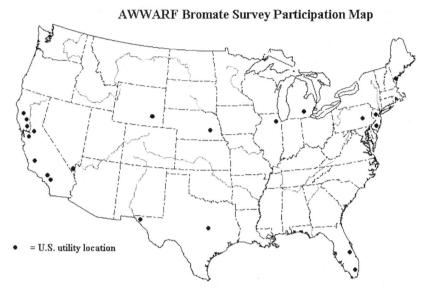

AWWARF Bromate Survey Participation Map

• = U.S. utility location

NOTE: 24 utilities total (21 U.S. and 3 French)

*Figure 3. Map showing the location of U.S. participants in the AWWARF Bromate Survey.*

## NOM Fractionation of the Three Waters

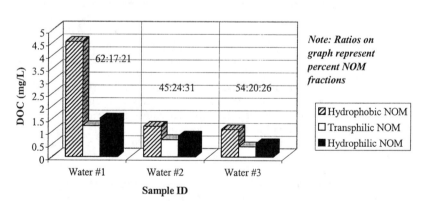

*Figure 4. NOM fractionation results for the three waters.*

Table II. Water Quality of Waters from Three Participating Utilities

| PART A: Water Quality of Bulk Water Parameter (Units) | Water #1 | #2 | #3 |
|---|---|---|---|
| Bromide, Br⁻ (µg/L) | 21 | 72 | 33 |
| Dissolved Organic Carbon, DOC (mg/L as C) | 9.4 | 2.7 | 1.9 |
| UV Absorbance @ 254 nm, $UVA_{254}$ (cm⁻¹) | 0.330 | 0.044 | 0.044 |
| Specific UVA, SUVA (L/mg-m) [a] | 3.5 | 1.6 | 2.3 |
| pH, at given temperature (standard units) | 7.3 @ 11 C | 8.1 @ 21 C | 7.9 @ 21 C |
| Ammonia-Nitrogen, $NH_3$-N (mg/L as N) | 0.20 | 0.06 | 0.07 |
| Carbonate Alkalinity (mg/L as $CaCO_3$) | 28 | 130 | 110 |
| PART B: Water Quality of Diluted Aliquots [b] Parameter (Units) | Water #1 | #2 | #3 |
| Percent Dilution | 68 | 63 | 47 |
| Bromide, Br⁻ (µg/L) | 6.7 | 26.6 | 17.5 |
| Dissolved Organic Carbon, DOC (mg/L as C) | 3.0 | 1.0 | 1.0 |
| UV Absorbance @ 254 nm, $UVA_{254}$ (cm⁻¹) | 0.106 | 0.016 | 0.023 |
| Specific UVA, SUVA (L/mg-m) [a] | 3.5 | 1.6 | 2.3 |
| pH, adjusted (standard units) | 7.0 | 7.0 | 7.0 |
| Ammonia-Nitrogen, $NH_3$-N (mg/L as N) | 0.06 | 0.02 | 0.04 |
| Carbonate Alkalinity (mg/L as $CaCO_3$) | 9.0 | 42 | 58 |

NOTE: [a] SUVA = $UVA_{254}$*100/DOC.
[b] Values listed in Part B are approximations based on volumetric dilutions of the bulk water.

One NOM characterization technique not included in Table III is the UV spectra results. The UV spectrum of each water, scanned from 200 to 400 nm, is presented in Figure 5. Arguably, every water has a unique UV "signature" based on the organic amount and character. This data is not particularly revealing but is shown for comparative reasons. The spectra shown in terms of differential UV spectra are much more informative. Figure 6 displays the differential UV spectra of the three waters. Components of NOM that are reactive with ozone will become oxidized, resulting in a decrease in the UV absorbance. On the differential UV spectrum graphs, locations of organic material with high ozone reactivity are indicated by a peak in the spectrum and the relative size of the peak is proportional to the degree of reactivity. Waters #2 and #3 had very similar peak locations (235 and 240 nm) as well as similar peak sizes. Water #1 appears to have the highest degree of reactivity, with a large peak near 210 nm. It must be noted that the $\Delta UVA_{254}$ values and differential UV spectra were derived using aliquots of a diluted water, as a consequence of the bench-scale ozonation process. Refer to Part B of Table II for the water quality of the diluted waters. With a DOC of 3 mg/L for water #1 and 1 mg/L for waters #2 and #3, it would be expected that water #1 would show the largest $\Delta UVA_{254}$ and UV differential

## UV Spectrum of the Three Waters
### 200 to 400 nm

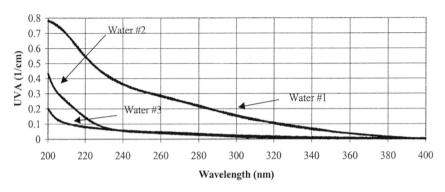

*Figure 5. Analysis of UV spectrum, 200 to 400 nm, for the three waters.*

## Differential UV Spectrum of the Three Waters
### 200 to 400 nm

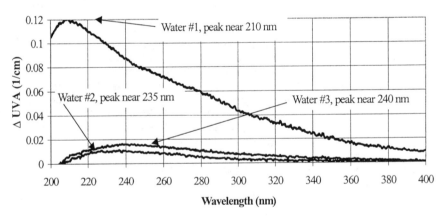

*Figure 6. Differential UV spectrum from 200 to 400 nm for the three waters.*

spectrum. However, when water #1 is normalized to a DOC level of 1 mg/L, it is found that water #1 still has the largest $\Delta UVA_{254}$ (0.0253 cm$^{-1}$ versus 0.0087 and 0.0144). Interpretation of UV spectrum analyses may be misleading due to possible interference of inorganic constituents (e.g., $NO_3^-$) in wavelengths less than 220 nm.

**Table III. NOM Characterization Results from Analyses on the Three Waters**

| NOM Characterization Measurement | Units | Water #1 | #2 | #3 |
|---|---|---|---|---|
| $\Delta UVA_{254}$ [a] | 1/cm | 0.0760 | 0.0087 | 0.0144 |
| Apparent Molecular Weight, MW [b] | Daltons | 1250 | 980 | 1240 |
| Hydrophobic NOM | Percent DOC | 62 | 45 | 54 |
| Transphilic NOM | Percent DOC | 17 | 24 | 20 |
| Hydrophilic NOM | Percent DOC | 21 | 31 | 26 |

NOTE: [a] $\Delta UVA_{254}$ = $UVA_{254}$ "before" ozonation – $UVA_{254}$ "after" ozonation, using diluted aliquots.
[b] Apparent molecular weight is the weight-averaged MW. See reference (*16*).

**Bench-Scale Ozonation**

Diluted versions of the three waters were subjected to bromate formation potential experiments. As described in the scope section, ozone decay curves were generated to yield an estimate of CT. An example of an ozone decay curve is shown in Figure 7. The curve is assumed to be linear from 0 to 1 minute (i.e., zero order) and then fit to a mathematical function until the ozone residual is 0.0 mg/L. For instance, the displayed decay curve was fit to an exponential function (overall, pseudo-first order), and the area under the curve resulted in an ozone exposure value of 10.6 mg-min/L. For this true-batch ozonation experiment, the corresponding value of bromate measured was 2.3 µg/L. The same ozone exposure versus bromate relationship was generated for the other waters, and the results can be found in Figure 8. Using the origin as an additional data point, the slope of the relationship from each water can be compared to determine which water is more susceptible to form bromate than the others. Keep in mind that these bromate values are levels in the final diluted mixture that is ozonated, which consists of sample, ozone stock, and phosphate buffer. Because of this dilution a new water quality matrix is created for the waters, which can be found in Table II, Part B. Water #3, with a slope of 0.69, showed the highest tendency to form bromate when compared to the other waters (0.61 for water #2 and 0.21 for water #1). For reference purposes, approximate ranges of *Giardia* inactivation and *Cryptosporidium* inactivation are displayed on the graph. The value of 0.5 mg-min/L for *Giardia* was selected because it is the required CT value to achieve a 1-log reduction of *Giardia* at 10 degrees C, as outlined in the SWTR. The range of 2.5 to 10 mg-min/L is used for *Cryptosporidium* in attempts to estimate what the "new" requirements may be (conservatively selected as 5 to 20 times the *Giardia* requirements) (*1,12*). Figure 8 can be potentially misleading, because it does not

## Ozone Decay Curve, Water #1

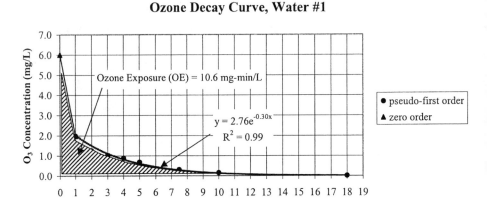

Figure 7. Example of an ozone decay curve to generate an estimate of ozone exposure.

## Ozone Exposure vs. Bromate:
## True-Batch Ozonation (2 mg O3 : 1 mg DOC)
## (k = slope of relationship)

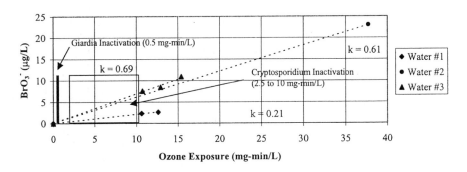

Figure 8. Ozone exposure versus bromate relationships derived from true-batch ozonation experiments with the three waters.

account for the difference in diluted source water bromide levels. In attempts to remedy this inequality, the same data are presented in terms of percent bromide conversion. Bromide conversion is simply the mass percentile of bromate (as bromide) in the ozonated water compared to the mass of bromide in the water before ozonation. These new relationships are presented in Figure 9. Another way to explain this new presentation format is as efficiency (percent) of bromate incorporation. Relative to each water's bromide concentration, waters #1 and #3 are equally as efficient in bromate formation, while water #2 is much less efficient. It appears that ozone is not as reactive in water #2, due to a lengthy ozone decay curve. Water #2 has the lowest SUVA value, which could attribute to the lack in bromate formation for a given ozone exposure.

**Scale-Up Comparison Study**

Upon receipt of the full- and pilot-scale data from the participating utility (#3), relationships of calculated CT versus bromate formation were developed. Combined with the bench-scale relationship, the three scales can be compared based on the relative slopes of the regression lines. Figure 10 shows the results of the scale-up comparison study. The full-scale ozone contactor was shown to form the most bromate at a given CT (a slope of 1.35); pilot-scale (0.49), and bench-scale continuous-flow (0.45) contactors followed. Basically, for a set CT value the bromate consequence is largest for the full-scale contactor and smallest for the continuous-flow bench-scale system. For the full- and pilot-scale relationships, the $t_{10}/t_{50}$ ratio is provided as information about the hydrodynamics of the contactors. Because an ideal plug-flow reactor (PFR) has a $t_{10}/t_{50}$ ratio of 1.0, it appears that the pilot-scale system more closely behaves like a PFR ($t_{10}/t_{50}$ of 0.87 based on tracer study information), and the full-scale contactor behaves more like a continuous stirred tank reactor (CSTR) with a $t_{10}/t_{50}$ of 0.49. In comparing the lab system with the higher scales, the continuous-flow system seems to agree favorably with the pilot-scale contactor. However, this system was shown to underestimate the bromate that the full-scale system demonstrated. Due to temperature control capability with the continuous-flow system, the experimentation performed with this system was conducted at 9 degrees C, as was the full- and pilot-scale tests. One positive finding is that for each type of ozone contactor, the individual relationships of calculated CT versus bromate were shown to be linear across the range of CT values used. This particular utility may have a challenge in complying with the established bromate standard at ambient pH if forced to operate with CT values at or above 8.0 mg-min/L.

# Conclusions

In summary:

- The survey indicated that there are some bromate issues with current ozonation practices at the full scale (which currently target *Giardia* inactivation). Future

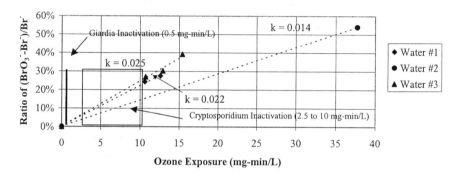

Figure 9. Ozone exposure versus bromate relationships for the three waters in terms of percent bromide conversion.

## Scale-Up Comparison Study: Ozone Exposure vs. Bromate
### (Water #3, k = slope of relationship)

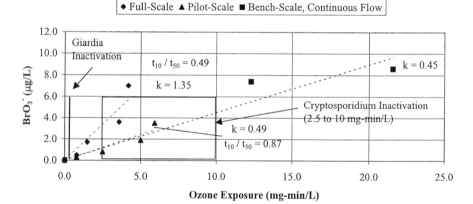

Figure 10. Ozone exposure versus bromate relationships for full-, pilot- and bench-scale ozone reactors resulting from the scale-up comparison

compliance may be difficult for some utilities when operating at disinfection levels capable of *Cryptosporidium* inactivation at ambient pH.

- Some components of NOM seem to exert a strong ozone demand (hydrophobic NOM). Waters #2 and #3 are similar in terms of NOM characterization analyses, but differ substantially in terms of bromate formation potentials. A much more rigorous data analysis effort is needed to assist in interpretation of the effects of NOM; conclusions are limited because the relationships are based on a collection of only three waters.
- Lab-scale derived relationships of CT (ozone exposure) versus bromate are shown to be linear and can also be compared with relationships derived at larger scales. For water #3, the continuous-flow bench-scale system was shown to be a good simulation of the pilot-scale contactor.

## Acknowledgments

The authors would like to thank the following groups that have helped and assisted with the research effort. The universities include the University of Illinois at Urbana-Champaign (Dr. Roger Minear and Dr. Benito Mariñas), the University of Toronto in Canada (Dr. Robert Andrews), and the University of Poitiers in France (Dr. Jean-Philippe Croué). Montgomery Watson has acted as the project managers. A special thanks is extended to the American Water Works Association Research Foundation (AWWARF) and project manager Kenan Ozekin for direction and support.

## References

1. Finch, G. R.; Black, E. K.; Gyürék, L.; Belosevic, M. *Ozone Disinfection of Giardia and* Cryptosporidium; AWWARF and AWWA, Denver, CO, 1994.
2. U.S. Environmental Protection Agency. National Primary Drinking Water Regulations; Disinfectants and Disinfection Byproducts; Proposed Rule, *Fed. Reg.* **1994**, *59* (145): 38668.
3. U.S. Environmental Protection Agency. National Primary Drinking Water Regulations; Disinfectants and Disinfection Byproducts; Final Rule, *Fed. Reg.* **1998**, *63* (241), 69390-69476.
4. U.S. Environmental Protection Agency. National Primary Drinking Water Regulations; Enhanced Surface Water Treatment Requirements; Proposed Rule, *Fed. Reg.* **1994**, *59* (145), 38832-38858.
5. Krasner, S. W.; Glaze, W. H.; Weinberg, H. S.; Daniel, P. A.; Najm, I. N. *Jour. AWWA* **1993**, *86*, 73-81.
6. Siddiqui, M. S.; Amy, G. L. *Jour. AWWA* **1993**, *86*, 63-72.
7. Najm, I. N.; Krasner, S. W. *Jour. AWWA* **1995**, *88*, 106-115.
8. Amy, G. L.; Westerhoff, P.; Minear, R. A.; Song, R. *Formation and Control of Brominated Ozone By-Products; Water Treatment* (90714); AWWARF, Denver, CO, 1997.

298

9.  Westerhoff, P.; Song, R.; Amy, G. L.; Minear, R. A. *Jour. AWWA* **1998**, *89*, 82-94.
10. Song, R.; Westerhoff, P.; Minear, R. A.; Amy, G. L. In *Water Disinfection and Natural Organic Matter: Characterization and Control*; Minear, R. A., Amy, G. L., Ed.; ACS Symposium Series 649; American Chemical Society: Washington, D.C., 1996; pp 322-349.
11. U.S. Environmental Protection Agency. National Primary Drinking Water Regulations; Disinfection; Turbidity, *Giardia lambia*, Viruses, Legionella, and Heterotrophic Bacteria; Final Rule, *Fed. Reg.* **1989**, *54*, 27486 (June 29, 1989).
12. Oppenheimer, J. A.; Aieta, E. M.; Jacangelo, J. G.; Najm, I. CT Requirements for Disinfection of *Cryptosporidium* in Natural Waters. Proc. AWWA Water Qual. Technol. Conf., Denver, CO, 1997.
13. Amy, G. L.; Siddiqui, M. S.; Zhai, W.; Debroux, J.; Odem, W. *Survey of Bromide in Drinking Water and Impacts on DBP Formation; Monitoring and Analysis* (90662); AWWARF, Denver, CO, 1994.
14. Aiken, G. R.; McKnight, D. M.; Thorn, K. A.; Thurman, E. M. *Org. Geochem.* **1992**, *18*, 567-573.
15. Echigo, S.; Minear, R. A.; Yamada, H. Ultra Low Bromate Detection in Drinking Water. Proc. ACS National Meeting, Anaheim, CA (unpublished, 1999).
16. Chin, Y. P.; Aiken, G. R.; O'Loughlin, E. *Environ. Sci. Technol.* **1994**, *28*, 1853-1858.
17. Edzwald, J. K.; Van Benschoten, J. E. Aluminum Coagulation of Natural Organic Matter. Proc. Fourth Int'l Gothenburg Symposium on Chemical Treatment, Madrid, Spain, Oct 1990.

Chapter 19

# Characterization and Comparison of Disinfection By-Products of Four Major Disinfectants

Xiangru Zhang[1], Shinya Echigo[1], Roger A. Minear[1], and Michael J. Plewa[2]

Departments of [1]Civil and Environmental Engineering and [2]Crop Sciences, University of Illinois at Urbana-Champaign, Urbana, IL 61801

Disinfection by-products (DBPs) generated from chlorination, chloramination, ozonation, and chlorine dioxide treatment were characterized and compared. DBPs examined included four trihalomethanes, nine haloacetic acids, four haloacetonitriles, two haloketones, chloropicrin, total organic halogen (TOX), total organic bromine (TOBr), total organic chlorine (TOCl), thirteen aldehydes, and bromate. The contributions of known DBPs to TOX, TOCl and TOBr formed from using different disinfectants were given. The reaction of humic substances with small amount of free chlorine in equilibrium with $NH_2Cl$ constitutes an important pathway for the formation of TOCl during chloramination. The yields of TOBr and total aldehydes produced from using each disinfectant were found to be related to the redox potential corresponding to each disinfectant. This work makes a step toward better decisions about which disinfectant poses the lowest risk to human health.

Many oxidants are being considered as alternatives to chlorine to reduce halogenated organic disinfection by-products (DBPs). The most popular of these alternative disinfectants are ozone, chlorine dioxide and chloramine. While the use of alternatives helps to minimize the trihalomethanes (THMs), other by-products will be produced. Singer et al. (*1*) and Richardson (*2*) have developed excellent summaries of DBPs in drinking water. There have been numerous publications on DBPs in drinking water since 1970s, however, most of the research focused on the characterization of one or two DBP groups resulting from using only one or two

**299**

disinfectants, which limited a comprehensive comparison of DBPs resulting from using different disinfectants. Only Lykins et al. (3) evaluated the DBPs resulting from use of several major disinfectants, but their attention was focused on a disinfectant contact time of 30 min, followed by the addition of chlorine.

Maintenance of a disinfectant residual to control bacterial concentrations throughout the distribution system can cause both increase or decrease in DBP concentrations (4-6). Thus, one focus of this research is to examine the DBPs formed after extended contact time in order to simulate tap water.

In this work, the DBPs resulting from a 5-day contact time with chlorination, chloramination, and chlorine dioxide treatment were characterized and compared. Ozone DBPs were also examined, but after a contact time of 30 min. DBPs examined included four THMs, nine haloacetic acids (HAAs), four haloacetonitriles (HANs), two haloketones (HKs), chloropicrin, total organic halogen (TOX), total organic bromine (TOBr), total organic chlorine (TOCl), 13 aldehydes, and bromate, which covers halogenated and nonhalogenated, organic and inorganic, specific and nonspecific parameters.

## Experimental Methods

Suwannee river fulvic acid (SRFA, International Humic Substances Society) was added to deionized, distilled water to simulate a water of 3.0 mg/L of total organic carbon. Two sets of samples were prepared, one containing 0.100 mg/L of NaBr as $Br^-$, one without $Br^-$.

Chlorination was performed in the model water (with/without $Br^-$) which was buffered with 0.50 mL of 0.25 M phosphate and small amount of HCl solution at pH 7.4. Stock solutions of sodium hypochlorite were prepared by the adsorption of high purity chlorine gas with 1 M NaOH solution, and standardized by the iodometric method (7). Sodium hypochlorite was added to the sample in 250-mL, glass-stoppered bottles at a dose of 4.5 mg/L as $Cl_2$.

Chloramination was performed in the model water (with/without $Br^-$) which was buffered with 0.50 mL of 0.25 M phosphate at pH 7.5. Monochloramine solutions were prepared just before use by reacting of ammonium chloride and sodium hypochlorite solutions in a chlorine-to-ammonia ratio of 0.8 mol/mol (8) to eliminate free chlorine. Monochloramine was added to the sample in 250-mL, glass-stoppered bottles at a dose of 5.0 mg/L as $Cl_2$.

Chlorine dioxide treatment was performed in the model water (with/without $Br^-$) which was buffered with 0.50 mL of 0.25 M phosphate at pH 7.5. Stock chlorine dioxide solution was prepared from the reaction between $H_2SO_4$ and $NaClO_2$. $ClO_2$ gas was purified by bubbling through a saturated solution of $NaClO_2$ prior to absorption in deionized, distilled water. The generated $ClO_2$ stock solution was found to be essentially free of chlorine (i.e., purity of 99% or greater). Concentration of $ClO_2$ in stock solution was measured spectrophotometrically (9). Chlorine dioxide was added to the sample in 250-mL, glass-stoppered bottles at a dose of 6.0 mg/L as $ClO_2$.

All the sample bottles were vigorously shaken for thorough mixing, then stored headspace-free at 25°C in the dark for 5 days. After 5 days, samples were collected headspace-free in 40- or 60-mL glass vials with polypropylene screw caps and Teflon-lined septa. The vials contained 0.2 or 0.3 mL of 0.2 N $Na_2S_2O_3$ solution to quench disinfectant residuals. Samples were stored at 5°C for no more than 2 days prior to analysis. At the time of sample collection, each of the treated waters was analyzed for disinfectant residuals by the DPD method (7). The results showed that after 5 days of contact time, the presence of disinfectant residuals were ensured at the following levels: 0.29 and 0.58 mg/L as $Cl_2$ in the chlorinated, 3.1 and 3.2 mg/L as $Cl_2$ in the chloraminated samples, 0.12 and 0.18 mg/L as $ClO_2$ in the chlorine dioxide treated samples, with and without bromide, respectively.

Ozonation was conducted in a batch reactor (a modified 500-mL graduated cylinder) containing the model water and 2 mM phosphate buffer at pH 7.5. The aqueous stock solution of ozone was injected rapidly below the water surface in the reactor at a dose of 6.0 mg/L. A time series of samples produced from the reactor were analyzed for ozone residual. The result shows that ozone disappeared in about 30 min. After 30 min, samples were collected headspace-free in 40- or 60-mL glass vials with polypropylene screw caps and Teflon-lined septa. Samples were stored at 5°C for no more than 2 days prior to analysis.

Gas chromatography was used for determination of HAAs, THMs, HANs, HKs and chloropicrin by using and slightly modifying EPA Methods 552.2, 551.1 and 501 (10-12). With the modified procedures, more than three series of DBPs were well separated with the same DB-1701, fused silica capillary column (30 m × 0.32 mm i.d., 0.25 μm film thickness) under two different programs. The program for the analysis of HAAs was: solvent, methyl tert-butyl ether; carrier gas, nitrogen; inlet pressure, 0.75 kg/cm$^2$; 35°C for 12 min, ramp to 135°C at 5°C/min, ramp to 220°C at 20°C/min. The program for the analysis of THMs, HANs, HKs and chloropicrin was: solvent, pentane; carrier gas, nitrogen; inlet pressure, 0.30 kg/cm$^2$; 35°C for 18 min, ramp to 145°C at 5°C/min, ramp to 220°C at 20°C/min.

The procedure for the determination of aldehydes included derivatization of aldehydes with o-(2,3,4,5,6-pentafluorobenzyl)-hydroxylamine hydrochloride followed by extraction with hexane and analysis by gas chromatography with an electron capture detector using a DB-5, fused silica capillary column (13).

An ion chromatograph, coupled with an anion column (AS9-HC, Dionex) and a guard column (AG9-HC, Dionex), was used to determine bromate concentration; a carbonate eluent (9 mM $Na_2CO_3$) and a 500 μL injection loop were used.

TOX was analyzed according to Standard Methods 5320B (14). TOCl and TOBr were measured by combining TOX analysis and ion chromatography (IC) protocols where the titration cell of the TOX analyzer is electrode-disabled and, instead of electrochemically titrating halide ions, is used as a collection reservoir for sample subsequently analyzed by IC for Cl⁻ and Br⁻ (15).

# Results and Discussion

*Disinfection By-Products Data*
Table I shows DBP data from the determination of treated samples.

*Bromide versus Non-Bromide Samples*
Compared with bromide-free samples, the concentrations of chlorinated DBPs in the bromide-containing samples decreased while the levels of mixed bromochloro- and bromo- species increased. With bromide in the water treated with chlorine, chloramine or chlorine dioxide, for instance, bromodichloromethane, chlorodibromomethane and bromoform levels increased with concurrent decrease of the concentration of chloroform; bromoacetic acid, bromochloroacetic acid, tribromoacetic acid levels increased with concurrent decrease of the concentration of chloroacetic acid, dichloroacetic acid or trichloroacetic acid; bromochloroacetonitrile or dibromoacetonitrile increased with concurrent decrease of concentration of dichloroacetonitrile. Minear and Bird (*16*) found that increasing bromide concentration, at a given chlorine dose increased the bromine-substituted THMs, particularly bromoform. Cowman and Singer (*17*) pointed that increasing bromide concentration gradually shifted HAA speciation to the mixed bromochloro species to the brominated species. Other chlorination by-products like haloacetonitriles, halopicrins and halonitromethanes were also affected with the same trend (*18, 19*). According to our data, it seems that the findings from chlorination studies can be extended to other disinfection processes like chloramination or chlorine dioxide treatment. This is because bromine produced from oxidizing bromide by disinfectants is much more reactive than chlorine in substitution reactions.

Now that the presence of bromide represents a typical raw water condition, only the bromide-containing samples are discussed in the following unless specified.

*Trihalomethanes*
Figure 1 shows the total amount of THMs from using different disinfectants. No significant THM levels were observed during the ozone and chlorine dioxide treatment. The only species detected in the ozonated sample was bromoform at a sub-ppb level of 0.17 µg/L. The concentrations of chloroform, bromodichloromethane, chlorodibromomethane and bromoform in the chlorine dioxide treated sample were 0.36, 0.76, 1.1, and 0.46 µg/L, respectively, which were so low and near the method detection limits that chlorine dioxide had been thought not to produce THMs (*20, 21*). The highest THM concentrations were present in the chlorinated sample with a total amount of 247 µg/L. The total THM concentration in the chloraminated sample was only 7.6 µg/L. Compared to the chlorination, chloramination decreases THM levels greatly; this was also observed by other researchers (*22*).

Table I. DBP data from using different disinfectants with/without Br⁻ (μg/L)

| DBPs | $NH_2Cl$ w/o[a] Br⁻ | $NH_2Cl$ w[b] Br⁻ | $ClO_2$ w/o Br⁻ | $ClO_2$ w Br⁻ | $Cl_2$ w/o Br⁻ | $Cl_2$ w Br⁻ | $O_3$ w/o Br⁻ | $O_3$ w Br⁻ |
|---|---|---|---|---|---|---|---|---|
| Bromate ion | <2.0[c] | <2.0 | <2.0 | <2.0 | <2.0 | <2.0 | <2.0 | 18.4 |
| Chloroform | 5.40 | 5.13 | 1.69 | 0.36 | 211 | 183 | <0.15 | <0.15 |
| Bromodichloromethane | <0.15 | 1.88 | 0.29 | 0.76 | 0.89 | 57.0 | <0.15 | <0.15 |
| Chlorodibromomethane | <0.15 | 0.47 | <0.15 | 1.06 | <0.15 | 7.37 | <0.15 | <0.15 |
| Bromoform | <0.15 | <0.15 | <0.15 | 0.46 | <0.15 | 0.26 | <0.15 | 0.17 |
| Chloroacetic acid | 2.46 | 2.07 | 0.48 | 1.72 | 2.15 | 1.71 | <0.48 | <0.48 |
| Bromoacetic acid | <0.33 | <0.33 | <0.33 | 0.65 | <0.33 | 2.90 | <0.33 | 1.21 |
| Dichloroacetic acid | 28.5 | 25.2 | 19.3 | 7.30 | 48.4 | 48.9 | <0.50 | <0.50 |
| Trichloroacetic acid | 0.36 | 0.33 | 0.36 | 0.24 | 59.5 | 51.0 | <0.17 | <0.17 |
| Bromochloroacetic acid | <0.33 | 3.91 | <0.33 | 8.89 | <0.33 | 9.60 | <0.33 | <0.33 |
| Bromodichloroacetic acid | <0.33 | <0.33 | <0.33 | <0.33 | <0.33 | <0.33 | <0.33 | <0.33 |
| Dibromoacetic acid | <0.16 | 0.64 | <0.16 | 13.2 | <0.16 | 2.21 | <0.16 | 0.43 |
| Chlorodibromoacetic acid | <0.83 | <0.83 | <0.83 | <0.83 | <0.83 | <0.83 | <0.83 | <0.83 |
| Tribromoacetic acid | <1.6 | <1.6 | <1.6 | 7.00 | <1.6 | 5.35 | <1.6 | <1.6 |
| Trichloroacetonitrile | <0.20 | <0.20 | <0.20 | <0.20 | <0.20 | <0.20 | <0.20 | <0.20 |
| Dichloroacetonitrile | 2.67 | 2.35 | <0.20 | <0.20 | <0.20 | <0.20 | <0.20 | <0.20 |
| Bromochloroacetonitrile | <0.20 | <0.20 | <0.20 | <0.20 | 0.29 | 2.79 | <0.20 | <0.20 |
| Dibromoacetonitrile | <0.20 | <0.20 | <0.20 | <0.20 | <0.20 | 0.37 | <0.20 | <0.20 |
| Dichloropropanone | <0.20 | <0.20 | <0.20 | 0.57 | <0.20 | <0.20 | <0.20 | <0.20 |
| Trichloropropanone | 0.45 | 1.46 | <0.20 | <0.20 | <0.20 | 2.12 | <0.20 | <0.20 |
| Chloropicrin | 1.12 | 1.77 | 0.63 | <0.20 | 3.52 | 4.28 | <0.20 | <0.20 |
| TOX (as Cl) | 116 | 155 | NA[d] | 61 | 568 | 572 | 1.6 | 6.3 |
| TOCl (as Cl) | NA | 143 | NA | 25 | NA | 534 | NA | 0.0 |
| TOBr (as Cl) | NA | 12 | NA | 36 | NA | 38 | NA | 6.3 |
| Formaldehyde | 1.34 | 1.34 | 3.42 | 1.45 | 3.60 | 3.64 | 12.0 | 8.03 |
| Acetaldehyde | 0.65 | 1.45 | 1.84 | 1.06 | 4.51 | 2.16 | 2.77 | 3.93 |
| Propionaldehyde | <1.0 | <1.0 | <1.0 | <1.0 | 1.00 | <1.0 | <1.0 | <1.0 |
| Butyraldehyde | <0.50 | <0.50 | 0.98 | <0.50 | <0.50 | <0.50 | <0.50 | <0.50 |
| Valeraldehyde | 0.69 | 0.62 | <0.50 | <0.50 | 0.66 | 0.70 | <0.50 | <0.50 |
| Hexanal | <1.0 | <1.0 | <1.0 | <1.0 | <1.0 | <1.0 | <1.0 | <1.0 |
| Heptanal | <1.0 | <1.0 | <1.0 | <1.0 | <1.0 | <1.0 | 1.72 | <1.0 |
| Octyl aldehyde | <1.0 | <1.0 | <1.0 | <1.0 | <1.0 | <1.0 | <1.0 | <1.0 |
| Benzaldehyde | <0.50 | <0.50 | <0.50 | <0.50 | <0.50 | <0.50 | <0.50 | <0.50 |
| Nonyl aldehyde | 0.66 | <0.50 | <0.50 | <0.50 | <0.50 | <0.50 | <0.50 | <0.50 |
| Decyl aldehyde | <1.0 | 1.57 | <1.0 | <1.0 | <1.0 | 3.28 | 1.07 | 1.00 |
| Glyoxal | 1.12 | 1.19 | 4.24 | 3.73 | 1.55 | 0.23 | 3.84 | 3.65 |
| Methyl glyoxal | 1.56 | 1.23 | 6.55 | 4.03 | 1.14 | 1.39 | 5.75 | 5.05 |

[a] w/o, without. [b] w, with. [c] <, below minimum reporting levels. [d] NA, not analyzed.

*Haloacetic acids*

Figure 2 shows the total amount of HAAs from using different disinfectants. The highest HAA concentrations were formed during chlorination, which produced trichloroacetic acid and dichloroacetic acid as the predominant species. Bromochloroacetic acid, tribromoacetic acid, bromoacetic acid and chloroacetic acid were also produced to some degree. Of interest is that in the chloraminated sample, dichloroacetic acid was the predominant species with a concentration of 25.2 µg/L much higher than that of trichloroacetic acid, 0.33 µg/L. When comparing the HAA speciation resulting from chlorination and chloramination, Cowman and Singer reported that in the chlorinated samples, the trihalogenated and dihalogenated HAAs are the predominant species, constituting about 95% of the total HAA concentration. In contrast to the speciation in chloraminated waters, the dihalogenated species were the principal species formed, while the trihalogenated species were the minor species (*17*). In the chlorine dioxide treated sample, dihalogenated species were also the principal species with concentrations relatively higher than those of trihalogenated species. The total amount of HAA species in the chlorine dioxide treated sample was close to that in the chloraminated sample but much lower than that in the chlorinated sample. The only species found in the ozonated sample were bromoacetic acid and dibromoacetic acid, with a total amount of 1.6 µg/L.

*Haloacetonitriles, Haloketones and Chloropicrin*

Concentrations of haloacetonitriles were relatively low and formed in only two processes: chlorination and chloramination, which produced 3.2 and 2.4 µg/L total haloacetonitriles, respectively. None of the HANs was detected in the ozone and chlorine dioxide treated samples. Only two haloketones, 1,1,1-trichloropropanone and 1,1-dichloropropanone were analyzed. The sum of concentrations of the two species was 2.1, 1.5 and 0.57 µg/L for chlorine, chloramine and chlorine dioxide treatment, respectively. Neither of the haloketones was detected in the ozonated sample. Chloropicrin was found in the chlorinated and chloraminated samples with concentrations of 4.3 and 1.8 µg/L, respectively. Chloropicrin was absent in the ozone and chlorine dioxide treated samples.

*Total Organic Halogen*

TOX levels formed from disinfectants are shown in Figure 3. The highest concentration of TOX, up to 572 µg/L, was present in the chlorinated sample. This number is quite close to Lykins' (*3*), 540 µg/L, under a similar amount of TOC, similar chlorine dose and contact time. Under our conditions, 191 µg TOX/mg TOC was produced by chlorination at a $Cl_2/C$ mass ratio of 1.5:1 and a 120 h contact time; Reckhow (*23*) reported a value of 191 µg TOX /mg TOC produced by chlorination of fulvic acid at a $Cl_2/C$ ratio of 5:1 for a period of 7 days. This demonstrates that our result is quite consistent with these studies. The concentration of TOX in the chlorine dioxide treated sample was 61 µg/L after a 5 days of contact time, i.e., 20 µg TOX was produced per mg TOC consumed. Lykins reported that with an average non disinfected influent concentration of 25 µg/L, average TOX concentration increased significantly after 30 min of chlorine dioxide contact time to 86 µg/L (*3*),

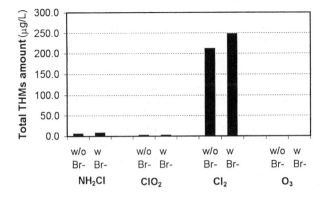

*Figure 1. Total amount of THMs in different samples*

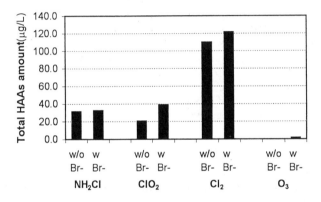

*Figure 2. Total amount of HAAs in different samples*

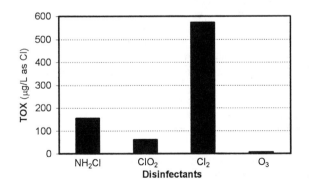

*Figure 3. TOX levels in different samples with bromide*

the net increase was 61 µg/L. TOX concentration in the chloraminated sample was 155 µg/L after a 5 days of contact time, in comparison to Lykins' result of 74 µg/L after 30 min of chloramination of similar raw water (3). TOX concentration in the ozonated sample was only 6.3 µg/L.

The percentage of TOX that can be accounted for by the known DBPs varied with disinfectants (Figure 4). In the chlorinated sample, nearly 50% of TOX could be accounted for by the known DBPs, while in the ozonated sample, only 8.3% of TOX could be represented by the known DBPs. For the chlorine dioxide and chloramine treated sample, this number was 28 and 17%, respectively. Stevens et al. (24), Glaze et al. (25), and Andrews et al. (26) studied the known fraction of TOX during chlorination, ozonation, and chlorine dioxide treatment, respectively. The results are in good agreement with the previously reported values.

In all the samples, total amount of THMs (TTHMs) and total amount of HAAs (THAAs) were the major contributors to TOX, but their contribution amounts varied with disinfectant. In the chlorinated sample, TTHMs and THAAs respectively accounted for 35.6 and 11.9% of TOX, the contribution of TTHMs was greater than that of THAAs. In the chloraminated sample, 3.9 and 10.8% of TOX were accounted for by TTHMs and THAAs, respectively; the contribution amount of TTHMs was less than that of THAAs. In the chlorine dioxide and ozone treated sample, the contribution amount of TTHMs was much lower than that of THAAs. The contribution amount of TTHMs and THAAs were 2.5 and 25.3% respectively for chlorine dioxide treatment, 1.1 and 7.1% respectively for ozonation. Of all the samples, the sum of HANs and HKs and chloropicrin accounted for 0-2.5% of TOX. Their contribution to TOX seems to be negligible compared to THMs and HAAs.

*Total Organic Chlorine*

The concentrations of TOCl in the chlorine, chloramine, chlorine dioxide and ozone treated samples were 534, 143, 25 and 0 µg/L, respectively (Figure 5). The free chlorine dose was highest in the chlorinated sample, thus the highest level of TOCl was produced in it. No free chlorine was present in the ozonated sample, and no TOCl was detected therein. TOCl level produced from chloramine or chlorine dioxide treatment fell between these extremes. It seems that the formation of TOCl is directly associated with free chlorine (HOCl/OCl⁻). Cowman and Singer (17) suggested that formation of HAAs from chloramination may be interpreted as a special case of chlorination with very low chlorine doses, assuming that HAA formation is occurring through the reaction of humic substances with small amounts of free chlorine in equilibrium with monochloramine. If TOCl formation from chloramin-ation is occurring through the reaction of humic substances with small amounts of free chlorine in equilibrium with monochloramine, the validity of the assumption needs to be examined.

The formation of monochloramine has been shown to be an elementary reaction: $HOCl + NH_3$ (aq) $\rightleftarrows NH_2Cl + H_2O$, with a rate constant of $5.1 \times 10^6$ L/mole-sec. The back reaction, decomposition of monochloramine to $NH_3$ and HOCl, is considerably slower, with a rate constant of $3 \times 10^{-5}$ sec$^{-1}$ (27). The ratio of forward

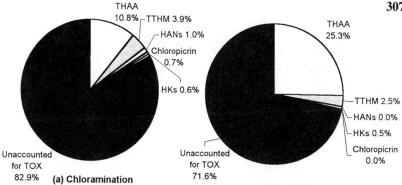

(a) Chloramination

(b) Chlorine dioxide treatment

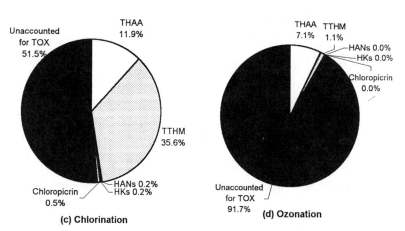

(c) Chlorination

(d) Ozonation

*Figure 4. Contribution of known DBPs to TOX formed from disinfection processes*

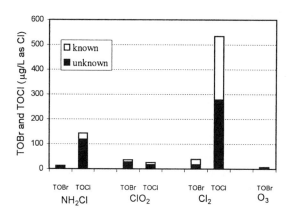

*Figure 5. TOBr and TOCl levels from using different disinfectants*

to back reaction rate constants yields an equilibrium constant (Keq) for monochloramine formation of $1.7 \times 10^{11}$ (28). Since the ammonium sulfate and sodium hypochlorite were added at a concentration of $8.80 \times 10^{-5}$ M and $7.04 \times 10^{-5}$ M as $Cl_2$, respectively, the initial concentration of free chlorine can be obtained as $3.02 \times 10^{-9}$ M as $Cl_2$. In the chlorinated sample, the initial concentration of free chlorine is $6.34 \times 10^{-5}$ M. The concentration of free chlorine in the chloraminated sample is only about 1/21000 of that in the chlorinated sample, while the concentration of TOCl in the chloraminated sample is more than 1/4 of that in the chlorinated sample. Compared to that in chlorination, it seems "unlikely" for the small amount of free chlorine in equilibrium with monochloramine during chloramination to form such an amount of TOCl. However, this small amount of free chlorine can react with SRFA and Br⁻, depleting the free chlorine concentration, and driving the equilibrium with $NH_2Cl$ to produce more free chlorine. For example, a reaction between free chlorine and a precursor site such as phenolic functionality (–ArOH) in SRFA is considered as follows:

$$NH_2Cl + H_2O \underset{k_2}{\overset{k_1}{\rightleftharpoons}} HOCl + NH_3 \text{ (aq)} \qquad HOCl + ArOH \overset{k_3}{\to} ClArOH + H_2O$$

where $k_3$ is about 65 L/mole-sec (29). The kinetics equations are given by:

$$d[HOCl]/dt = k_1[NH_2Cl] - k_2[HOCl][NH_3] - k_3[HOCl][ArOH] \qquad (1)$$

$$d[ClArOH]/dt = k_3[HOCl][ArOH] \qquad (2)$$

Assume an acid ($NH_4^+$) and its conjugate base ($NH_3$) can reach equilibrium instantaneously:

$$[NH_3] = \{ k_a / ( k_a + [H^+] ) \} \times ( [NH_4^+ + NH_3] ) = \alpha_{HOCl} C_{T, NH3} \qquad (3)$$

The initial concentrations of monochloramine and total ammonia can be calculated according to Keq before the addition of SRFA: $[NH_2Cl]_0 = 7.04 \times 10^{-5}$ M, $C_{T, NH3} = 1.76 \times 10^{-5}$ M. The initial concentration of phenolic functionality can be estimated according to the elementary composition of fulvic acid and the percentage of oxygen in this group (28):

$$[ArOH]_0 = 3 \text{ mg/L as C} \times ( \text{44-50\% as O} / \text{40-50\% as C} ) \times 14\% \text{ of O in ArOH}$$
$$= 0.42 \text{ mg/L ArOH as O} = 2.6 \times 10^{-5} \text{ M}$$

Since $k_2$ and $k_3$ are much larger than $k_1$, the initial concentration of free chlorine and concentration of free chlorine are very small. The consumption of $[ArOH]_0$ should be less than 10% based on experimental results. Therefore, the mass balances for the above reactions are given by:

$$[NH_2Cl]_0 = [NH_2Cl] + ( C_{T,OCl} - C_{T,OCl,0} ) + [ClArOH]$$
$$\cong [NH_2Cl] + [ClArOH] \tag{4}$$

$$C_{T, NH3,0} = C_{T, NH3} - ( [NH_2Cl]_0 - [NH_2Cl] ) = C_{T, NH3} - [ClArOH] \tag{5}$$

$$[ArOH]_0 = [ArOH] + [ClArOH] \{ [ArOH] \tag{6}$$

Assume HOCl reacts essentially as fast as it is produced, so that $d[HOCl]/dt = 0$. The steady-state concentration of HOCl is given from Eq. (1) by:

$$[HOCl] = k_1[NH_2Cl]/\{ k_2[NH_3] + k_3[ArOH] \} \tag{7}$$

Substitute Eqs. (3), (4), (5), (6), (7) into Eq. (2):

$$d[ClArOH]/dt = \{ k_1k_3[ArOH]_0 ( [NH_2Cl]_0 - [ClArOH] )\} / \{ k_2\alpha_{NH3}$$
$$\times ( C_{T, NH3,0} + [ClArOH] ) + k_3[ArOH]_0 \} \tag{8}$$

Integrate Eq.(8) to give the concentration of ClArOH as a function of time:

$$\{ k_3[ArOH]_0 + k_2\alpha_{NH3} ( [NH_2Cl]_0 + C_{T, NH3,0} ) \} \ln \{ [NH_2Cl]_0 / ( [NH_2Cl]_0 -$$
$$[ClArOH] ) \} - k_2\alpha_{NH3} [ClArOH] = k_1k_3[ArOH]_0 \times t \tag{9}$$

Substitute the initial concentrations, the rate constants, and the contact time (432,000 sec) into Eq. (9), the concentration of ClArOH can be obtained as $1.05 \times 10^{-6}$ M, or 37 $\mu g/L$ as Cl, which can account for about 26% of TOCl (143 $\mu g/L$) formed during chloramination. Considering of other competing precursors in SRFA and further chlorination of ClArOH, it appears that the reaction of humic substances with small amount of free chlorine in equilibrium with $NH_2Cl$ constitutes an important pathway, if it is not the only one, for the formation of TOCl during chloramination.

The percentage of TOCl that can be accounted for by the known chlorine-containing DBPs varied with disinfectants (Figure 5). In the chlorinated sample, the known chlorine-containing DBPs constituted nearly 50% of TOCl, while in the chloramine and chlorine dioxide treated sample, this number was relatively low—17 and 31%, respectively.

*Total Organic Bromine*

The concentration of TOBr in the chlorine, chlorine dioxide, chloramine and ozone treated samples were 38, 36, 12 and 6.3 $\mu g/L$ as Cl, respectively; or 86, 81, 30 and 14 $\mu g/L$ as Br, respectively (Figure 5, the initial amount of $Br^-$ was 100 $\mu g/L$). TOBr levels in the chlorine and chlorine dioxide treated samples are quite close to each other but are much higher than those in the chloramine and ozone treated samples. The differences of TOBr formation from different disinfectants can be explained by the differences of $HOBr/OBr^-$ formation, which are directly related to the redox potentials (Table II). Since the concentration of HOBr is proportional to the total concentration of HOBr and OBr⁻ at a fixed pH, the discussion is focused on HOBr only.

**Table II. Electrode potentials corresponding to different disinfectants**

| Half-reaction | $E_0$ (v)(30) | Assumption | E (v) |
|---|---|---|---|
| $O_3(g) + 2H^+ + 2e^- \rightleftharpoons O_2 (g) + H_2O$ | +2.07 | | ≅+2.07 |
| $HOCl + H^+ + 2e^- \rightleftharpoons H_2O + Cl^-$ | +1.49 | [Cl⁻]/[HOCl]=0.1 | +1.29 |
| $ClO_2 + e^- \rightleftharpoons ClO_2^-$ | +1.15 | [ClO₂⁻]/[ClO₂]=0.1 | +1.21 |
| $NH_2Cl + H_2O + 2e^- \rightleftharpoons Cl^- + NH_3 + OH^-$ | +0.75 | [Cl⁻]/[NH₂Cl]=0.1 | +1.13 |
| $HOBr + H^+ + 2e^- \rightleftharpoons H_2O + Br^-$ | +1.33 | [Br⁻]/[HOBr]=10 | +1.08 |

Under the experimental conditions, $E_{HOCl/Cl^-}$ and $E_{ClO2/ClO2^-}$ were higher than $E_{HOBr/Br^-}$, HOCl and ClO₂ can readily oxidize the small quantity of Br⁻ to HOBr, which led to the similar amount of TOBr produced during chlorine and chlorine dioxide treatment. $E_{NH2Cl/Cl^-}$ was just a little higher than $E_{HOBr/Br^-}$, NH₂Cl cannot oxidize all the Br⁻ to HOBr; $E_{O3/O2}$ were much higher than $E_{HOBr/Br^-}$, O₃ can oxidize Br⁻ to HOBr, and further to $BrO_2^-$ and $BrO_3^-$, moreover, O₃ can oxidize and diminish the organic-DBP precursors. Accordingly, both disinfectants resulted in smaller amount of HOBr formation, and hence smaller amount of TOBr formation. Further consideration of the pathway/mechanism shows that the oxidation of Br⁻ to HOBr by ClO₂ or NH₂Cl may be through an intermediate — HOCl. Wajon et al. (31) reported that when phenol (a functional group in humic substances) is present, ClO₂ adds to the phenoxyl radical para to the oxygen, and p-benzoquinone is formed with concomitant release of HOCl. Ni et al. (32) confirmed that HOCl is an important intermediate formed from the reaction between ClO₂ and phenolic compounds.

The percentage of TOBr that can be accounted for by the known bromine-containg DBPs also varied with disinfectants. In the chlorinated sample, the known bromine-containing DBPs represented over 60% of TOBr; in comparison to the chlorine dioxide, chloramine and ozone treated samples, only 26, 14 and 8.2% of TOBr can be accounted for, respectively.

*Aldehydes*

In the ozonated samples, formaldehyde, acetaldehyde, glyoxal and methyl glyoxal were the predominant of aldehydes. These four species constituted 87-93% of the total measured aldehydes, which is in good agreement with the finding of Paode et al. (33): 75-99% of the total measured aldehydes was accounted for by the four species. Heptanal and decanal also occurred in the ozonated sample, but their individual concentrations were very low (less than 2 μg/L).

Formaldehyde, acetaldehyde, glyoxal and methyl glyoxal were also found to be present in each of the chlorine, chloramine and chlorine dioxide treated samples, typically at concentrations substantially higher than the other species. Pentanal and decanal were detected at relatively significant concentrations in the chloramine treated sample. In the chlorinated sample, the concentrations of formaldehyde and acetaldehyde were 3.6 and 2.2 µg/L, respectively. Krasner et al. (19) first demonstrated that formaldehyde and acetaldehyde were produced in the chlorinated waters at concentrations of 3.5 and 2.6 µg/L, respectively.

The total measured concentrations of aldehydes produced upon disinfection are shown in Figure 6. Ozonation produced the highest level of total measured aldehydes, chloramination produced the lowest level of total measured aldehydes, chlorine dioxide and chlorine treatment produced levels falling between them. The formation of aldehydes may be primarily due to the process of oxidation of SRFA, which depends on the oxidizing power of oxidants. Under the experimental conditions, the redox potentials corresponding to each disinfectant are also shown in Figure 6. Assuming that most of the formed aldehydes are not transformed to organic acids, the difference in redox potentials may be a good explanation for the difference of aldehyde yields.

Another thing of interest was that during ozone, chlorine or chlorine dioxide treatment, the total aldehyde amounts from bromide-containing samples were somewhat lower than that from bromide-free samples; except for during chloramination, the total aldehyde amount from bromide-containing sample was quite close to that from bromide-free sample (Figure 6). The possible cause for the small decline in the total aldehyde levels in the bromide-containing sample is that bromide consumes some of the disinfectants and results in a small reduction in oxidizing power.

*Bromate*

The concentration of bromate in the ozonated sample was 18.4 µg/L. The concentrations of bromate in the chlorine, chloramine and chlorine dioxide treated samples were below the minimum reporting level.

## Conclusions

Disinfection by-products from chlorination, chloramination, ozonation, and chlorine dioxide treatment with/without bromide were characterized and compared. Chlorination produced the largest amounts of organic halogenated DBPs (indicated by TTHMs, THAAs, TOX, TOCl, TOBr etc.). Chloramination and chlorine dioxide treatment reduced the organic halogenated DBPs greatly compared with chlorination. Ozonation produced the lowest levels of organic halogenated DBPs but the highest levels of aldehydes and bromate.

The contributions of known DBPs to TOX, TOCl and TOBr formed using different disinfectants were in the range of 8.2-60%.

It appears that the reaction of humic substances with small amounts of free chlorine in equilibrium with $NH_2Cl$ constitutes an important pathway, if it is not the only one, for the formation of TOCl during chloramination.

312

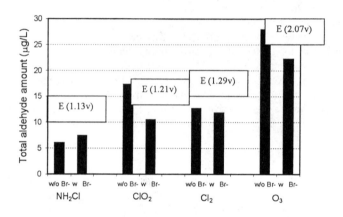

*Figure 6. Total aldehyde amounts from using different disinfectants*

The yields of TOBr and total aldehydes produced from using each disinfectant were found to be directly related to the redox potential corresponding to each disinfectant.

## Acknowledgments

The authors would like to acknowledge the funding for the studies from the AWWA Research Foundation (Grant No. 554), the USEPA (Grant No. R825956-01) and the USGS (Grant No. INT HQ 96-GR02668-2, through the University of Illinois Water Research Center).

## References

1. Singer, P. C. *J. Environ. Eng.* **1994**, *120* (4), 727-744.
2. Richardson, S. D. In *Encyclopedia of Environmental Analysis and Remediation;* John Wiley & Sons, Inc.: New York, NY, 1998; pp 1398-1421.
3. Lykins, B. W. Jr.; Koffskey, W. E.; Patterson, K. S. *J. Environ. Eng.* **1995**, *120* (4), 745-757.
4. Norton, C. C.; LeChevallier, M. W. *J. Am. Water Works Assoc.* **1997**, *89* (7), 66-77.
5. LeBel, G. L.; Benoit, F. M.; Williams, D. T. *Chemosphere*, **1997**, *34* (11), 2301-2317.
6. Carson, M.; Hardy, D. *J. Am. Water Works Assoc.* **1998**, *90* (2), 95-106.
7. *Standard Methods for the Examination of Water and Wastewater*; Eaton, A. D.; Clesceri, L. S.; Greenberg, A. E., Eds.; APHA, AWWA, and WEF: Washington, DC, 1995; pp 436-447.
8. Yoon, J.; Jenson, J. N. In *Disinfection By-Products in Water Treatment: The Chemistry of Their Formation and Control;* Minear, R. A.; Amy, G. L., Eds.; CRC Press, Inc.: Boca Raton, FL, 1996; pp 351-361.
9. Karpel Vel Leitner, N.; Laat, J. D.; Dore, M.; Suty, H. In *Disinfection By-Products in Water Treatment: The Chemistry of Their Formation and Control;* Minear, R. A.; Amy, G. L., Eds.; CRC Press, Inc.: Boca Raton, FL, 1996; pp 393-407.
10. USEPA. *Methods for the Determination of Organic Compounds in Drinking Water*: Supplement 3. EPA600R95131. 1995.
11. USEPA. *Methods for the Determination of Organic Compounds in Drinking Water*: Supplement 2. EPA600R92129. 1992.
12. USEPA. *Methods for the Determination of Organic Compounds in Drinking Water*: Supplement 1. EPA600490020. 1990.
13. Sclimenti, M. J.; Krasner, S. W.; Glaze, W. H.; Weinberg, H. S. *Ozone Disinfection By-Products: Optimization of the PFBHA Derivatization Method for the Analysis of Aldehydes*; AWWA WQTC: San Diego, CA, 1990.

314

14. *Standard Methods for the Examination of Water and Wastewater;* Clesceri, L. S.; Greenberg, A. E.; Trussell, R. R. Eds.; American Public Health Association: Washington, DC, 1989.
15. Echigo, S.; Zhang, X.; Minear, R. A.; Plewa, M. J. In *Division of Environmental Chemistry Preprints of Extended Abstracts*; 217th ACS National Meeting: Anaheim, CA, 1999; Vol. 39 (1), pp 254-256.
16. Minear, R. A.; Bird, J. D. In *Water Chlorination: Environmental Impact and Health Effects;* Jolley, R. L., et al., Eds.; Ann Arbor Science: Ann Arbor, MI, 1980; Vol. 3, pp 151-160.
17. Cowman, G. A.; Singer, P. C. *Environ. Sci. Technol.* **1996**, *30* (1), 16-24.
18. Tribaud, H.; Delaat, J.; Dore, M. *Wat. Res.* **1988**, *22*, 381-390.
19. Krasner, S. W.; McGuire, M. J.; Jacangelo, J. G.; Patania, N. L.; Reagan, K. M.; Aieta, M. E. *J. Am. Water Works Assoc.* **1989**, *81*, 41-53.
20. Werdehoff, K. S.; Singer, P. C. *J. Am. Water Works Assoc.* **1987**, *79* (9), 107-113.
21. Ben Amor, H.; De Laat, J.; Dore, M. *Environ. Technol. Lett.* **1988**, *9*, 1105-1108.
22. Norman, T. S.; Harms, L. L.; Looyenga, R. W. *J. Am. Water Works Assoc.* **1980**, *72* (3), 176-180.
23. Reckhow, D. A.; Singer, C.; Malcolm, R. L. *J. Am. Water Works Assoc.* **1990**, *82*, 173-180.
24. Stevens, A. A.; Moore, L. A.; Slocum, C. J.; Smith, B. L.; Seeger, D. R.; Ireland, J. C. In *Water Chlorination: Chemistry, Environmental Impact and Health Effects*; Lewis Publishers: Chelsea, MI, 1989; Vol. 6, pp 579-604.
25. Glaze, W. H.; Weinberg, H. S.; Cavanagh, J. E. *J. Am. Water Works Assoc.* **1993**, *85* (1), 96-103.
26. Andrews, R. C.; Ferguson, M. J. In *Disinfection By-Products in Water Treatment: The Chemistry of Their Formation and Control;* Minear, R. A.; Amy, G. L., Eds.; CRC Press, Inc.: Boca Raton, FL, 1996; pp 17-55.
27. Morris, J. C. In *Principles and Applications of Water Chemistry*; Faust, S. D.; Hunter, J. V., Eds.; John Wiley: New York, NY, 1967.
28. Snoeyink, V. L.; Jenkins, D. *Water Chemistry*; John Wiley & Sons: New York, NY, 1980.
29. Lee, G. F. In *Principals and Applications of Water Chemistry*; Faust, S. D.; Hunter, J. V., Eds.; John Wiley: New York, NY, 1967.
30. Glaze, W. H. In *Water Quality and Treatment*; American Water Works Association, Ed.; McGraw-Hill, Inc.: New York, NY, 1990; pp 750-751.
31. Wajon, J. E.; Rosenblatt, D. H.; Burrows E. P. *Environ. Sci. Technol.* **1982**, *16*, 396-402.
32. Ni, Y.; Shen, X.; Vanheiningen, A. *J. Wood Chem. Technol.* **1994**, *14* (2), 243-262.
33. Paode, R. D.; Amy G. L.; Krasner, S. W.; Summers, R. S.; Rice, E. W. *J. Am. Water Works Assoc.* **1997**, *89* (6), 79-93.

# DISINFECTION BY-PRODUCT METHODS DEVELOPMENT

Chapter 20

# Ultra-Low Bromate Detection in Drinking Water: A Post-Column Derivatization Method without Anion Suppressors for Bromate Analysis

Shinya Echigo[1], Roger A. Minear[1], and Harumi Yamada[2]

[1]Department of Civil and Environmental Engineering, University of Illinois at Urbana-Champaign, Urbana, IL 61801
[2]Research Center for Environmental Quality Control, Kyoto University, Otsu, Japan

A modified post-column derivatization method for sub-ppb bromate detection, based on the reaction system introduced by Weinberg and Yamada has been developed. The Weinberg and Yamada reaction system converts bromate ion to tribromide ion of high absorbance at 267 nm and requires two additional anion suppressors to supply stable concentrations of derivatizing reagents (i.e., HBr and $HNO_2$). To eliminate this added expense, the same reaction conditions are achieved by simply pumping $H_2SO_4$ and the mixture of NaBr and $NaNO_2$ to the reaction coil without the anion suppressors. The PQL (0.38 µg/L) of this method is comparable to that of a configuration with suppressors. This modification is expected to reduce the cost of the reaction system considerably and to contribute to more reliable monitoring of low bromate concentrations.

## Introduction

Bromate, a probable human carcinogen (*1*), has been gathering public concern as a disinfection by-product (DBP) (*2-5*) and has been regulated at 10 µg/L 1998 in the United States since December of as part of the Information Collection Rule (*6*). This level is above the $10^{-4}$ cancer risk level (*1*) and has been set based on current analytical limitations with conventional ion-chromatographic analysis (*7,8*). Thus, for better risk management and monitoring of bromate, it is important to develop a simple and more sensitive method for this compound.

During the past few years, several novel approaches have been introduced for low level bromate detection (*9-17*) Those methods employ inductively coupled plasma mass spectrometry (9-11), electrospray ion mass spectrometry (12), or post-column derivatizing systems coupled with ion chromatography (7, 13-17), for example. While those methods achieve detection limits below 1.0 µg/L, they also possess some drawbacks. Inductively coupled plasma mass spectrometry and electrospray ion mass spectrometry are not feasible for routine analysis at each water agency due to their initial costs. Thus, in terms of the initial cost, the methods with post-column derivatization and UV detection are more promising. Among those methods, the one which uses a mixture of nitric acid, potassium bromide, and the o-dianisidine is the simplest one (15). However, o-dianisidine used in this method is a probable carcinogen (15). That is, this method requires special handling of the waste from the system. On the other hand, other post-column systems use safer reagents, but those systems are more complicated and require additional cost for anion suppressors for stable delivery of reagents (7) or a longer flow cell path length (15 mm) for better sensitivity (17).

Considering these limitations, we chose the reaction scheme introduced by Weinberg and Yamada (7,14) because of its relative simplicity and laboratory safety considerations, and modified their reaction system in order to achieve the same performance with a simpler configuration.

Their method is based on the reactions as follows (18-21):

$$BrO_3^- + 5Br^- + 6H^+ \xrightarrow{\quad HNO_2 \quad} 3Br_2 + 3H_2O \qquad (1)$$
$$Br_2 + Br^- \xrightarrow{\qquad\qquad} Br_3^- \qquad (2)$$

The bromate concentration is determined by measuring the absorbance at 267 nm since $Br_3^-$ (tribromide ion) has a high absorption coefficient (40900 $M^{-1} \cdot cm^{-1}$ at 267 nm) (21). To achieve these reactions in a post-column system, Weinberg and Yamada developed a system as shown in Figure 1 (7). In their system, two anion suppressors (i.e., cation exchange membranes) were employed to achieve stable bromide and nitrite concentrations under acidic conditions in the process, because the mixture of HBr and $HNO_2$ is not stable (13), and cannot be stored in a reagent feed vessel. However, it is expected that stable bromide and nitrite concentrations under acidic conditions can be achieved by a reaction system without anion suppressors as shown in Figure 2. That is, 3.0 N $H_2SO_4$ and the mixture of 1.0 M NaBr and 0.145 mM $NaNO_2$ are mixed directly in the system with a gradient pump or two flow injection pumps. The configuration with suppressors costs about $4,000 (two suppressors and one flow injection pump), and the configuration without suppressors also costs about $4,000 (two flow injection pumps). Also, a laboratory is less likely to possess more than one anion suppressor in advance than to possess two extra pumps (or an extra gradient pump). Hence, the development of this method expands the choice of configuration for the ion-chromatographic system, which may reduce the total cost for the system.

This chapter consists of three main parts. First, with a discussion on the effect of reagent concentrations and comparison with the performance of a configuration with anion suppressors (Method 1), the performance of direct acid injection configuration (Method 2) is examined. (Note: the conditions for the separation of bromate are different from those by Weinberg and Yamada [7].) Second, Method 2 is also compared with a conductivity method that is almost identical to the USEPA method 300.1. Lastly, some practical aspects of Method 2 (i.e., interference by a natural water matrix) are discussed.

# Experimental

## Reagents

All the reagents used in this study were of analytical grade (Fisher, Pittsburgh, PA), and standard solutions, eluent, and post-column reaction reagents were prepared with Millipore quality water (Millipore, Bedford, MA).

## Analytical Methods

*Method 1 (with Anion Suppressors for in situ Generation of HBr and HNO₃) (7,14)*
A Dionex DX-300 ion chromatograph system with a UV detector (model UVIS-200, Linear) was used (see Figure 1). The ion-chromatographic conditions used for separation of oxyhalides were as follows: analytical column, IonPac® AS9-HC (4 × 250 mm, Dionex, Sunnyvale, CA); guard column IonPac® AG9-HC (4 × 50 mm, Dionex); eluent, 9 mM $Na_2CO_3$; eluent flow rate, 1.0 mL/min; sample loop size, 500 μL. The post-column reaction conditions were as follows: post-column derivatization reagent, 0.5 M NaBr solution containing 0.145 mM of $NO_2^-$ (this mixture is stable if it is not in acidic condition) (7), reagent flow rate, 1.0 mL/min; acidification of the reagent, two anion suppressor (ASRS-Ultra, 4 mm, Dionex) with 1.5 N $H_2SO_4$; acid flow rate, 3.0 mL/min; reaction coil dimensions, 0.25 mm i.d., 2.0 m length; reaction temperature, 68 °C; UV detector path length, 6 mm; UV detector wave length, 267 nm; UV detector signal output full-scale, 1.0 absorbance unit (AU). Though the post-column reaction conditions are similar to those reported (7, 14), some conditions were different from the ones previously reported (7, 14). These changes were made based on the results of preliminary experiments for optimization. The conditions for the separation of bromate were the same as USEPA method 300.1 except for a reduced eluent flow rate. It is also of note that the conductivity detector was disconnected from the reaction system, because preliminary experiments suggested that bypassing the conductivity detector slightly improved the baseline stability due to a simpler hydraulic condition. However, this does not imply the conductivity detector is incompatible with the post column detection system.

*Method 2 (Direct Acid Injection without Anion Suppressors)*
The ion-chromatographic conditions were exactly the same as those given in the above section on Method 1. In Method 2, $H_2SO_4$ was directly mixed with the post-column reaction reagent instead of acidification by way of anion suppressors (see Figures 1 and 2). The post-column reaction conditions for this scheme were as follows: post-column derivatization reagent, 1.0 M NaBr solution containing 0.145

mM of $NO_2^-$; reagent flow rate, 1.0 mL/min; acidification, 3.0 N $H_2SO_4$. The reagent concentrations were determined considering dilution effect and were based on preliminary experiments. The effect of reagent concentrations will be discussed in later sections. These reagents were delivered by a gradient pump (GMP-2, Dionex) with 50/50 delivery. To reduce the effect of rapid heat production by mixing $H_2SO_4$ and the mixture of $NaNO_2$ and NaBr and to enhance the mixing, Teflon® tube (2 mm id, 2 m length) was installed between the first mixing-T and the second mixing-T. The dimensions of the reaction coil and the reaction temperature were the same as for Method 1. For comparison with Method 1, the UV detector path length was chosen as 6 mm. For other experiments, the flow cell path length was changed to 10 mm to enhance the sensitivity of the method. Other detector conditions were UV detector wave length, 267 nm; UV detector signal output full-scale, 1.0 absorbance unit (AU). The conductivity detector was not connected.

*Conductivity Method*
   An analysis with a conductivity detector was also performed following the USEPA Method 300.1 (*8*) with a minor modification (i.e., flow rate was reduced to 1.0 mL/min). The ion-chromatographic conditions were identical to those shown in the above sections. The conditions for the conductivity detector were as follows: detector, PED-2 (Dionex); suppressor, ASRS-Ultra (Dionex); suppression mode, external water mode.

**Test Sample Preparation**

*Standard Solutions*
   Standard solutions were prepared by diluting a 1.0 mg/mL bromate stock solution with Millipore quality water. This stock solution was also prepared in our laboratory from $KBrO_3$ (Fisher, Pittsburgh, PA). Also, a set of spiked samples was prepared with a bottled mineral water to see the effect of the natural water matrix. This bottled water (pH 7.2) was reported by the manufacturer to contain chloride (4.5 mg/L) and bicarbonate (357 mg/L) ions.

*Chlorination and Chloramination*
   Recently, bromate formation was found not only during ozonation but also during chlorination (*22,23*), although the concentration is relatively low. As an example of the application of this method, chlorination and chloramination experiments were performed both with and without natural organic matter (NOM). NaOCl stock solution was prepared by dissolving chlorine gas into 1 N NaOH solution. The experimental conditions were as follows: pH, 7.4 (adjusted with concentrated HCl); bicarbonate 2 mM; bromide ion, 400 µg/L; incubation time, 6 days; NOM source, Suwannee River Fulvic Acid (International Humic Substance Society, St. Paul, MN); NOM addition, 0 or 3 mg/L as C; chlorine dose, 5.0 mg/L as $Cl_2$; chloramine dose, 5.0 mg/L as $Cl_2$ without free chlorine (*24*); chlorine-to-ammonia ratio, 0.8 mol/mol (4 mg $Cl_2$/mg $NH_3$-N); temperature, 25 °C. Samples were prepared in 100-ml, glass-stopped bottles. After mixing all the reagents by

shaking vigorously, the samples were stored in the dark headspace-free. The bottles were shaken every 24 hours during the experiment.

# Results and Discussion

### Comparison between Method 1 and Method 2

Table I compares the practical quantification limits (PQL) of the two methods. The PQL of Method 1 was 0.30 μg/L, and that of Method 2 was 0.38 μg/L. The PQL of Method 2 is slightly higher than that of Method 1 (one possible reason for this is higher background absorbance due to the presence of sulfate [by 3.4 mAU]). However, both of the PQLs are below 0.5 μg/L (i.e., $10^{-5}$ cancer risk level) (*1*), and Method 2 is considered to be comparable to Method 1.

**Table I. Comparison of Quantification Limits (with a 6 mm flow cell).**

|  | Standard concentration (μg/L) | SD (μg/L) | MDL[a] (μg/L) | PQL (μg/L)[b] | Peak/noise ratio |
|---|---|---|---|---|---|
| Method 1 | 1.0 | 0.048 | 0.15 | 0.30 | 7.1 |
| Method 2 | 1.0 | 0.063 | 0.19 | 0.38 | 7.3 |

[a]MDL (method detection limit) = SD × $t_{s,99\%}$ =3.14 × SD for n=7.

[b]PQL (practical quantification limit) = 2 × MDL.

Figure 3 shows chromatograms of standard solutions by Method 2. Distinct peaks were obtained for both 1.0 μg/L and 0.4 μg/L. Also, with a 10 mm flow cell, Method 2 achieved a PQL of 0. 26 μg/L (peak/noise ratio for 1.0 μg/L = 9.8), while Method 1 achieved 0.22 μg/L with a 10 mm flow cell (peak/noise ratio for 1.0 μg/L = 10.3).

### Effects of Reagent Concentrations

*Effect of NaBr Concentration*
Figure 4 shows the effect of NaBr concentration. For 0.5 M, the peak area decreased about 20% compared with those for 1.0 M and 2.0 M. For 2.0 M, however, the baseline became noisier compared with that for 1.0 M. Thus, the optimum concentration was considered to be around 1.0 M. The optimal range of NaBr for Method 2 was higher that for Method 1 (0.5 M).

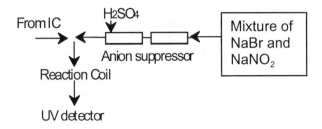

Figure 1. *Schematic diagram of Weinberg and Yamada method (7).*

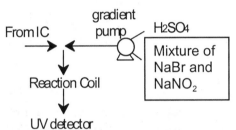

Figure 2. *Schematic diagram of a modified post-column reaction system for bromate analysis.*

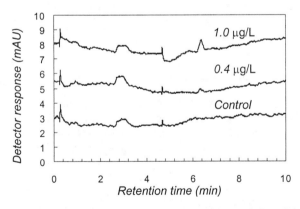

Figure 3. *Chromatograms of bromate ion by Method 2. Cell path length, 6 mm; NaBr, 1.0 M; NaNO₂, 0.145 m; retention time of bromate, 6.3 min.*

*Effect of NaNO₂ Concentration*

Figure 5 shows the effect of nitrite ion concentration for 1.0 M NaBr. For NaNO₂ concentrations between 0.0725 mM to 0.295 mM, no significant change was observed in terms of peak area. For 0.0363 mM, the peak area slightly decreased (about 10%), and the baseline became noisy for 0.590 mM (this data point is not shown in the figure). Hence, this suggests that NaNO₂ concentration has to be maintained from 0.1 to 0.3 mM. This tendency (i.e., constant response in a certain range of concentration) was also observed for Method 1.

*Effect of H₂SO₄ Concentration*

Figure 6 shows the effect of $H_2SO_4$ concentration with 1.0 M NaBr and 0.145 mM NaNO₂. It was found that increasing $H_2SO_4$ concentration increases the peak area in the range between 0.01 N and 3.0 N. A higher concentration might be applicable, but 3.0 N of $H_2SO_4$ appears to be close to a response plateau.

With the above results, the optimal concentrations of NaBr, NaNO₂, and $H_2SO_4$ for the modified system with a 10 mm flow cell were determined as 1.0 M, 0.145 mM (10 mg/L as NO₂), and 3.0 N (1.5 M), respectively. Unless otherwise noted, these conditions were used with a 10 mm cell in the studies discussed below.

## Standard Curves for Method 2

Figure 7 shows the relationship between peak area and standard concentration in the range between 0.2 and 2.0 µg/L. A linear relationship was obtained. In addition, Figure 8 shows the standard curve for higher concentration (between 1.25 to 50 µg/L). Although a bromate concentration higher than 2.0 µg/L can be determined by a conventional conductivity analysis, this linear relationship shows that Method 2 also covers a part of the working range of conductivity analysis sufficiently, if the concentration is less than 50 µg/L (i.e., Method 2 is applicable to assess whether a water meets the current regulation [6]).

## Comparison with a Conductivity Method

Figures 9 (conductivity) and 10 (Method 2) compare the peak signals for bromate by Method 2 and a conductivity method based on USEPA method 300.1. In a bottled water, though the peaks for 1.25 µg/L of bromate can been seen by both methods, it is obvious that Method 2 achieves better baseline and peak response. The recovery was calculated as 98.3 % (n=4) for Method 2 compared with standard solution. It is of note that for conductivity analysis the sample was treated by On-Guard Ag/H pretreatment cartridges (Dionex), while for Method 2 the sample was pretreated only by On-Guard H cartridge. Method 2 is not interfered with by the chloride ion. Method 2 was also tested with both On-Guard Ag and H pretreatment cartridges, but the recovery did not change significantly with a recovery 97.1 %

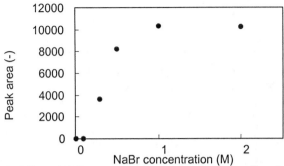

Figure 4. Effect of NaBr concentration on the peak area. Standard concentration, 1.0 μg/L; cell path length, 10 mm; NaNO₂, 0.145 mM; H₂SO₄, 3.0 N.

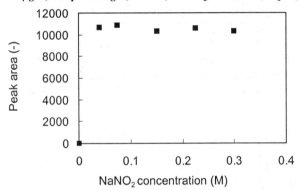

Figure 5. Effect of NaNO₂ concentration on the peak area. Standard concentration, 1.0 μg/L; cell path length, 10 mm; NaBr, 1.0 M; H₂SO₄, 3.0 N.

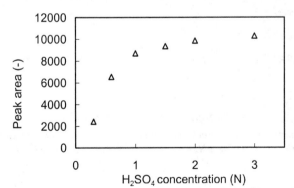

Figure 6. Effect of H₂SO₄ concentration on the peak area. Standard concentration, 1.0 μg/L; cell path length, 10 mm; NaBr, 1.0 M; NaNO₂, 0.145 mM.

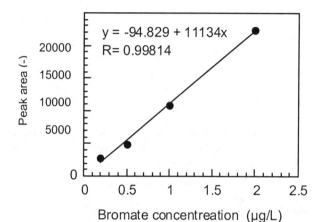

*Figure 7. Relationship between bromate concentration and peak area (0.2–2.0 µg/L). Cell path length, 10 mm; NaBr, 1.0 M; NaNO₂, 0.145 mM; H₂SO₄, 3.0 N.*

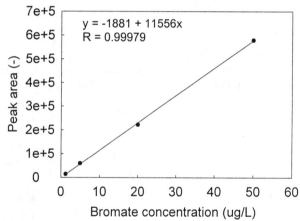

*Figure 8. Relationship between bromate concentrtion and peak area (1.25–50.0 µg/L). Cell path length, 10 mm; NaBr, 1.0 M; NaNO₂, 0.145 mM; H₂SO₄, 3.0 N.*

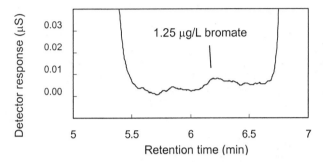

*Figure 9. Chromatograms of bromate in a natural water matrix by conductivity detection. Pretreatment, On-Guard Ag and H pretreatment cartridges.*

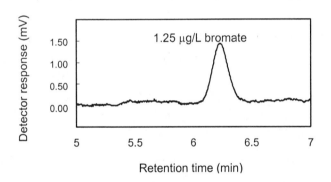

*Figure 10. Chromatograms of bromate in a natural water matrix by the modified post-column method. Pretreatment, On-Guard H pretreatment cartridge.*

(n=4). The reason for the use of the On-Guard H cartridge is explained in the next section.

### Effect of Bicarbonate

Figure 11 shows the effect of the pretreatment by an On-Guard H pretreatment for a sample that contains 5 mM of sodium bicarbonate. In this figure, it can be seen that the presence of bicarbonate deformed the peak shape. Also, excess concentration of carbonate or bicarbonate can change the elution time of bromate, since excess carbonate or bicarbonate lead to a localized change of eluent concentration. Thus, it is recommended that carbonate and bicarbonate be removed from samples with high alkalinity by an On-Guard H pretreatment cartridge, even for the analysis by post-column reaction system.

### Bromate Formation during Chlorination and Chloramination

Table II shows the results for the bromate formation during chlorination and chloramination. During chlorination, bromate was formed both with and without (23) NOM. Also, it was found that the presence of NOM suppresses the formation of bromate during chlorination, though some bromate is still produced with NOM. However, no significant bromate formation was observed in chloramination.

**Table II. Bromate Formation during Chlorination and Chloramination.**

|  | Without NOM addition[a] | With NOM addition (3.0 mg/L) |
|---|---|---|
| Chlorination (5.0 mg/L as $Cl_2$) | 1.56[b] | 0.41[b] |
| Chloramination (5.0 mg/L as $Cl_2$) | trace[c] | trace[c] |

[a] In 2 mM carbonate buffer solution.

[b] Units are μg/L.

[c] Small peak below PQL of Method 2.

## Conclusion

In this chapter, a modified post-column derivatization method for sub-ppb bromate detection based on the reaction system introduced by Weinberg and Yamada (7) has been developed. This system does not require anion suppressors or highly

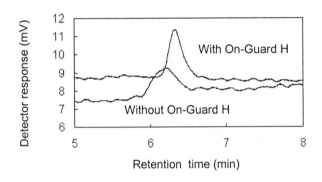

*Figure 11. Effect of bicarbonate ion for the modified method. Concentrations in the sample: bicarbonate, 5 mM; bromate, 2.5 μg/L.*

toxic chemicals for its operation, while achieving a PQL below 0.4 µg/L without interference by chloride ion. For real samples, it is recommended to employ an On-Guard H cartridge to remove carbonate to obtain better peak shape. The PQL of this method is comparable to that by a configuration with suppressors. The successful direct addition of $H_2SO_4$ into the mixture of NaBr, and $NaNO_2$ without the anion suppressors has shown the potential to achieve a simple, cheap, and safe system for low level bromate quantification.

## Acknowledgments

This study was funded by the American Water Works Research Foundation (Grants 554 and 493), and the United States Environmental Protection Agency (Grant R825956-01) and the United States Geological Survey (Grant INT AQ 96-6R02668-2 through the University of Illinois Water Research Center). The authors express their deepest appreciation to Dr. Peter Jackson of Dionex Corp (Sunnyvale, CA) for supplying the extra anion suppressors.

## References

1.  *IARC Monographs on the Evaluation of Carcinogenic Risk to Humans*; IARC Press: Lyon, France, 1986; Vol. 40, p 207.
2.  von Gunten, U.; Hoigné, J. *Environ. Sci. Technol.* **1994**, *28*, 1234-1242.
3.  Ozekin, K.; Westerhoff, P.; Amy, G. L.; Siddiqui, M. *J. Environ. Engin.* **1998**, *124*, 456-461.
4.  Haag, W. R.; Hoigné, J. *Environ. Sci. Technol.* **1983**, *17*, 261-267.
5.  Gordon, G.; Emmert, G.; Bubnis, B. Bromate Ion Formation in Water When Chlorine Dioxide is Photolysed in the Presence of Bromide Ion; *Proceedings, AWWA Water Quality Technology Conference*, San Diego, CA, **1996**, 1985-1992.
6.  *Fed. Reg.* **1998**, *63* (241), 69389-69476.
7.  Weinberg, H.; Yamada, H. *Anal. Chem.* **1998**, *70*, 1-6.
8.  U.S. Environmental Protection Agency *USEPA Method 300.1, Revision 1.0: Methods for the Determination of Inorganic Anions in Drinking Water by Ion Chromatography*; USEPA: Cincinnati, OH, 1997.
9.  Yamanaka, M.; Sakai, T.; Kumagai, H.; Inoue, Y. *J. Chromatography A* **1997**, *789*, 259-265.
10. Creed, J. T.; Magnuson, M. L.; Pfaff, J. D.; Brockhoff, C. *J. Chromatography A* **1997**, *753*, 261-267.
11. Nowak, M.; Seubert, A. *Anal. Chim. Acta* **1998**, *359*, 193-204.
12. Charles, L.; Pepin, D.; Casetta, B. *Anal. Chem.* **1996**, *68*, 2554-2558.
13. Inoue, Y.; Sakai, T.; Kumagai, H.; Hanaoka, Y. *Anal. Chim. Acta* **1997**, *346*, 299-350.
14. Weinberg, H.; Yamada, H.; Joyce, R. *J. Chromatography A* **1998**, *804*, 137-142.

15. Wagner, H. P.; Alig, A. E.; Pepich, B. P.; Frebis, C. P.; Hautman, D. P.; Munch, D. A Study of Ion Chromatographic Methods for Trace Level Bromate Analysis in Drinking Water Comparing the Selective Anion Concentration (SAC) Method, U.S. EPA 300.1, and a Post-Column Reagent Procedure; *Proceedings, AWWA Water Quality Technology Conference*, San Diego, CA, **1996**, p 4D-1.

16. Gordon, G.; Bubnis, B. P.; Sweetin, D.; Kuo, C. Y. *Ozone. Sci. Engin.* **1994**, *16*, 79-89.

17. Walters, B. D.; Gordon, G.; Bubnis, B. *Anal. Chem.* **1997**, *69*, 4275-4277.

18. Hunter, G.; Goldspink, A. A. *Analyst* **1954**, *79*, 467-478.

19. Faniele, G. *Gazzeta* **1960**, *90*, 1585-1596.

20. Chiu, G.; Eubanks, R. D. *Mikrochim. Acta* **1989**, *2*, 145-148.

21. Wang, T. X.; Kelley, M. D.; Cooper, J. N.; Beckwith, R. C.; Margerum, D. W. *Inorg. Chem.* **1994**, *33*, 5872-5878.

22. Macalady, D. L.; Carpenter, J. H., Moore, C. A. *Science* **1977**, *195*, 4284-4286.

23. Yamada, H.; Sulaiman, M.; Matsui, S. The Formation of Bromate Ion as By-Products During Chlorination; *Proceedings, International Ozone Association Conference*, Sydney, Australia, 1996, pp I-100 – I-105.

24. Yoon J.; Jenson J. N. *Disinfection By-Products in Water Treatment: The Chemistry of Their Formation and Control*; Minear, R. A.; Amy, G. L., Eds.; CRC Press Inc.: Boca Raton, FL, 1996, pp 351-361.

Chapter 21

# Differentiation of Total Organic Brominated and Chlorinated Compounds in Total Organic Halide Measurement: A New Approach with an Ion-Chromatographic Technique

Shinya Echigo[1], Xiangru Zhang[1], Roger A. Minear[1], and Michael J. Plewa[2]

Departments of [1]Civil and Environmental Engineering and [2]Crop Science, University of Illinois at Urbana-Champaign, Urbana, IL 61801

A new method to differentiate total organic chlorinated compounds (TOCl) and total organic brominated compounds (TOBr) has been developed. In this method, HBr and HCl, which are contained in the off-gas from the TOX combustion furnace and are equivalent to TOBr and TOCl in samples, are collected in water with a bubble diffuser instead of being titrated in a micro coulometric cell. Then the concentrations of bromide and chloride ions are determined by ion chromatography. For standard compounds such as 2,4,6-trichlorophenol (100 μg/L as Cl), recoveries of more than 75% were obtained from the full process, including carbon adsorption. Interference with chloride ion quantification by carbon dioxide gas, an auxiliary gas for the combustion furnace, was a barrier to earlier attempts of this method. This interference can be eliminated without loss of HBr and HCl by sparging the solutions with nitrogen gas. These differentiated values are expected to be used as new bulk indices of chlorinated and brominated disinfection by-products.

## Introduction

Despite extensive efforts to identify unknown disinfection by-products (DBPs), a significant fraction of DBPs is still unidentified mainly due to the complexity and diversity of natural organic matter (NOM) in source water which reacts with disinfectants to form organic DBPs during drinking water treatment and distribution (*1*). However, halogenated DBPs have been considered to be one major contributor to human health risk of drinking water in addition to oxidation DBPs, such as aldehydes (*1*). Thus, in addition to determining the concentrations of individual compounds, total organic halide (TOX) measurements have played an important role in estimating the extent of total halogenated DBPs and their risks (*2, 3*).

Chlorinated compounds, however, are not the only group of halogenated organic compounds which are formed during disinfection. In the presence of bromide or iodide ions in source waters, not only chlorinated organic compounds (TOCl) but also brominated (TOBr) and iodinated organic compounds (TOI) are formed during chlorination (*4-6*). Though each of these three groups of DBPs contains the compounds which may have carcinogenic potential, it has been suggested that brominated species are more toxic than chlorinated ones (*7-9*). Thus, it is expected that better understanding of the formation characteristics of DBPs and better evaluation of the toxicity of drinking waters will be achieved, if those fractions are separately evaluated. Iodinated DBPs are generally rare.

In a conventional TOX measurement, unfortunately, these three fractions (i.e., chlorinated, brominated, and iodinated compounds) cannot be differentiated, since the silver coulometric titration cell is insensitive to the difference between bromide, chloride, and iodide ions (*10*). Thus, it is necessary to develop an alternative detection component to evaluate the three halogenated DBP fractions separately.

One recent approach successfully differentiated extractable organic chlorine (EOCl), extractable organic bromine (EOBr), and extractable organic iodine (EOI) fractions in a sediment sample by neutron activation analysis (*11*). While the same principle would be applicable to aqueous samples, the instrumental facility that is required for this analysis is not available at each water agency for the monitoring of halogenated DBPs, in part due to its cost. That is, the use of this method is limited only to research applications. For routine monitoring application, the system has to be simple and inexpensive.

The problem with the conventional method is the differentiation of specific halide ions. All the halogen atoms on organic DBPs are converted to chloride, bromide, and iodide ions correspondingly during combustion. The simplest way to differentiate these three ions is ion chromatography.

Based on the above consideration, this study develops a method to differentiate total organic chlorinated compounds (TOCl) and total organic brominated compounds (TOBr). This new method employs an ion chromatograph system to quantify bromide and chloride separately, instead of silver coulometric titration. We focused on only the differentiation of TOBr and TOCl, but the same principle is applicable to TOI differentiation.

This chapter consists of two main parts. In the first part, the method development is introduced with the results of preliminary experiments. In the second part, chlorination, chloramination, and chlorine dioxide treatments were conducted with a model raw water in the presence of bromide ion, and the distributions of TOCl and TOBr compared, as an example of the application of this method.

# Experimental

### Reagents

All the reagents used in this study were of ACS-reagent grade and purchased from Fisher (Pittsburgh, PA) unless otherwise noted. Bromoform, dichloroacetic acid (Aldrich, Milwaukee, WI), and dibromoacetic acid (Aldrich) and 2,4,6-trichlorophenol (Aldrich) were selected as test compounds. Standard solutions of these compounds were freshly prepared on each day of each experiment. These standards, eluent for ion chromatography, and other dilution of solutions were prepared with Millipore quality water (Millipore, Bedford, MA).

### Procedure

The analytical procedure for this new TOCl and TOBr differentiation method consists of three main steps. First, the target compounds are adsorbed onto granular activated carbon with an adsorption module (AD-3, Dohrmann, Cincinnati, OH) (*10*). Then the carbon is combusted in a conventional furnace for TOX measurement with a single boat inlet module with collection of halide ions from the off gas adsorbed in the water in the test tube. The last step is ion-chromatographic analysis. The details of the latter two steps are given below.

*Combustion Component*

The furnace (S-300, Dohrmann, Cincinnati, OH) outlet is connected to a test tube instead of the titration cell so that HCl and HBr in the exit gas from the furnace is trapped in 10 mL of Millipore quality water in the test tube through a bubble diffuser (see Figure 1). Oxygen and carbon dioxide were used as reactant and auxiliary gas, respectively (*10*). A slightly lower gas flow rate (125 mL/min) than the 150 mL/min flow rate suggested in the operation manual was used (*12*) to minimize the splashing of the sample solution during the absorption step. Furnace temperature was controlled at 800 °C during operation. Also, preliminary results suggested that 20 minutes is sufficient for the combustion of each sample. After combustion and absorption, the gas transfer line was flushed with 3 mL of Millipore quality water to remove condensed bromide and chloride on the gas exit tube, and this water was added to the 10 mL of distilled water used for absorption. Then, the volume of water in the test tube was adjusted to 15 mL before ion-chromatographic analysis.

*Detection Component*

The ion chromatograph used was an ordinary system (DX-300, Dionex, Sunnyvale, CA) which can differentiate between chloride ion and bromide ion with a conductivity detector (PED-2, Dionex). In this study, an IonPac® AS9-HC column (Dionex) was employed as the analytical column, which was protected by an IonPac®

AG9-HC column (Dionex). As the eluent, 9 mM sodium carbonate was used. For suppression, an ASRS-Ultra suppressor (Dionex) was used with an external water mode configuration. Before the sample injection, each sample was sparged with nitrogen gas for 5 min at 5 psi to remove excess of carbon dioxide.

## Chlorination, Chloramination, and Chlorine Dioxide Treatment

*Sample Preparation*

To demonstrate the application of the above method, a simulated water sample was treated with three different disinfectants: chlorination, chloramination, and chlorine dioxide. The doses of disinfectants were 4.5 mg/L as $Cl_2$, 5.0 mg/L as $Cl_2$, and 6.0 mg/L as $Cl_2$, respectively. Stock solutions of sodium hypochlorite were prepared by the adsorption of high purity chlorine gas with 1 N NaOH solution. The monochloramine solution was prepared by reacting ammonium chloride and sodium hypochlorite solutions in a chlorine-to-ammonia ratio of 0.8 mol/mol to eliminate the potential for free chlorine (*13*). The stock chlorine dioxide solution was prepared from the reaction between $H_2SO_4$ and $NaClO_2$ (*14*). The model water consists of sodium bromide (100 µg/L as Br⁻), Suwannee River fulvic acid (3.0 mg/L as dissolved organic carbon) (International Humic Substance Society, St. Paul, MN), phosphate buffer (0.5 mM, adjusted to pH7.5). The contact time of the disinfectant with the model water was 5 days, during which period the samples were stored in 250–mL glass-stopped bottles in the dark at ambient temperature (25°C) (*14*). Conventional TOX analysis was also conducted with a DX-20 TOX analyzer (Dohrmann) to confirm the results of analysis.

*Analysis of Trihalomethanes (THMs) and Haloacetic Acids (HAAs)*

HAA and THM concentrations were also determined for the samples described in the above section by using and slightly modifying EPA Methods 552.2, 551.1 and 501 (i.e., gas chromatography with an electron capture detector ) (*14-17*). DB-1701, fused silica capillary column (30 m length, 0.32 mm i.d., 0.25 µm film thickness) was employed under two different programs. The program for the analysis of HAAs was: solvent, methyl *tert*-butyl ether; carrier gas, nitrogen; inlet pressure, 0.75 kg/cm²; 35°C for 12 min, ramp to 135°C at 5°C/min, ramp to 220°C at 20°C/min. The program for the analysis of THMs, and HANs was: solvent, pentane; carrier gas, nitrogen; inlet pressure, 0.30 kg/cm²; 35°C for 18 min, ramp to 145°C at 5°C/min, ramp to 220°C at 20°C/min.

## Results and Discussion

### Method Development

*Interference by Carbon Dioxide*

Preliminary experiments were conducted under various conditions. One important finding from these experiments is that an unknown peak (see Figure 2), which is suspected to be carbon dioxide (the auxiliary gas), overlaps the chloride ion peak (retention time 7.8 min). The interference by this peak for the quantification of chloride (i.e., TOCl) is removed by sparging the sample solution with nitrogen gas for 5 minutes (see Figure 3). The loss of chloride and bromide ions during this pre-treatment is estimated to be less than 5% for the standard solutions (100 μg/L as Cl, n=4).

Table I shows the recoveries for standard compounds which were directly injected to the sample boat of the TOX analyzer to assess the adsorption efficiency of HCl and HBr in the off-gas. Compared with the recovery in conventional analysis (e.g., 92% for 2,4,6-trichlorophenol), the recoveries by this TOCl and TOBr separation method was considered to be sufficient for our purpose.

**Table I. Recoveries of TOBr and TOCl without Carbon Adsorption.**

| Compound[*] | Recovery (%) | Standard Deviation (% for n=4) |
|---|---|---|
| dichloroacetic acid | 88.0 | 4.0 |
| dibromoacetic acid | 84.5 | 3.7 |
| bromoform | 90.2 | 3.1 |
| 2,4,6-trichlorophenol | 91.6 | 2.8 |

[*] The amount of standard compounds injected corresponds to that in 100 mL of standard solutions (100 μg/L).

*Effects of Heater Tape and Transfer Line Flushing*

For the analysis of real samples (i.e., the samples processed by adsorption onto activated carbon), the recoveries were greatly improved with the addition of heater tape to the transfer line and flushing it with water. This is because heater tape and flushing prevent the condensation of water onto the tube walls with subsequent loss of chloride and bromide ions. Greater condensation was observed for adsorbed samples because the carbon column contains more water than directly injected concentrated standard solutions. For example, with the heater tape and flushing, the recoveries for dibromoacetic acid and dichloroacetic acid were improved from 23.3% to 76.4% and from 24.1% to 83.6%, respectively (see Figure 4).

The recoveries of standards with carbon adsorption were summarized in Table II. From these results, the recoveries for TOCl and TOBr were determined as 84 and 80% on average, respectively.

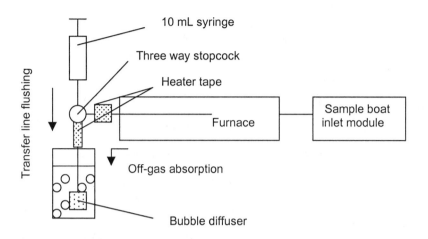

*Figure 1. Schematic diagram of TOBr and TOCl absorption system.*

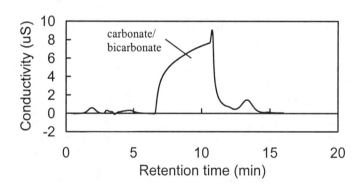

*Figure 2. Interference of ion-chromatographic chloride detection by carbon dioxide.*

**Table II. Recoveries of TOBr and TOCl with Adsorption Process.**

| Compound* | Recovery (%)# | Standard Deviation (% for n=4) |
|---|---|---|
| dichloroacetic acid | 83.6 | 6.8 |
| dibromoacetic acid | 76.4 | 7.2 |
| bromoform | 82.6 | 5.4 |
| 2,4,6-trichlorophenol | 84.0 | 5.9 |

* Sample volume, 100 mL; standard solutions, 100 μg/L.

# All the analyses were performed with heater tape and flow train flushing. Recoveries were calculated

*Working Range*

The working concentration of TOBr and TOCl depends on the sample volume. For example, if the sample volume is 100 mL, TOCl and TOBr levels higher than 20 μg/L as Cl or Br are easily determined with a standard deviation less than 14% (n=4). For, lower concentrations, larger sample volumes are recommended for reliable ion-chromatographic analysis.

## Comparison of TOCl/TOBr Distributions During Chlorination, Chloramination, and Chlorine Dioxide Treatment

Table III summarizes the concentrations and distributions of TOCl and TOBr formed during chlorination, chloramination, and chlorine dioxide treatment of a model water containing Suwannee River fulvic acid and bromide ion. The TOX levels after chlorination, chloramination, and chlorine dioxide treatment were 583, 155, and 61 μg/L, respectively. The corresponding percent of TOBr in the TOX was found to be 6.7, 7.8, and 59.0 (Table III and Figure 5). In Table III, TOCl and TOBr values were calculated based on chloride ion and bromide ion concentration recovered by ion chromatograph and the recoveries determined in the previous section. The TOX values measured by a conventional method and the ones by our method matched well within an error of 9%. Although these errors are sufficiently small, the fractions of organic chlorine and organic bromine in TOX can be also defined based on the ratio of TOBr and TOCl with the TOX values measured by a conventional method as: $TOX_{Cl} = TOX \times TOCl / (TOCl + TOBr)$ and $TOX_{Br} = TOX \times TOBr / (TOCl + TOBr)$, respectively. Table IV shows the distribution of organic chlorine and organic bromine if the latter definition is applied.

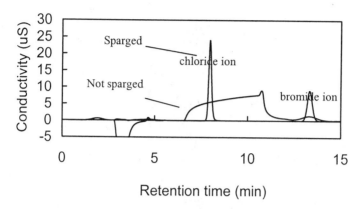

*Figure 3. Effect of nitrogen sparging on ion-chromatographic analysis for bromide and chloride in TOCl/TOBr differentiation.*

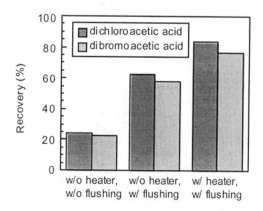

*Figure 4. Effect of heater tape and transfer line flushing on recoveries of TOCl and TOBr.*

338

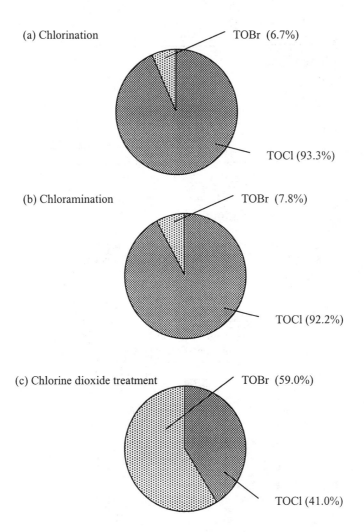

*Figure 5. Distributions of TOCl and TOBr formed during (a) chlorination, (b) chloramination, and (c) chlorine dioxide treatment.*

Table III. Comparison between Conventional TOX Analysis and the
TOCl/TOBr Differentiation Method.

| Disinfectant | Dose (mg/L as Cl₂) | TOX* (μg/L as Cl) | TOCl# (μg/L as Cl) | TOBr# (μg/L as Cl) | TOCl# + TOBr# (μg/L) |
|---|---|---|---|---|---|
| Chlorine | 4.5 | 583 | 511 | 36 | 547 |
| Chloramine | 5.0 | 155 | 131 | 11 | 142 |
| Chlorine dioxide | 6.0 | 61 | 23 | 33 | 56 |

*Determined by conventional TOX analysis (10).

#The subscript m indicates the value was calculated by the peak area of chloride or bromide ion with the recoveries (84% for chloride and 80% bromide).

Table IV. Distribution of Organic Chlorine and Organic Bromine during
Chlorination, Chloramination, and Chlorine Dioxide Treatment.

| Disinfectant | Dose (mg/L as Cl₂) | TOX (μg/L as Cl) | TOX$_{Cl}$ (μg/L as Cl) | TOX$_{Br}$ (μg/L as Cl) |
|---|---|---|---|---|
| Chlorine | 4.5 | 583 | 544 | 39 |
| Chloramine | 5.0 | 155 | 143 | 12 |
| Chlorine dioxide | 6.0 | 61 | 25 | 36 |

It is of note that the percentage of TOBr for chlorine dioxide treatment was the highest of the three, and the TOBr level formed was larger than that during chloramination, although total TOX (TOBr [as Cl] + TOCl) during chlorine dioxide treatment was the lowest of the three. One possible mechanism of TOBr formation during chlorine dioxide treatment is the HOCl formation by the reaction of chlorine dioxide and phenolic compounds (18,19) followed by the reaction of HOCl with bromide ion. Also, chloramination of Suwannee River fulvic acid produced a relatively high TOX. Similar observations (i.e., organic halogen formation during chloramination at a level around 100 μg/L) have been reported (20-22), though it is well know that chloramination produces much less THMs and HAAs compared with chlorination (*23,24*). Some fraction of TOX formation during chloramination is attributed to the presence of a very low level of free chlorine that is in equilibrium with chloramine. An estimation based on kinetics of the reaction between chlorine and phenolic compounds accounts for at least 25% of TOX formation during chloramination of fulvic acid (*14*).

Table V compares the chlorine-to-bromine ratios (mol/mol) in THMs, HAAs, and TOX. The lower Cl/Br ratio for chlorine dioxide treatment discussed above with respect to TOX is found to be consistent with the Cl/Br ratios in THMs and HAAs.

**Table V. Distribution of TOCl and TOBr during Chlorination, Chloramination, and Chlorine Dioxide Treatment.**

| Disinfectant | Dose (mg/L as Cl₂) | Cl/Br ratio in THMs (mol/mol) | Cl/Br ratio in HAAs (mol/mol) | Cl/Br ratio in TOX (mol/mol) |
|---|---|---|---|---|
| Chlorine | 4.5 | 12.6 | 11.5 | 13.9 |
| Chloramine | 5.0 | 9.6 | 15.4 | 11.8 |
| Chlorine dioxide | 6.0 | 1.2 | 0.74 | 0.69 |

## Summary

In this study, a new method to differentiate between TOCl and TOBr has been developed with a combustion furnace and an ion chromatograph. In this method, HBr and HCl, which are contained in the off-gas from a furnace for TOX measurement and are equivalent to TOBr and TOCl in samples, are collected in water with a bubble diffuser instead of being titrated in a micro coulometric cell. The concentrations of bromide ion and chloride ion were quantified separately by ion chromatography. For standards, recoveries of more than 75% were obtained from the full process, including the carbon adsorption process. Interference with chloride ion quantification by carbon dioxide gas, an auxiliary gas for the furnace, was eliminated without loss of HBr and HCl by sparging the solutions with nitrogen gas. These differentiated values are expected to be used as new bulk indices of chlorinated and brominated disinfection by-products.

A demonstration was also performed to suggest that different disinfectants yield different TOCl/TOBr distributions. Notably, for this demonstration, chlorine dioxide treatment shows a higher TOBr percentage compared to both chloramination and chlorination; however, the overall TOX level was significantly less than for the disinfectants tested.

## Acknowledgments

This study was a part of research projects funded by the United States Environmental Protection Agency (Grant R825956-01), the United States Geological Survey (Grant INT AQ 96-6R02668-2 through the University of Illinois Water Research Center), and American Water Works Association Research Foundation (Grant 554).

## References

1. Weinberg, H. S.; Krasner, S. W.; Richardson, S. D.; Sangaiah, R.; Singer, P. C. Current Research into the Occurrence of New Disinfection By-Products in Drinking Water; *Extended Abstracts of 217th ACS National Meeting, Division of Environmental Chemistry*, Anaheim, CA, *39* (1), **1999**, pp 265-267.
2. Andrews, R. C.; Ferguson, M. In *Disinfection By-Products in Water Treatment*; Minear R. A., Amy G. L., Eds.; Lewis Publishers: New York, 1995; pp 17-55.
3. Stevens, A. A.; Dressman, R. C.; Sorrel, R. K.; Braa, H. J. *J. Am. Water Works Assoc.* **1985**, *77* (4), 146-154.
4. Krasner, S. W.; Sclimenti, M. J.; Chinn, R.; Chowdhury, Z. K.; Owen, D. M. In *Disinfection By-Products in Water Treatment*; Minear R. A.; Amy G. L., Eds.; Lewis Publishers: New York, 1995; pp 59-90.
5. Zhang, X.; Echigo, S.; Minear, R. A.; Plewa, M. J. Characterization and Comparison of Disinfection By-Products from Using Four Major Disinfectants; *Extended Abstracts of 217th ACS National Meeting, Division of Environmental Chemistry*, Anaheim, CA, *39* (1), **1999**, pp 251-254.
6. Glaze, W. H.; Weinberg, H. S.; Cavanagh, J. E. *J. Am. Water Works Assoc.* **1993**, *85* (1), 96-104.
7. Plewa, M. J.; Kargaliouglu, Y.; McMillan, B. J.; Minear, R. A. A New Assessment of the Cytotoxicity and Genotoxicity of Drinking Water Disinfection By-Products; *Extended Abstracts of 217th ACS National Meeting, Division of Environmental Chemistry*, Anaheim, CA, *39* (1), **1999**, pp 198-201.
8. Richard, B. J. Carcinogenic Properties of Brominated Haloacetates; *Workshop Report on Disinfection By-Products in Drinking Water: Critical Issues in Health Effects Research*, Chapel Hill, NC, 1995, p 29.
9. Bull, R. J.; Birnbaum, L. S.; Cantor, K. P.; Rose, J. B.; Butterwoth, B. E.; Pegram, R.; Tuomisto, J. *Fundamental. Appl. Toxicol.* **1995**, *28*, 155-166.
10. *Standard Methods for the Examination of Water and Wastewater, 19th Ed.*; Eaton, A. D.; Clesceri, L. S.; Greenberg, A. E., Eds.; APHA; AWWA; WEF: Washington, DC, 1995, pp 5-22 – 5-27.
11. Kannan, K.; Kawano, M.; Kashima, Y.; Matsui, M.; Giesy, J. P. *Environ. Sci. Technol.* **1999**, *33*, 1004-1008.
12. Song, R. *S-300 Furnace Instruction Manual*, University of Illinois at Urbana-Champaign, unpublished, 1992.

13. Yoon J.; Jenson J. N. In *Disinfection By-Products in Water Treatment: The Chemistry of Their Formation and Control*; Minear, R. A.; Amy, G. L., Eds.; CRC Press Inc., Boca Raton, FL, 1996, pp 351-361.

14. Zhang, X.; Echigo, S.; Minear, R. A.; Plewa, M. J. Characterization and Comparison of Disinfection By-Products of Four Major Disinfectants. In *Natural Organic Matter and Disinfection By-Products: Characterization and Control in Drinking Water*; ACS Symposium Series; American Chemical Society: Clarendon Hills, IL (in press).

15. USEPA. *Methods for the Determination of Organic Compounds in Drinking Water: Supplement 3*. EPA600R95131. 1995.

16. USEPA. *Methods for the Determination of Organic Compounds in Drinking Water: Supplement 2*. EPA600R92129. 1992.

17. USEPA. *Methods for the Determination of Organic Compounds in Drinking Water: Supplement 1*. EPA600490020. 1990.

18. Wajon, J. E.; Rosenblatt D. E.; Burrows E. P. *Environ. Sci. Technol.* 1982, *16*, 396-402.

19. Ni, Y.; Shen, X.; Vanheiningen, A. *J. Wood Chem. Technol.* 1994, *14* (2), 243-262.

20. Krasner, S. W.; Symons, J. M.; Speitel Jr. G. E.; Diehl, A. C.; Hwang, C. J.; Xia, R.; Barrett, S. Effect of Water Quality Parameters on DBP Formation During Chlorination; *Proceedings, American Water Works Association 1996 Annual Conference (Water Quality)*, Toronto, Ontario, 1996, pp 601-628.

21. Cowman G. A.; Singer, P. C. *Environ. Sci. Technol.* 1996, *30*, 16-24.

22. Krasner, S. W.; McGuire, M. J.; Jacangelo, J. G.; Patania, N. L.; Reagan, K. M.; Aieta, E. M. *J. Am. Water Works Assoc.* 1989, *81* (8), 41-53.

23. Norman, T. S.; Harms, L. L.; Looyenga, R. W. *J. Am. Water Works Assoc.* 1980, *72* (3), 176-180.

24. Cowman G. A.; Singer, P. C. *Environ. Sci. Technol.* 1996, *30*, 16-24.

Chapter 22

# Quantification of Nine Haloacetic Acids Using Gas Chromatography with Electron Capture Detection

Katherine S. Brophy, Howard S. Weinberg, and Philip C. Singer

Department of Environmental Sciences and Engineering, University of North Carolina, Chapel Hill, NC 27599

The EPA promulgated Stage 1 of the Disinfectant/Disinfection By-Product Rule in December 1998. Under this rule, five of the nine bromine- and chlorine-containing haloacetic acids (monochloro-acetic acid [MCAA], dichloroacetic acid [DCAA], trichloroacetic acid [TCAA], monobromoacetic acid [MBAA], and dibromoacetic acid [DBAA]) were regulated for the first time. Research has shown, however, that the four unregulated species may in fact be formed at significant levels in certain types of chlorinated waters. Hence, future EPA regulations may include these species. This research was aimed at adapting current methodologies to establish a more reliable, sensitive method that can quantify all nine HAA species.

Stage 1 of the Disinfectant/Disinfection By-Products (D/DBP) rule was promulgated in December 1998. (*1*). Stage 1 set the Maximum Contaminant Level (MCL) for five haloacetic acids (HAA5; monochloroacetic acid [MCAA], dichloroacetic acid [DCAA], trichloroacetic acid [TCAA], monobromoacetic acid [MBAA], and dibromoacetic acid [DBAA]) at 60 μg/L. Studies have shown, however, that the four haloacetic acid (HAA) species that are not regulated under this rule (bromochloroacetic acid [BCAA], bromodichloroacetic acid [BDCAA], dibromochloroacetic acid [DBCAA], and tribromoacetic acid [TBAA]) may in fact contribute significantly to total HAA concentrations, particularly in high bromide waters. For this reason, Stage 2 of the D/DBP rule may include MCLs for all nine bromine- and chlorine-containing HAA species. Studies performed in our laboratory indicated that Standard Method 6251B in its current form (*2*), which uses

© 2000 American Chemical Society    **343**

diazomethane as a derivatizing agent to measure HAA6 (HAA5 plus BCAA), could not effectively measure the remaining three HAA species (HAA3). The method exhibited poor derivatization efficiency for HAA3, and poor precision between duplicate samples. EPA Method 552.2 (3), which uses acidic methanol as the derivatizing agent, was also evaluated in our laboratory. While this method was reproducible for HAA3, the derivatization efficiency was low, with resulting inadequate quantitation limits. Due to the need for a more reliable, sensitive method for the quantification of all nine bromine- and chlorine-containing HAA species, this research was aimed at adapting current methodologies so that they could be used for analysis of HAA9 in our laboratory.

## Materials and Methods

Haloacetic acids were analyzed using modified versions of both EPA Method 552.2 (3) and Standard Method 6251B (2). Twenty-milliliter aliquots of water were extracted at pH < 2 into 4 mL of methyl $t$-butyl ether (MtBE) containing internal standard. The extraction was facilitated by the addition of approximately 12 g of anhydrous sodium sulfate, which had been previously baked at 400°C for a minimum of 4 h. The extracted acids were derivatized using either acidic methanol (EPA Method 552.2) or diazomethane (Standard Method 6251B), to form their methyl ester derivatives. The derivatized acids were analyzed on an HP-5890 Series II gas chromatograph equipped with an electron capture detector (Hewlett-Packard, San Fernando, CA). Helium (99.999% purity) was used as the carrier gas and nitrogen (99.999% purity) was used as the make-up gas. Gas chromatographic conditions were as follows: Column: DB-1701 fused silica capillary (J&W Scientific, Folsom, CA), 30 m length, 0.25 mm internal diameter, 1 μm film thickness. Temperature program: Initial = 35°C for 10 min, ramp to 75°C at 5°C/min and hold for 15 min, ramp to 100°C at 5°C/min and hold for 5 min, ramp to 135°C at 5°C/min and hold for 10 min. Injector: temperature = 180°C, injection volume = 2 μL, split valve opened at 0.5 min. Detector: temperature = 280-300°C. Gas flow: helium = 1-1.5 mL/min, nitrogen = 40-60 mL/min.

High purity HAA standards were obtained from Supelco, Inc. (Bellefonte, PA). HAA6 were purchased as a mixture containing 2 mg/mL of each species dissolved in MtBE, whereas the HAA3 standards were purchased as neat standards. Single component stock solutions of the HAA3 species were prepared in MtBE at concentrations ranging from 4-7 mg/mL. Haloester standards were also purchased as a mixture of HAA6 in MtBE and HAA3 as neat standards. The surrogate recovery standard, 2,3-dibromopropionic acid, was purchased from Supelco, Inc. as a single component stock solution (1 mg/mL) in MtBE. The internal standard, 1,2,3-trichloropropane, was purchased as a neat standard from Aldrich Chemical Co. (Milwaukee, WI). Stock solutions of the internal standard were prepared in MtBE at concentrations ranging from 4-5 mg/mL. All stock solutions were stored in a freezer at –15°C and replaced every three months, or upon evidence of degradation. Standard degradation was recognized by a decrease in detector response for the standard at a given concentration over time, and/or the appearance of extraneous chromatographic peaks when the pure standard was analyzed.

pH–adjusted solutions were prepared using either a phosphate buffer (pH 7) or a borate buffer (*10*). The buffer solutions were prepared as described in the *CRC Handbook of Chemistry and Physics* (*4*) by adding 0.1 molar sodium hydroxide to 50 mL of 0.1 molar potassium dihydrogen phosphate (pH 7) or 0.025 molar borate (pH 10). Five mL of buffer was added per liter of water, and 0.1 molar sodium hydroxide or 0.1 molar hydrochloric acid was used to achieve the target pH, if necessary. All synthetic waters were prepared using deionized, organic-free laboratory-grade water (LGW) (Dracor Inc., Durham, NC).

The toxicity and carcinogenicity of each reagent used in these methods have not been precisely defined. Each chemical should be treated as a potential health hazard. Certain individuals may experience adverse effects upon exposure to the extraction solvent, MtBE, via skin contact or inhalation of vapors. Protective clothing and gloves should be used when handling MtBE, and the solvent should be handled in a fume hood or glove box. Diazomethane is a toxic and explosive gas. Diazomethane should be stored in the freezer at -15°C when not in use. The gas should not be allowed to come in contact with ground glass joints. Protective clothing and gloves should be used when handling diazomethane, and diazomethane should only be handled in a fume hood or glove box.

## Method Development

Current applications of Standard Method 6251B (*2*) and EPA Method 552.2 (*3*) in our laboratory yielded near 100% recovery from water for HAA6 and near 100% conversion from acid to ester during derivatization, permitting a practical quantitation limit (PQL) for each species of 1 µg/L. The PQL is defined as the concentration at which a consistent, linear chromatographic response is observed as compared to a standard calibration curve. In examining the efficiencies of the two derivatizing agents in esterifying the HAA3 species, however, the percent conversions from acid to ester were found to be much lower. Conversion efficiencies were determined by comparing calibration curves built from ester standards to curves built from acid standards that had been extracted from water and derivatized using either diazomethane or acidic methanol. Values on the abscissa scale of these curves reflect the theoretical ester concentration of the derivatized acid standards assuming 100% extraction and conversion efficiency. The ordinate axis represents the relative area of the analyte as measured by the ratio of the analyte peak area to internal standard. Since the latter may change with each fresh batch of extracting solvent, relative areas for a specific analyte concentration may change with each new batch extraction or between experiments. The conversion efficiencies of diazomethane and acidic methanol in esterifying HAA3 were compared in order to decide which method to pursue in terms of optimization for HAA3 analysis.

Examples of the conversion efficiencies of the two derivatizing agents in esterifying HAA3 are shown in Figures 1-4. While the conversion efficiency is good for BDCAA (near 100%), the conversion efficiencies for DBCAA and TBAA are

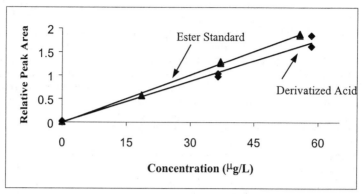

*Figure 1. Average derivatization efficiency for BDCAA*

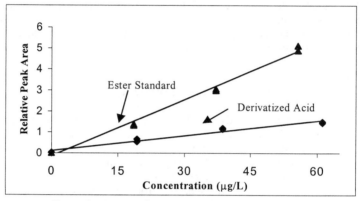

*Figure 2. Average derivatization efficiency for DBCAA*

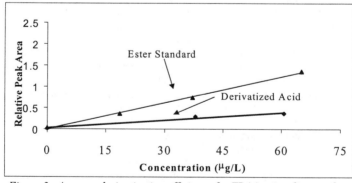

*Figure 3. Average derivatization efficiency for TBAA using diazomethane*

low, approximately 25% and 30%, respectively. The conversion efficiencies of both agents were similar for BDCAA and DBCAA, but diazomethane was slightly better than acidic methanol at converting TBAA to its corresponding ester (Figure 3 versus Figure 4). This may be due to the fact that TBAA can undergo partial decarboxylation to form bromoform upon exposure to methanol (5), thereby lowering conversion to ester of this species by acidic methanol. Figures 2 and 3 demonstrate that the derivatizing agents were less than effective in quantitatively esterifying DBCAA and TBAA, respectively. Due to the limitations of acidic methanol in esterifying TBAA, as well as the laborious and time-consuming nature of the acidic methanol process, it was decided to pursue the diazomethane method in terms of optimization for HAA3.

An examination of each step in the procedure revealed a detrimental impact on the esterification of HAA3 caused by the presence of water co-extracted with the analytes into the MtBE prior to derivatization. Little is known regarding the relative reaction rates of HAA6, HAA3, and water with diazomethane. A possible explanation for the detrimental impact of water on HAA3 esterification is that the bulky molecular size and structure of the HAA3 species as compared to the HAA6 species result in slower rates of esterification for HAA3. Since water reacts with diazomethane to form methanol and nitrogen gas, dissolved water in the extracts exhibits a demand on diazomethane when the derivatizing agent is added. As a result, the diazomethane may become depleted to the extent that esterification of HAA3 becomes kinetically hindered before 100% conversion has occurred. Kinetic studies on the rates of esterification for HAA6, HAA3, and water were beyond the scope of this research, thus this hypothesis cannot be supported by kinetic data at this time.

Various approaches towards removing dissolved water from the MtBE extracts were considered, mainly focusing on treating the MtBE with a drying agent prior to derivatization. Standard Method 6251B (2) suggests the option of filtering the extracts through acidified sodium sulfate ($Na_2SO_4$) to remove water before esterification. A simplified version of this drying step was evaluated for HAA3 in our laboratory, using anhydrous magnesium sulfate ($MgSO_4$) as the drying agent. Disposable glass Pasteur pipettes were fitted with glass wool, and approximately one gram of $MgSO_4$ was placed on top of each glass wool plug. The MtBE extracts were filtered through the salt, and 2 mL of each sample was collected in 2-mL volumetric flasks for derivatization.

The results of this experiment for DBCAA and TBAA are shown in Figure 5 and Figure 6, respectively. As can be seen, poor recovery was obtained for these species, most likely due to adsorption of the analytes onto the surface of the glass wool. For this reason, an alternative approach was evaluated to remove water from the MtBE. Rather than filtering the extracts through the salt, approximately 100 mg of $MgSO_4$ was added directly to each extract prior to addition of diazomethane. This method of water removal allowed near 100% esterification of HAA3, as illustrated in Figures 7-9.

A drawback to the addition of the drying step to the procedure was the creation of a peak due to a reaction between sulfate and diazomethane. A similar peak, tentatively identified as dimethyl sulfide, is observed using EPA Method 552.2 (3). When the modified diazomethane method was evaluated in terms of HAA9, it was found

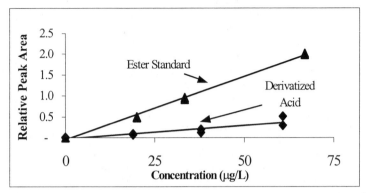

*Figure 4. Average derivatization efficiency for TBAA using acidic methanol*

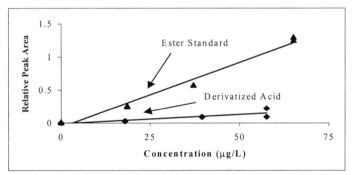

*Figure 5. Derivatization efficiency for DBCAA when filtered through MgSO₄*

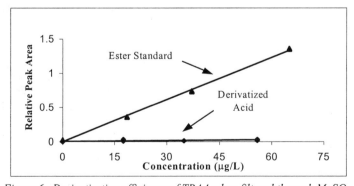

*Figure 6. Derivatization efficiency of TBAA when filtered through MgSO₄*

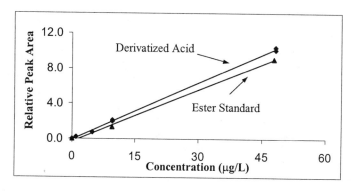

*Figure 7. Derivatization efficiency of BDCAA — MgSO₄ added to extract*

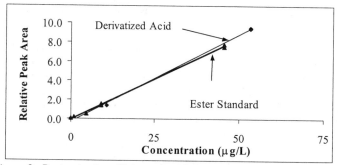

*Figure 8. Derivatization efficiency for DBCAA — MgSO₄ added to extract*

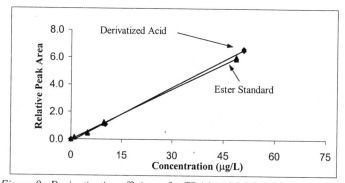

*Figure 9. Derivatization efficiency for TBAA — MgSO₄ added to extract*

that this peak co-eluted with BCAA when using a DB1701 column. At concentrations above 5 µg/L of BCAA, resolution between the two peaks allowed quantitation of BCAA. Below 5 µg/L, however, the BCAA peak became engulfed by the dimethyl sulfide peak. In order to rectify this problem, several alternative capillary columns were considered. It was found that use of a DB-1 column (30 m, 0.25 mm, 1 µm) allowed resolution of all nine HAA species, as well as the surrogate and internal standards. The temperature program using the DB-1 column was as follows: Initial temperature = 37°C, hold for 21 min, ramp to 136°C at 5°C/min, hold for 3 min, ramp to 250°C at 20°C/min, hold for 3 min. Using the adapted method with MgSO$_4$ added to the derivatization flask and the DB-1 column for separation, all nine HAA species could be consistently quantified, with a PQL of 1 µg/L for all nine species.

## Removal of Chlorine

Five chlorine-quenching agents were evaluated to determine the best choice for removal of chlorine while stabilizing levels of formed HAA9: ammonium chloride (NH$_4$Cl), ammonium sulfate ((NH$_4$)$_2$SO$_4$), sodium sulfite (Na$_2$SO$_3$), sodium thiosulfate (Na$_2$S$_2$O$_3$), and sodium meta-arsenite (NaAsO$_2$). EPA Method 552.2 (*3*) and Standard Method 6251B (*2*) both call for the use of NH$_4$Cl. However, all five quenching agents have been used in the literature to quench chlorine for disinfection by-product analysis. The quenching agents were evaluated on a mixture of HAA9 prepared in LGW in 40-mL vials at temperature conditions simulating what might occur during sample collection in the field and at two pH values (7 and 10). Approximately 10 mg of quenching agent was added to each 40-mL vial (except the control), making a final concentration of approximately 250 mg/L. The results of the evaluation for the sum of HAA9 are illustrated in Figures 10 and 11, along with the conditions employed, and indicate that either NH$_4$Cl, (NH$_4$)$_2$SO$_4$ or NaAsO$_2$ could be used as a quenching agent without affecting overall HAA9 stability. In examining the impacts of the quenching agents on individual HAA species, (NH$_4$)$_2$SO$_4$ appeared to least impact HAA stability and was therefore chosen as the quenching agent to be used during field sample collection.

## Quality Assurance

The adapted method for HAA analysis was quality assured by conducting split sampling analyses with two other laboratories each using their own quality assured method. Lab #2 uses EPA Method 552.2 (*3*) for HAA9 analysis, and Lab #3 uses Standard Method 6251B (*2*). The water used in these tests was collected from a drinking water utility treating low-bromide surface water. Upon receipt of a settled water sample at the UNC laboratories, the water was spiked with 300 µg/L bromide then chlorinated with 7 mg/L Cl$_2$ (Sample A). Residual chlorine was quenched using approximately 250 mg/L ammonium sulfate, and aliquots of Sample A were adjusted to pH 6 and pH 10 and spiked with approximately 10 µg/L of each HAA9 species

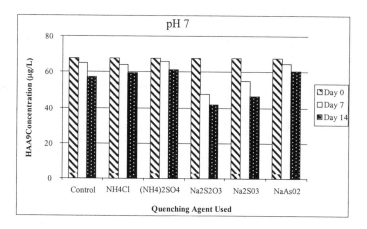

*Figure 10. Impact of quenching agents on HAA stability—24 h at 15°C, 2 weeks at 4°C*

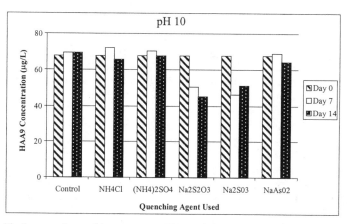

*Figure 11. Impact of quenching agents on HAA stability—24 h at 15°C, 2 weeks at 4°C*

(Samples B and C, respectively). Each sample had 10 mg/L of sodium azide, evaluated elsewhere (6), added prior to shipping to the participating laboratories to prevent biodegradation of the haloacetic acids prior to analysis. The split sampling results are illustrated in Table I and illustrate a few trends of note. When using acidic methanol, the calculated level of MCAA in Sample A is more than double that obtained by diazomethane and appears to be divergent from the true value for Sample C. Acidic methanol appears to demonstrate an underestimation for CDBAA and TBAA at low levels (Sample A), while the existing standard method overestimates BDCAA. This inconsistency among the established methods, especially at lower levels of HAAs, illustrates the need to establish a standard methodology for HAA9 analysis so that all laboratories analyzing for HAA9 can achieve consistent, accurate results.

The adapted method was used to determine spike recoveries for HAA9 in several different matrices in order to ascertain that matrix effects would not prevent accurate quantitation of all nine species. One spike recovery study was performed on water collected from the point-of-entry (POE) to the distribution system at a water utility which treats low-bromide surface water from the Mississippi River. The results for the spike recovery are shown in Table II, demonstrating good spike recoveries for all species and indicating that accurate results can be obtained using the adapted methodology. Table II also indicates that the currently unregulated HAA3 species account for over 10% of the total HAA9 concentrations, even in a low bromide-containing water. This further demonstrates the need for a more reliable methodology for the quantitation of all nine HAA species that can be easily deployed in the drinking water industry.

The method detection limit (MDL) was also evaluated using the adapted methodology. A solution was prepared containing 1 µg/L of each HAA species in water, and seven replicates were extracted and derivatized using the adapted methodology. The MDL was determined by multiplying the standard deviation between the seven replicates for each species by the student's t-value at 99% confidence and n-1 degrees of freedom (3.143 for seven replicates). The results of this study are shown in Table III. For all HAA9 species, the relative standard deviation was less than or equal to 10%. This demonstrates the ability of the adapted methodology to provide accurate, precise results, even at low levels of HAA9.

## Conclusions

This research resulted in the establishment of an analytical method that can more reliably quantify all nine bromine- and chlorine-containing HAA species in chlorinated waters at levels as low as 1 µg/L. By removing water from MtBE extracts with $MgSO_4$ prior to derivatization, nearly 100% conversion of each acid to its corresponding ester can be achieved. Use of a DB-1 capillary column permits complete resolutions of all nine HAAs, the surrogate recovery standard, and the internal standard. Samples collected in the field are quenched of residual chlorine with $(NH_4)_2SO_4$, which does not impact the levels of HAA9 present in the range of pH 7-10.

**Table I. Split Sampling for HAA9 Analysis**

| Sample | MCAA | MBAA | DCAA | BCAA | TCAA | DBAA | BDCAA | CDBAA | TBAA |
|---|---|---|---|---|---|---|---|---|---|
| A -UNC | 2.01 | BDL* | 3.4 | 6.79 | 3.18 | 6.09 | 10.9 | 11.7 | 4.7 |
| A -Lab #2 | 5.5 | 1.7 | 5.9 | 9.1 | 3.7 | 9.1 | 11.4 | 10.0 | 3.2 |
| A -Lab #3 | BDL** | 2.2 | 4.9 | 9.0 | 4.6 | 10 | 15 | 12 | 4.7 |
| Average | 3.76 | 1.95 | 4.73 | 8.30 | 3.83 | 8.40 | 12.4 | 11.2 | 4.20 |
| St. Dev. | 2.47 | 0.35 | 1.26 | 1.31 | 0.72 | 2.05 | 2.24 | 1.08 | 0.87 |
| % RSD | 66% | 18% | 27% | 16% | 19% | 24% | 18% | 10% | 21% |
| B -UNC | 11.2 | 12.1 | 14.2 | 17.7 | 14.1 | 16.7 | 21.7 | 21.1 | 16.9 |
| B -Lab #2 | 12 | 12.6 | 17.2 | 19.8 | 15.2 | 18.4 | 22.3 | 19.4 | 16.6 |
| B -Lab #3 | 11 | 13 | 15 | 20 | 16 | 21 | 27 | 23 | 19 |
| Average | 11.4 | 12.6 | 15.5 | 19.2 | 15.1 | 18.7 | 23.7 | 21.2 | 17.5 |
| St. Dev. | 0.53 | 0.45 | 1.55 | 1.27 | 0.95 | 2.17 | 2.90 | 1.80 | 1.31 |
| % RSD | 5% | 4% | 10% | 7% | 6% | 12% | 12% | 9% | 7% |
| C -UNC | 12.4 | 11.7 | 14.0 | 17.4 | 13.7 | 16.3 | 21.3 | 20.7 | 16.5 |
| C -Lab #2 | 8.9 | 12.9 | 18.5 | 20.3 | 15.6 | 18.6 | 22.5 | 19.5 | 16.6 |
| C -Lab #3 | 12 | 12 | 16 | 20 | 16 | 19 | 27 | 24 | 19 |
| Average | 11.1 | 12.2 | 16.2 | 19.2 | 15.1 | 18.0 | 23.6 | 21.4 | 17.4 |
| St. Dev. | 1.92 | 0.62 | 2.25 | 1.59 | 1.23 | 1.46 | 3.00 | 2.33 | 1.42 |
| % RSD | 17% | 5% | 14% | 8% | 8% | 8% | 13% | 11% | 8% |

\* Detection limit = 1.00 µg/L
\**Detection limit = 2.00 µg/L

**Table II. Spike Recoveries for HAA9 in Sample Matrix**

| Sample | Concentration (µg/L) | | | | | | | | |
|---|---|---|---|---|---|---|---|---|---|
| | MCAA | MBAA | DCAA | BCAA | TCAA | DBAA | BDCAA | DBCAA | TBAA |
| POE | < 1.00 | 1.01 | 18.0 | 1.66 | 10.6 | < 1.00 | 1.35 | 2.53 | < 1.00 |
| POE+Spike | 10.1 | 11.3 | 26.3 | 12.6 | 21.2 | 10.6 | 12.9 | 12.8 | 11.4 |
| Spike Conc. | 10.0 | 10.0 | 10.0 | 10.0 | 10.0 | 10.0 | 10.7 | 10.4 | 10.4 |
| % Spike Recovery | 99% | 103% | 83% | 109% | 106% | 106% | 108% | 99% | 107% |

**Table III. Method Detection Limits for HAA9 Using Adapted Methodology**

| Analyte | Standard Conc. Added (μg/L) | Mean Measured Conc. (μg/L) | Standard Deviation (μg/L) | Relative Std. Dev. % | Experimental MDL (μg/L) |
|---|---|---|---|---|---|
| ClAA | 1.00 | 1.00 | 0.081 | 8% | 0.255 |
| BrAA | 1.00 | 0.923 | 0.036 | 4% | 0.113 |
| Cl₂AA | 1.00 | 0.932 | 0.046 | 5% | 0.146 |
| BrClAA | 1.00 | 0.929 | 0.031 | 3% | 0.096 |
| Cl₃AA | 1.00 | 1.048 | 0.107 | 10% | 0.337 |
| Br₂AA | 1.00 | 1.01 | 0.080 | 8% | 0.252 |
| BrCl₂AA | 1.07 | 1.56 | 0.110 | 7% | 0.344 |
| Br₂ClAA | 1.04 | 1.08 | 0.095 | 9% | 0.300 |
| Br₃AA | 0.945 | 0.958 | 0.061 | 6% | 0.191 |

The adapted methodology was applied to a low-bromide surface water to analyze for HAA9. The method was quality assured by analyzing spike recoveries within different sample matrices. HAA9 analysis of this water revealed that DBCAA and BDCAA, which are currently unregulated, were present at significant levels within the water, accounting for 12.4% of the total HAA concentrations. This finding in a low-bromide water illustrates the importance of quantifying all nine HAA species, as analysis of HAA5 may significantly underestimate total HAA concentrations. The impact can be expected to be much greater for waters that contain high levels of bromide, e.g. >100 μg/L.

## Acknowledgments

This paper represents a part of an on-going study into the formation, occurrence, stability, and dominance of haloacetic acids and trihalomethanes in treated drinking water. The research is funded by the American Water Works Association Research Foundation.

## Literature Cited

1. U.S. Environmental Protection Agency. National Primary Drinking Water Regulations: Disinfectants and Disinfection Byproducts. *Fed. Reg.* **1998**, *63* (241), 69389.

2. Eaton, A. D., Clesceri, L. S., Greenberg, A. E., Eds. *Standard Methods for the Examination of Water and Wastewater*, 19th Edition. APHA: Washington, DC, 1995; pp 6-67 to 6-76.

3. Munch, D. J.; Munch, J. W.; Pawlecki, A. M. *Method 552.2: Determination of Haloacetic Acids and Dalapon in Drinking Water by Liquid-Liquid Extraction, Derivatization and Gas Chromatography with Electron Capture Detection.* U.S. Environmental Protection Agency: Cincinnati, OH, 1995; pp 552.2-1 to 552.2-34.

4. Lide, D. R., Ed. *CRC Handbook of Chemistry and Physics,* 76th Edition. CRC Press: Boca Raton, FL, 1995; p 8-42.

5. Peters, R. J. B.; Erkelens, C.; de Leer, E. W. B.; de Galan, L. *Wat. Res.* **1991,** *25,* 473-477.

6. Brophy, K. S. M.S. Thesis, University of North Carolina, Chapel Hill, NC, 1999.

Chapter 23

# Development of a Capillary Electrophoresis Method for Haloacetic Acids

Yuefeng Xie[1], Haojiang (Joe) Zhou[1], and Joseph P. Romano[2]

[1]Environmental Engineering Programs, Pennsylvania State University at Harrisburg, Middletown, PA 17033
[2]Waters Corporation, 34 Maple Street, Milford, MA 01757

An innovative analytical method was developed for directly determining haloacetic acids in drinking water samples without sample derivatization using capillary electrophoresis. All nine haloacetic acids, including monohaloacetic acids, dihaloacetic acids, and trihaloacetic acids, were analyzed. The detection limits for monohaloacetic acids and dihaloacetic acids were less than 3 μg/L, and ranged from 4.0 to 7.0 μg/L for trihaloacetic acids. The spiking recovery for eight haloacetic acids except tribromoacetic acid was in the range of 74.8% to 107%. The newly developed method for haloacetic acid analysis consists of liquid-liquid extraction and capillary electrophoresis with direct UV detection. The extraction and back-extraction took 30 to 45 min to process and the total run time for capillary electrophoresis was 12 min. Further study is needed to improve the peak asymmetry for trihaloacetic acids, method sensitivity, and to simplify or automate the sample extraction process.

## Introduction

Since chloroform was first identified in chlorinated waters, many disinfection by-products (DBPs) have been identified in finished drinking water. In addition to trihalomethanes (THMs), haloacetic acids (HAAs) are another group of DBPs that are commonly detected in chlorinated drinking waters (1-3). Due to their potential human health risks and widespread occurrence in chlorinated drinking water (1-5), the United States Environmental Protection Agency (USEPA) set a new maximum contaminant level (MCL) at 60 μg/L for five HAAs, including monochloroacetic acid (MCAA), monobromoacetic acid (MBAA), dichloroacetic acid (DCAA), dibromoacetic acid (DBAA), and trichloroacetic acid (TCAA) under the Stage 1 Disinfectants/DBP Rule promulgated in 1998 (6). In addition to these five HAAs,

**356**

there are four more un-regulated HAAs, including bromochloroacetic acid (BCAA), bromodichloroacetic acid (BDCAA), dibromochloroacetic acid (DBCAA), and tribromoacetic acid (TBAA) (7-8). These nine haloacetic acids could also be divided into three groups including monohaloacetic acids, dihaloacetic acids, and trihaloacetic acids. Several analytical methods for determining HAAs in drinking water have been developed in recent years (9-11). Currently, three analytical methods are approved for HAA monitoring under the Information Collection Rule (12). They are EPA Methods 552.1 and 552.2 and Standard Method 6251B (13-15). All these methods require sample extraction and methylation prior to gas chromatography-electron capture detection. For sample methylation, Standard Method 6251B uses diazomethane which is generally prepared with a potent carcinogen 1-methyl-3-nitro-1-nitrosoguanidine (MNNG) (15). Due to the toxicity of MNNG and the hazardous nature of diazomethane, there is concern about using these chemicals, especially in water utility laboratories. EPA Methods 552.1 and 552.2 use acidic methanol which is heated at a high temperature (e.g., 50°C) (13-14). There is also concern about incomplete methylation of trihaloacetic acids using the acidic methanol procedure (16). Therefore, an analytical method which is capable of directly determining HAAs without sample methylation will significantly benefit the drinking water industry.

Capillary ion electrophoresis is an emerging separation technology for analysis of low-molecular-weight inorganic and organic ions. Capillary electrophoresis (CE) offers many advantages for determining anions and cations in water and wastewater samples, including high separation efficiency, short analysis time, and minimal sample requirement (17). In 1997, Xie and Romano explored the use of CE for directly analyzing mg/L level of HAAs without methylation (18). The results indicated that all nine HAAs were separated in less than 10 min with a phosphate electrolyte solution. However, the concentrations of HAAs in the sample were approximately 100 times greater than typical concentrations in chlorinated drinking water. Song and Budde also explored the application of capillary electrophoresis/mass spectrometry for HAA analysis (19). Only synthetic samples were analyzed in their study.

The objective of the present study is to develop a CE method for directly determining μg/L level of HAAs in drinking water. Liquid-liquid extraction will be used to concentrate the samples and remove the interfering matrix ions (e.g., chloride and sulfate). The method detection limit, spiking recovery and other method performance will be evaluated with synthetic and field samples.

## Experimental Methods

All aqueous standards and synthetic samples were prepared with Alpha-Q water (Millipore Corporation, Bedford, MA) and a commercially available HAA standard (Absolute Standards, Inc., New Haven, CT). In addition to nine HAAs, the standard also contained dichloropropionic acid (dalapon), a chlorinated herbicide. The commercial standard was prepared at a concentration of 100 μg/mL for each HAA and dalapon with methyl tert-butyl ether (MTBE). Field samples were collected at Hershey Water Treatment Plant which is owned and operated by Pennsylvania

American Water Company. Before the extraction, ammonium chloride was added to all field samples to quench free residual chlorine. As shown in Figure 1, a 200 mL sample was then transferred into a 500 mL separatory funnel. After adding 6.1 μg of internal standard, difluoroacetic acid (DFAA), 10 mL of concentrated sulfuric acid and 50 g of sodium sulfate, the sample was extracted with 20 mL of MTBE for three min. The extract was then dried with anhydrate sodium sulfate and concentrated to approximately 1 mL under a nitrogen flow in a 50°C water bath. Finally, the concentrated extract was back-extracted with 0.2 mL of 2.5 mM phosphate solution (pH = 9). After separation, the aqueous phase was submitted for CE analysis.

The instrument employed in this study was a Waters Quanta 4000 Capillary Electrophoresis System. The instrument configuration included a negative power supply, a mercury lamp, a Waters Accu-Sep polyamide-coated fused-silica capillary (60 cm x 75 μm I.D.), an ultraviolet (UV) absorbance detector, and a Waters Millennium Chromatography Manager data process system, as shown in Figure 2. The analytical method used phosphate electrolyte solution and direct UV detection at 185 nm. The phosphate electrolyte solution contained 12.5 mM sodium dihydrogen phosphate, 12.5 mM disodium hydrogen phosphate, and 1.0 mM tetradecyltrimethyl ammonium hydroxide (TTAOH). The pH of the solution was adjusted to 8.0 with sodium hydroxide. Other detailed operating conditions are shown in Table I.

**Table I. Operating Conditions of Capillary Electrophoresis for HAA Analysis**

| Parameters | Settings |
|---|---|
| Electrolyte | 25 mM phosphate + 1.0 mM TTAOH (pH = 8.0) |
| Capillary | 75 μm (I.D.) x 375 μm (O.D.) x 60 cm (length) |
| Temperature | Ambient (20°C) |
| Power supply | Negative |
| Voltage | 15 kV |
| Current | 50μA (constant current) |
| Sampling | Hydrostatic for 30 seconds (37 nL) |
| Detection | Direct UV at 185 nm |
| Quantitation | Time corrected peak area |

For each HAA, a standard curve ranging from 4 μg/L to 100 μg/L was created with aqueous standards. To determine the method detection limit (MDL) for each HAA and dalapon, seven replicates of a 4-μg/L HAA aqueous standard were analyzed. The MDL was calculated as 3.14 times the standard deviation of replicate samples. The spiking recovery was determined by analyzing a field sample spiked with 0, 10, 20, 30, and 40 μg/L of HAAs. The mean recovery was equal to the slope of the plot of the measured concentration versus the spiked concentration.

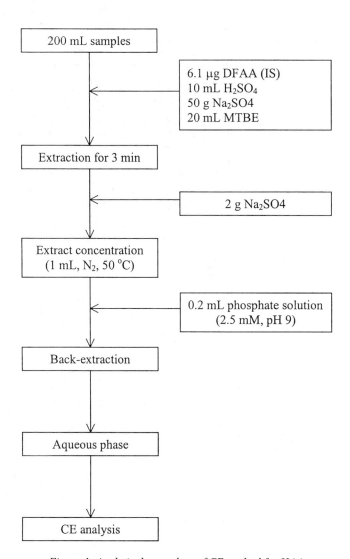

*Figure 1. Analytical procedure of CE method for HAAs*

## Results and Discussion

A typical electropherogram for HAA standards is shown in Figure 3a. The internal standard (DFAA), nine HAAs, and dalapon were well separated in 10 min. Figure 3b is the electropherogram for a spiked field sample. Little interference was observed in the field sample. Inorganic anions, including chloride and sulfate which cause interference in CE analysis, were removed during the initial solvent extraction and nonpolar or less polar organics were removed during the back-extraction. For all four trihaloacetic acids, however, tailing peaks were observed. This may be because the mobility of the electrolyte was faster than that of trihaloacetic acids. The mobility of the electrolyte could be decreased by lowering the pH of the electrolyte solution (17). Peak asymmetry could also be improved by increasing the ionic strength of the electrolyte or addition of organic solvents (e.g., acetonitrile) to the electrolyte (17). As shown in Figure 3, in contrast to other chromatographic techniques, the migration times for the sample HAA were different in the different samples or different runs. This is a common problem with CE that could pose a problem for peak identification. This migration time change could be overcome by using a marker and relative migration times.

The standard curves for nine HAAs and dalapon were obtained in the range of 4 to 100 µg/L. Three typical standard curves for MCAA, DCAA and TCAA are shown in Figures 4a, 5a and 6a. The linear regression coefficients ($r^2$) for all ten standard curves, including one for dalapon, are shown in Table II. The calculated MDLs for HAAs and dalapon are also shown in Table II. These MDLs were determined using seven 4-µg/L replicates. For dalapon, mono- and dihaloacetic acids, the MDLs were less than 3 µg/L. For trihaloacetic acids, the MDLs ranged from 4.0 to 7.7 µg/L. Again, the high MDLs for trihaloacetic acids were partially because of the asymmetrical peak shape of trihaloacetic acids.

### Table II. Method Performance Data for HAAs

| HAAs | Regression Coefficients ($r^2$) for Standard Curves | Method Detection Limits (µg/L) | Spiking Recovery |
|---|---|---|---|
| MCAA | 0.999 | 1.4 | 101% |
| MBAA | 0.999 | 2.6 | 92.4% |
| DCAA | 0.999 | 1.3 | 97.9% |
| BCAA | 0.993 | 1.5 | 105% |
| DBAA | 0.992 | 2.9 | 99.2% |
| Dalapon | 0.997 | 1.9 | 100% |
| TCAA | 0.999 | 4.4 | 78.3% |
| BDCAA | 1.000 | 5.9 | 74.8% |
| DBCAA | 0.991 | 4.0 | 107% |
| TBAA | 0.991 | 7.7 | - |

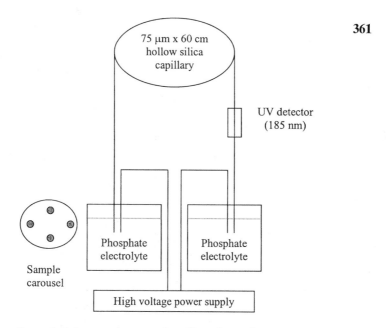

*Figure 2. Schematic diagram of capillary electrophoresis system*

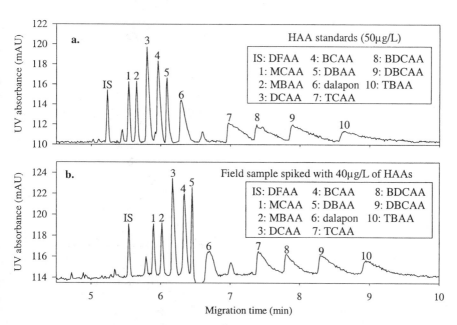

*Figure 3. Typical electropherograms of HAA samples*

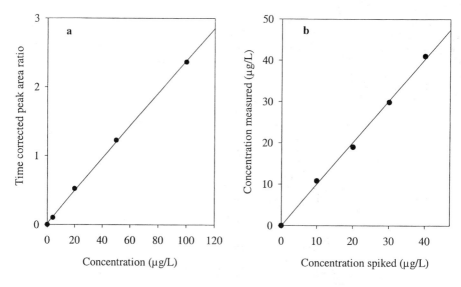

*Figure 4. The calibration curve (a) and spiking recovery (b) for MCAA*

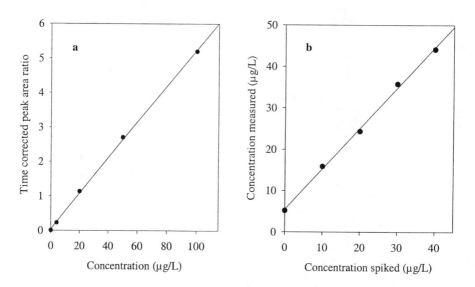

*Figure 5. The calibration curve (a) and spiking recovery (b) for DCAA*

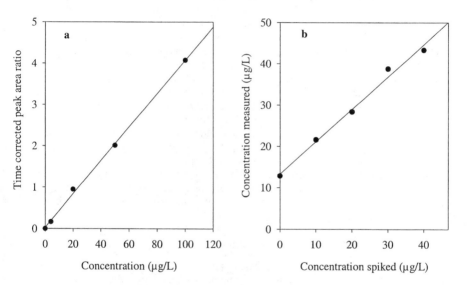

*Figure 6. The calibration curve (a) and spiking recovery (b) for TCAA*

The spiking recovery was determined using a field water sample with five spiking levels. Figures 4b, 5b, and 6b illustrate the mean spiking recoveries for MCAA, DCAA, and TCAA, which equal the slope of the plots of the measured concentration versus the spiked concentration. Except for tribromoacetic acid, the spiking recoveries for other HAAs and dalapon, as listed in Table II, were well above the 70-130% recovery criterion set by USEPA (13-14). The background concentrations of MCAA, DCAA, and TCAA in the field sample were nondetectable, 5.2 μg/L, and 12.8 μg/L, respectively. These results were similar to the results of samples collected in the same system and in same season and analyzed in the water utility laboratory using EPA Method 552.2. The spiking recovery for tribomoacetic acid was not determined due to an interfering peak.

Due to the relatively low sensitivity of direct UV detection, water samples were concentrated approximately 1000 times using liquid-liquid extraction and back extraction. The extraction procedure also removed interfering ions including chloride and sulfate. However, the extraction and back-extraction took approximately 30 to 45 min to process. The authors suggest that further study is needed to simplify the sample concentration process or automate the extraction process. Recent development in solid phase extraction materials could make solid phase extraction an alternative sample extraction technique. The sensitivity of the CE detection could also be improved by using enhanced UV detection cells (e.g., bubble cells or Z-cells) (20), or using other type of detectors (e.g., fluorescence or mass spectrometry detectors) (19). However, these options could increase the complexity of the instrument and the cost of analyses. On-capillary concentration, or electromigration sample introduction, is another option to concentrate the sample and increase the method sensitivity. However, concentration variation of interfering anions including chloride and sulfate may significantly affect reproducibility of this technique. Liquid-liquid extraction and back extraction may be needed in conjunction with on-capillary concentration.

## Conclusions

1. The newly developed CE method is capable of directly determining μg/L levels of HAAs in drinking water samples without sample derivatization.
2. The detection limits were less than 3 μg/L for monohaloacetic acids and dihaloacetic acids, and ranged from 4.0 to 7.0 μg/L for trihaloacetic acids. The spiking recovery for eight haloacetic acids except tribromoacetic acid was in the range of 74.8% to 107%.
3. Further study is needed to improve the peak asymmetry, method sensitivity, and simplify or automate the sample extraction process.

## Acknowledgments

The authors express their thanks to the Waters Corporation, for providing the capillary electrophoresis system under the project, and the Pennsylvania American Water Company, for assisting with the field sample collection. Thanks also go to Jim Krol and Mark Benvenuti of the Waters Corporation for their technical assistance. Haojiang (Joe) Zhou was supported by the Environmental Pollution Control assistantship at Pennsylvania State University.

## Literature Cited

1. Uden, P. C.; Miller, J. W. *J. Am. Water Works Assoc.* **1983**, *75(10)*, 524-527.
2. Krasner, S. W.; McGuire, M. J.; Jacangelo, J. G.; Patania, N. L.; Reagan, K. M.; Aieta, E. M. *J. Am. Water Works Assoc.* **1989**, *81(8)*, 41-53.
3. Reckhow, D. A.; Singer, P. C. *J. Am. Water Works Assoc.* **1984**, *76(4)*, 151-157.
4. Bull, R. J.; Sanchez, I. M.; Nelson, M. A.; Larson, J. L.; Lansing, A. J. *J. Toxicol.* **1990**, *63*, 341-359.
5. DeAngelo, A. B.; Daniel, F. B.; Stober, J. A.; Olson, G. R. *Fundam. Appl. Toxicol.* **1991**, *16*, 337-347.
6. United States Environmental Protection Agency *Fed. Reg.* **1998**, *63*, 69389-69476.
7. Xie, Y.; Rajan, R. V.; Reckhow, D. A. *Org. Mass Spectrom.* **1992**, *27*, 807-810.
8. Cowman, G. A.; Singer, P. C. *Environ. Sci. Technol.* **1996**, *30*, 16-24.
9. Chinn, R.; Krasner, S. W. Simplified Technique for the Measurement of Halogenated Organic Acids in Drinking Water by Electron Capture Gas Chromatography. *Proceedings*, 1989 Pacific Conference on Chemistry and Spectroscopy, Pasadena, CA, 1989.
10. Barth, R. C.; Fair, P. S. *J. Am. Water Works Assoc.* **1992**, *84(11)*, 94-98.
11. Xie, Y.; Reckhow, D. A.; Springborg, D. C. *J. Am. Water Works Assoc.* **1998**, *90(4)*, 131-138.
12. United States Environmental Protection Agency. *Fed. Reg.* **1996**, *61*, 24353-24388.
13. *Method 552.1: Determination of Haloacetic Acids and Dalapon in Drinking Water by Ion-Exchange Liquid-Solid Extraction and Gas Chromatography with an Electron Capture Detector.* Environmental Monitoring and System Laboratory, U.S. Environmental Protection Agency: Cincinnati, OH, 1992.
14. *Method 552.2: Determination of Haloacetic Acids and Dalapon in Drinking Water by Liquid-Liquid Extraction, Derivatization, and Gas Chromatography with Electron Capture Detection.* National Exposure Research Laboratory, U.S. Environmental Protection Agency: Cincinnati, OH, 1995.
15. Clesceri, L. S., Greenberg, A. E., Eaton, A. D., Eds. *Standard Methods for the Examination of Water and Wastewater*; 20th ed. American Public Health Associ-ation, American Water Works Association, Water Environment Federation: Washington, DC, 1998; pp 6/42–6/51.
16. Xie, Y. Unpublished results, 1999.
17. Krol, J.; Benvenuti, M.; Romano, J. P. *Ion Analysis Methods for IC and CIA® and Practical Aspects of Capillary Ion Analysis Theory.* Waters Corporation: Milford, MA, 1998.
18. Xie, Y.; Romano, J. P. Application of Capillary Ion Electrophoresis for Drinking Water Analysis. *Proceedings*, AWWA Water Quality Technology Conference, 1997. American Water Works Association: Denver, CO, 1997.
19. Song, X.; Budde, W. Disinfection By-Products by Capillary Electrophoresis/ Electrospray/Mass Spectrometry. *Proceedings*, 45th ASMS Conference on Mass Spectrometry and Allied Topics, Palm Springs, CA, 1997.
20. Hozalski, R. M. Personal communication, 1998.

Chapter 24

# Environmental Applications of Novel, Highly Enhancing Substrates for Surface Enhanced Raman Scattering

**Lin He, Shawn P. Mulvaney, Sarah K. St. Angelo, Bonnie E. Baker, and Michael J. Natan**

**Department of Chemistry, The Pennsylvania State University, University Park, PA 16802**

Surface enhanced Raman scattering (SERS) has the potential to be an ultrasensitive technique for environmental analysis. Highly enhancing SERS substrates assembled by evaporation of a thin Ag film over a Ag-coated, colloidal Au submonolayer have been developed. Optimal particle size and coverage was determined by combinatorial methods. Atomic force microscopy data and optical spectra have elucidated the island-like, discrete particle nature of the substrate surface. SERS data on 1,3,5-tri(pyridine)triazine and terbutryn illustrate possible applications.

Increased concern about the fate of pollutants in the biosphere mandates improvements in detection technology, both in ease of data acquisition and measurement sensitivity. Surface enhanced Raman scattering (SERS) is a technique for acquiring Raman spectra of analyte molecules in close proximity to roughened noble metal surfaces. Though largely untested, the notion of using Raman spectroscopy in environmental analysis is attractive for several reasons. First, Raman spectra comprise fingerprint-like collections of vibrational data that, like infrared spectra, are molecule-specific (*1, 2*). Second, $H_2O$ is a poor Raman scatterer, simplifying measurements in "nature's solvent". Third, recent advances in instrumentation (*3, 4*) have made Raman spectroscopy both inexpensive and quite sensitive. Nevertheless, for detection of species present at the sub-microgram level, the additional signal available through SERS enhancement is required.

SERS experiments typically involve adsorption of an analyte in aqueous solution onto a SERS-active substrate (followed by acquisition of the Raman spectra via laser excitation). There are a limited number of applications of SERS in analytical

[1]Corresponding author.

chemistry. This is due to the fact that, despite numerous approaches to substrate fabrication (*5, 6*), the requisite combination of reproducibility, ease of manufacture, low cost, simplicity of use, durability and enhancement factor have not yet been attained.

Our research group has been focused on development of SERS as a technique for environmental analysis. Our efforts have been concentrated in two areas. The first is the integration of SERS detection into conventional capillary electrophoretic separations (*7*), with an eye toward analysis of disinfection byproducts. Results in this area will be described elsewhere (*8*). The second, which builds on our considerable experience with colloidal metal nanoparticles (*9-11*), is fabrication of substrates that meet the criteria delineated above. Aspects of our progress in this area are described herein; a more detailed study will be presented elsewhere (*12*).

# Experimental

<u>Materials</u>. Glass slides used were Fisher Premium Microscope Slides. 18.2 M$\Omega$ H$_2$O was distilled through a Barnstead Nanopure system. 3-Mercaptopropylmethyl-dimethoxysilane (MPMDMS) and 3-Mercaptotrimethoxysilane (MPTMS) were obtained from United Chemical. HAuCl$_4$·3H$_2$O, AgNO$_3$, trisodium citrate dihydrate, NaOH, *trans*-1,2-bis(4-pyridyl)ethylene (BPE), and N,N-dimethyl-4-nitrosoaniline (p-NDMA) were obtained from Aldrich. BPE was recrystallized in CH$_3$OH several times before use. A stock 10 mM BPE solution was then prepared in a fresh H$_2$O:CH$_3$OH (9:1) solvent solution. Dilutions in H$_2$O were made as needed. 12-nm colloidal Au was prepared from HAuCl$_4$·3H$_2$O reduced by citrate, as previously described. All references to "12-nm colloidal Au" particles in this text were nearly spherical, with a standard deviation less than 10%. Multiple preparations were used in this work. For each, nanoparticles were sized by transmission electron micrographs (>500 particles per image) using NIH Image software. LI Ag solution purchased from Nanoprobes was used in the electroless plating process and used as directed. Concentrated HCl, HNO$_3$, H$_2$SO$_4$, and 30% H$_2$O$_2$ were purchased from J. T. Baker Inc. Spectrophotometric grade CH$_3$OH was obtained from EM science. Ag (99.99%) and Au (99.99%) used for thermal evaporation were obtained from Johnson-Matthey Corp.

<u>3-Layer Substrate Preparation</u>. Glass slides were cut and cleaned as previously reported or through successively sonicating for 20 minutes in water, methanol, and acetone. Glass slides were functionalized in a 5-10% solution of MPMDMS in methanol for one hour. Copious rinsing with methanol and then water removed any unbound silane. Functionalized substrates were coated with a 12-nm colloidal Au solution for 90, 95, or 140 s, depending on colloid concentration and coverage desired. Substrates were then rinsed with water and partially dried under an Ar stream. Further exposure to LI Ag solution for 24 minutes produced a 2-layer SERS-active substrate. The first two layers of combinatorial substrates were assembled as previously reported. The third layer of non-continuous Ag film was thermally evaporated on top of two-layer substrates. Optical spectra were taken after each step to monitor the surface quality.

<u>Dislodgment of Particles from the Substrate.</u> Dislodging the particles from the slides was accomplished through sonication for 60 min. either in $H_2O$ or a NaOH solution (pH ca. 12). One slide was sonicated in a given solution at a given time. The concentration of the dislodged particles in solution was increased by sonication of three substrates in any given solution. Optical data was recorded for both the slide and solution after each step. The dislodged particle solutions were then centrifuged for 15 min. at 11.5 x $10^3$ rpm.

<u>Instrumentation.</u> *Film Evaporation:* Ag films were thermally evaporated in an Edwards Auto 306 evaporation system. Metal deposition occurred at a pressure of 2 x $10^{-6}$ mbar and at various deposition rates, with constant sample rotation to ensure even evaporation. All substrates were used immediately after evaporation.

*Surface Enhanced Raman Scattering:* SERS spectra were obtained on a Solution 633 micro-Raman spectrometer purchased from Detection Limit, Inc. Excitation was provided by a 30-mW, 632.8-nm HeNe laser with a 7 cm$^{-1}$ bandpass. All spectra were taken with a 3-mm focal-length objective (~5-μm diameter spot size). Spectra were collected by a CCD detector with TE-cooling system around -8.5 °C. The system was operated and monitored by Labview software on a Monorail laptop (Monorail Corp.). SERS response was optimized by manually adjusting the height of the probe that was attached to a translation stage. Spectral analysis was handled with either Grams 32 or Igor Pro software packages.

*Atomic Force Microscopy:* All AFM images were measured on a Digital Nanoscope IIIa instrument (Digital Instruments, CA) in the tapping mode. Images were captured using Nanoscope IIIa version software. Samples coated with evaporated Ag film were measured directly, while colloidal Au and 2-layer substrates were dried under an Ar stream before measurement.

## Results and Discussion

**Substrate Morphology:**

Many materials can give rise to the enhancement effect in SERS experiments; however, the most often employed are roughened noble metal films of Ag or Au (*2*). Scheme 1 is a cartoon depicting the assembly of a 3-layer, SERS-active substrate from Au and Ag. First, monodisperse colloidal Au particles are immobilized on a silanized glass support.

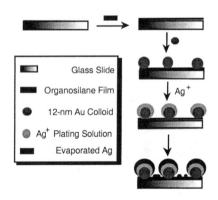

| | |
|---|---|
| ▭ | Glass Slide |
| ▬ | Organosilane Film |
| ● | 12-nm Au Colloid |
| ● | $Ag^+$ Plating Solution |
| ▬ | Evaporated Ag |

*Scheme 1. 3-Layer Substrate Fabrication.*

Film integrity is monitored using UV-visible absorbance spectroscopy, with quality control dependent upon the shape and height of the Au plasmon band (*11*). Suitable substrates are then treated with a commercial $Ag^+$ plating solution, yielding a Ag-clad, colloidal Au microarray (2-layer substrate). A corresponding shift in the optical properties, namely the rise of a shoulder around 620 nm attributed to inter-particle coupling, is seen after the Ag cladding is introduced. Finally, a discontinuous film of Ag is evaporated over the 2-layer substrate giving rise to the final 3-layer geometry. Optical spectra for 3-layer substrates demonstrate a more pronounced shoulder due to increased particle coupling on the densely-covered surface. No special concerns beyond practicing basic laboratory safety need be considered when preparing these substrates.

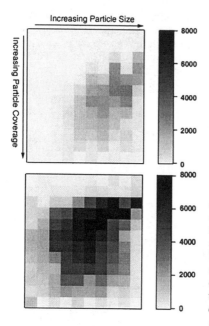

*Figure 1. Combinatorial examination of 2-layer (top) and 3-layer (bottom) substrates with varying particle coverage and size. SERS intensity (cps) is plotted for the 1198 $cm^{-1}$ peak of BPE as a function of the sample position. Shaded boxes represent a 0.2 $cm^2$ area. Acquisition parameters: 20 mW of 632.8 nm excitation, 2-s integration, 8 mm focal length, integration range = 1170 $cm^{-1}$ to 1220 $cm^{-1}$, reference range = 1120 $cm^{-1}$ to 1170 $cm^{-1}$.*

The general combinatorial approach to assembling 2-layer substrate was described in the previously published report (*10*). Briefly, this method comprises sequential, time-controlled translation of substrates first into colloidal Au then a solution of $Ag^+$. The translations produce gradients in particle coverage and particle size, respectively. Those nanostructures giving rise to highest SERS enhancement factors are elaborated via a positional map of SERS intensities for adsorbed analyte. Figure 1 (top) shows one such map for the 1198 $cm^{-1}$ peak of BPE. Note that the most active region of the surface is not that containing the highest density of large particles. A similar map can be generated after uniform gas-phase deposition of a Ag film (Figure 1, bottom). It is interesting to note that evaporation of 20 nm of Ag, previously determined to be optimal (*12*), changes the physical location of the most enhancing region of the surface, as well as increasing by at least a factor of two the overall SERS enhancement. Higher SERS intensity observed at the slightly smaller Au colloidal particle coverage can be explained as the result of the formation of small Ag islands between the larger particles.

The nanoscale surface morphology of a 3-layer substrate can be seen in atomic force micrographs illustrated in Figure 2. The 1 μm × 1 μm image (left) shows the island-like nature

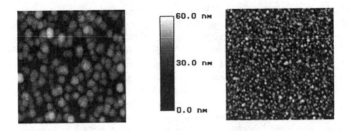

*Figure 2. Representative AFM images of 3-Layer substrates; (left) 1 μm × 1 μm image, (right) 3 μm × 3 μm image.*

of the metal film. Large surface features surrounded by small island-like features are clearly seen and assigned to Ag-clad, colloidal Au particles and inter-particle, evaporated Ag islands, respectively. Regularity of surface structure can be seen in the 3 μm × 3 μm image (right). This regularity helps explain the repeatability in SERS response observed in spectra taken at various places on the substrate surface. The surface architecture depicted in these representative micrographs has been confirmed with field emission scanning electron microscopy.

The island-like surface features, which are responsible for the additional enhancement effect of 3-layer substrates, appear to be discrete particles. An effective method for tracking this characteristic is by monitoring the metal plasmon band with UV-visible spectroscopy. The upper panel of Figure 3 shows the UV-visible spectra of 12-nm colloid (the first layer) in various states of relation to an MPTMS modified glass substrate. The trace a corresponds to the surface confined colloidal Au, while the other traces show colloidal Au dislodged from and remaining on the substrate

*Figure 3. (Upper) UV-visible spectra of 12-nm colloidal Au in solution and affixed to a MPTMS coated glass support both before and after 1 hr of sonication. (Lower) UV-visible spectra of Ag-clad, Au particles in solution and affixed to a MPTMS coated glass support both before and after 1 hr of sonication.*

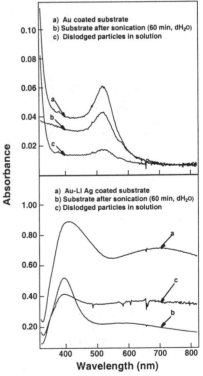

surface after 60 minutes of sonication. A significant percentage of the colloid has been dislodged from the substrate surface with sonication. The two post-sonication spectra demonstrate no shift in the plasmon band's $\lambda_{max}$, indicating that sonication neither altered the particle size nor induced aggregation. The consistent position of the $\lambda_{max}$ also indicates that the particles are discrete and do not interact significantly on the glass surface nor when they are in solution. Finally, the sum of the absorbance maxima for the dislodged and the remaining surface confined particles approximately equals that of the original substrate, further implying little or no particle fragmentation or agglomeration during sonication. The lower panel of Figure 3 shows UV-visible spectra that follows the dislodgment of Ag-clad, colloidal Au particles from a 2-layer substrate; the blue shifted plasmon band is attributed to the Ag-cladding, while the second band arises from inter-particle coupling as mentioned earlier. Again, sonication proves to be a viable means of dislodging these particles. However, the position of the plasmon band's $\lambda_{max}$ for both post-sonication spectra is blue-shifted with respect to the spectrum of the original substrate. This can be attributed to a lower degree of inter-particle interaction after the sonication treatment. Interestingly, this particular particle geometry would be difficult to recreate by solution synthesis; the portion of the Au particle affixed to the surface should not be coated by the Ag plating solution. However, at this time we have no proof discrete Au and Ag domains remain on the dislodged particles. In fact, the blue shift in the plasmon band could be attributed to the completion of the Ag-cladding.

**Composition:**

The enhancement effect of substrates is not only morphology dependent, as demonstrated in Figures 2 and 3, but it depends critically on the material composition of the metal film. It is widely reported in the SERS literature that the coinage metals, Cu, Au, and Ag have the highest enhancement factors, with the latter giving rise to the largest enhancement. This is demonstrated experimentally in Figure 4. SERS spectra are shown for 0.5 mM p-NDMA on a 3-layer substrate, a 2-layer substrate, and a 3-layer substrate where the evaporated Ag layer is replaced with an evaporated Au layer of the same thickness. The data indicate that the two substrates with an outer Ag-cladding (either 2-layer or 3-layer) demonstrate a significant enhancement advantage over the Au clad substrate. This trend has been confirmed with several SERS active molecules.

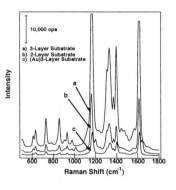

*Figure 4. SERS spectra of 0.5 mM p-NDMA on a Ag coated, 3-layer substrate, 2-layer substrate, and a Au coated, 3-layer substrate (12 nm colloidal Au, LI Ag layer, 20 nm evaporated Au). Acquisition parameters: 12 mW of 632.8 nm excitation, 1-s integration time, 3 mm focal length, and 20 µL samples.*

While a pure Ag film might in theory lead to a more enhancing SERS substrate, it is not practical to construct 3-layer substrates without the colloidal Au monolayer. Monodisperse colloidal Au, readily synthesized with a standard deviation of less than 10%, provides a highly regular foundation for construction of 3-layer substrates. This regularity is not easily attained with colloidal Ag because of the extreme difficulty of producing Ag sols with a standard deviation less than 10%. A 3-layer substrate with a polydispersed colloidal Ag foundation may lead to some surface sites with a greater SERS enhancement than 3-layer substrates, but will simultaneously negate the ability to prepare reproducible substrates, as well as to compare data taken at different spots on the substrate. Thus, by taking advantage of the physical properties of colloidal Au and the enhancement properties of the Ag-cladding, 3-layer substrates exhibit great promise in SERS application.

**Application:**

Measurements were conducted to evaluate the analytical merit of the newly developed SERS substrates using 1,3,5-tri(pyridine)triazine, a common environmental waste compound (Figure 5). Only 5 seconds of integration time are needed to collect a well-resolved SERS spectrum of a 60 pmols deposited on a 3-layer substrate. While a peak at 1000 cm$^{-1}$ was barely noticeable when the amount of deposited compound decreased to 60 fmols, a more discernible band can be observed by increasing the integration time. Analyte detection is further demonstrated with terbutryn, a commonly used pesticide. As can be seen in Figure 6, a well defined spectrum can be

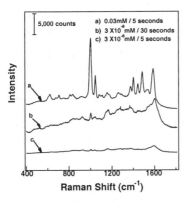

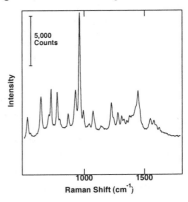

*Figure 5. SERS spectra of 1,3,5-tri(pyridine)triazine on a 3-layer substrate. Acquisition parameters: 20 mW of 632.8 nm excitation, 3 mm focal length, analyte concentration and integration time indicated on the graph.*

*Figure 6. SERS spectrum of terbutryn on a 3-layer substrate. Acquisition parameters: 12 mW of 632.8 nm excitation, 30-s integration time, 3 mm focal length, and 20 µL sample.*

gathered at a low excitation power with less than a minute of integration time when using 3-layer substrates. These two applications illustrate the promising future of 3-layer substrates in ultrasensitive detection.

## Conclusions

These data have shown that 3-layer substrates comprise viable substrates for environmental SERS analyses. Evaporation of a 20-nm thick Ag coating onto a combinatorially optimized two-layer architecture leads to increased SERS enhancement. The nanometer-scale morphology leading to increased enhancement was elaborated by atomic force microscopy. UV-vis spectroscopy was used to show how the particles comprising the discontinuous film could be dislodged from the substrate, leading to a loss in interparticle coupling. Finally, SERS spectra of representative analytes of environmental interest show that the requisite sensitivity can be obtained. Accordingly, our efforts in this area are continuing.

## Acknowledgements

NSF (CHE-9627338), NIH (DK48784-02), USDA (96-35102-3840) and EPA (R825363-01-0) are gratefully appreciated for support.

## References

1. Gerrard, D. L. *Anal. Chem.* **1994**, *66*, 547R-557R.
2. Creighton, J. A. *Surface Enhanced Raman Spectroscopy*; Neagle, W. and Randell, D. R., Eds. The Royal Society of Chemistry: Cambridge, UK, 1990, pp 13-26.
3. Asher, S. A.; Munro, C. H.; Chi, Z. *Laser Focus World* **1997**, *33*, 99-109.
4. Lyon, L. A.; Keating, C. D.; Fox, A. P.; Baker, B. E.; He, L.; Nicewarner, S. R.; Mulvaney, S. P.; Natan, M. J. *Anal. Chem.*, **1998**, *70*, 341R-361R.
5. Maya, L.; Vallet, C. E.; Lee, Y. H. *J. Vac. Sci. Technol, A* **1997**, *15*, 238-242.
6. Hulteen, J. C.; Van Duyne, R. P. *J. Vac. Sci. Technol., A* **1995**, *13*, 1553-1558.
7. Engelhardt, H.; Beck, W.; Kohr, J.; Schmitt, T. *Angew. Chem. Int. Ed. Engl.* **1993**, *32*, 629-649.
8. He, L.; Natan, M. J. *in preparation.*
9. Freeman, R. G.; Grabar, K. G.; Allison, K. A.; Bright, R. M.; Davis, J. A.; Guthrie, A. P.; Hommer, M. B.; Jackson, M. A.; Smith, P. C.; Walter, D. G.; Natan, M. J. *Science* **1995**, *267*, 1629-1632.
10. Baker, B. E.; Kline, N. J.; Treado, P. J.; Natan, M. J. *J. Am. Chem. Soc.* **1996**, *118*, 8721-8722.
11. Grabar, K. G.; Brown, K. R.; Keating, C. D.; Stranick, S. J.; Tang, S.-L.; Natan, M. J. *Anal. Chem.* **1997**, *69*, 471-477.
12. He, L.; Mulvaney, S. P.; Natan, M. J. *in preparation.*

# Chapter 25

# Identification of Polar Drinking Water Disinfection By-Products Using Liquid Chromatography–Math Spectometry

**Susan D. Richardson, Tashia V. Caughran, Thomas Poiger[1], Yingbo Guo[2], and F. Gene Crumley**

**National Exposure Research Laboratory, U.S. Environmental Protection Agency, Athens, GA 30605**

A qualitative method using 2,4-dinitrophenylhydrazine (DNPH) derivatization followed by analysis with liquid chromatography (LC)/negative ion-electrospray mass spectrometry (MS) was developed for identifying polar aldehydes and ketones in ozonated drinking water. This method offers advantages over the currently accepted method using pentafluorobenzylhydroxylamine (PFBHA) derivatization and gas chromatography/mass spectrometry (GC/MS) analysis, in that it allows for the detection of highly polar carbonyl compounds (with multiple polar substituents) and produces mass spectra and chromatographic behavior that can be used to distinguish between aldehydes and ketones in ozonated water. Results for many polar-substituted aldehyde and ketone standards are presented, as well as the identification of polar disinfection by-products (DBPs) in ozonated drinking water from full-scale plants and laboratory-scale ozonations of humic acid.

Although chlorine has been used to disinfect drinking water for approximately 100 years, there have been concerns raised over its use due to the formation of some potentially hazardous by-products. Because of these concerns, alternative disinfectants are being explored. Ozone is one of the most popular alternatives, as it is effective against resistant microorganisms, and it does not form the chlorine-containing by-products that are of concern. However, there is still much not known about the disinfection by-products (DBPs) formed by ozone. Ozone DBPs that have been reported to-date include aldehydes, ketones, keto-aldehydes, carboxylic acids, aldo-acids,

[1]Current address: Swiss Federal Institute for Fruit Growing, Horticulture and Viticulture, Schloss, CH–8820 Wädenswil, Switzerland.
[2]Current address: Metropolitan Water District of Southern California, La Verne, CA 91750–3399.

hydroxy-acids, alcohols, esters, and alkanes (*1-12*). The major uncertainty relative to ozone DBPs is the highly polar by-products that are believed to be present but have not been identified due to the difficulty in extracting polar compounds from water. A few polar DBPs (mostly low molecular weight aldehydes) have been identified using derivatization with pentafluorobenzylhydroxylamine (PFBHA) followed by gas chromatography/mass spectrometry (GC/MS) analysis. PFBHA imparts a nonpolar character to carbonyl-containing molecules, allowing them to be extracted from water and analyzed. This procedure is effective for those compounds that are amenable to gas chromatography and electron ionization (EI). However, for highly polar compounds, such as those with multiple hydroxyl or carbonyl groups, this method is not always effective, due to the absence of molecular ions or to poor chromatography. Recently, researchers have proposed double-derivatization/GC/MS methods for some of these highly polar chemicals. One such method employs methylation following reaction with PFBHA (*13*); another method involves silation with bis(trimethylsilyl)trifluoroacetamide (BSTFA) (*14*) following reaction with PFBHA. The first method has been shown to be useful for aldo- and keto-acids, and the second method is effective for hydroxy-ketones and aldehydes.

The focus of this paper involves a method using derivatization with 2,4-dinitrophenylhydrazine (DNPH), followed by analysis using negative-ion electrospray liquid chromatography/mass spectrometry (LC/MS) (Figure 1). By using LC/MS instead of GC/MS, detection of highly polar compounds (with multiple polar substituents) is possible, as LC/MS is well suited for polar compounds. Because the DNPH derivatives are not pre-charged, they can be easily concentrated onto an ordinary C18 cartridge. This allows for both improved detection and removal of salts from the sample (salts are not retained on the C18 phase). In addition, although these derivatives are not pre-charged, they readily form negative ions in the electrospray interface, due to the acidity of the NH group.

DNPH                    Acetaldehyde                    Hydrazone Derivative

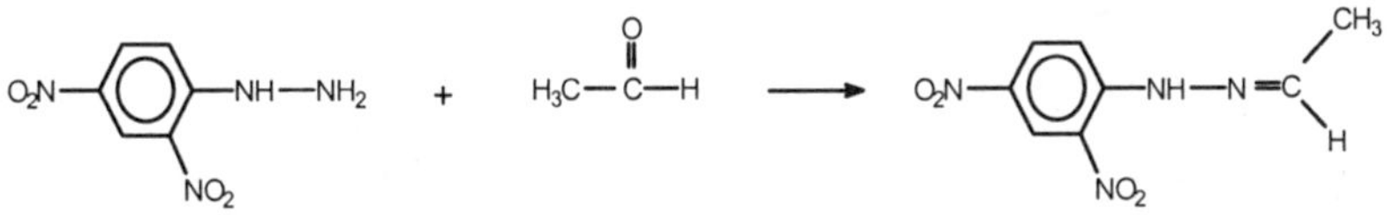

*Figure 1. DNPH derivatization procedure.*

The most common application of DNPH derivatization has been for the analysis of low molecular weight aldehydes and ketones in air (*15-18*). Air monitoring programs in both the United States and Europe employ DNPH derivatization followed by analysis with LC and UV detection (*15, 19-21*). Recently, two groups have used atmospheric pressure chemical ionization (APCI)-MS with DNPH derivatization to analyze carbonyl compounds in air (*17-18*). Because this DNPH derivatization has shown promise for analyzing polar carbonyls, it was investigated with electrospray LC/MS with the hope of identifying new, polar DBPs in ozonated drinking water.

# Experimental

## DNPH Derivatizations

Derivatizations were carried out by a modified procedure similar to that published by Grosjean and Grosjean (*15*). One liter of drinking water was first acidified to pH 2 with 2 mL of HCl, to which was added 4 mL of a 5 mg/mL solution of DNPH in acetonitrile, which was allowed to react for either 1 or 4 hr. For most compounds, a 1-hr reaction time was sufficient; however, di-carbonyl compounds required a longer derivatization time (4 hr) to produce more complete derivatization. A reaction time of 16 hr was also investigated, but was not found to increase the levels of derivatives beyond the 1 or 4 hr reaction time. Following the reaction, 100 mg of sodium phosphate was added. Samples were then adjusted to pH 5 with 1 M sodium hydroxide and extracted onto a C-18 Empore disk (3M Corp.) that had been preconditioned with acetonitrile and distilled water. The pH adjustment was made to minimize acid degradation of the Empore disk and the LC column and electrospray source to be used for the analysis. The Empore disk was then rinsed with 20 mL of purified water, dried under vacuum, and eluted with 20 mL of acetonitrile. Evaporation of the acetonitrile revealed bright yellow-orange crystals that were re-dissolved in a mixture of 1:1 acetonitrile/water to a volume of 750 μL. This 50:50 mixture of acetonitrile/water was used so that the solvent composition of the sample injected would match the initial composition of the LC gradient used (50:50 acetonitrile/water).

## LC/MS Analyses

LC/MS analyses were accomplished using a Hewlett Packard 1050 LC coupled with a VG Platform quadrupole mass spectrometer. Electrospray ionization was used with a flow rate of 0.3 mL /min and a source temperature of 160°C. Scans were performed over a *m/z* range of 30 to 700 Da at unit resolution, and the cone voltage was alternated between 15V and 30V to alternately obtain mass spectra containing mostly unfragmented molecular ions (15V) and collisionally induced dissociation (CID) spectra (30V). The CID spectra produced contained many structurally useful fragment ions. Samples (50 μL) were injected onto a Supelco Supelcosil C18 LC column (5 μm particle size, 150 x 2.1 mm i.d.), which was eluted using a gradient of 50:50 acetonitrile/water to 98:2 acetonitrile/water over 60 min. Diode array UV spectra (200 to 450 nm) were recorded simultaneously with mass spectra using a Hewlett Packard 1050 LC. Parent ion MS/MS analyses were performed at Research Triangle Institute on a PE Sciex API 365 triple quadrupole mass spectrometer.

## Purification of Water for Analysis of Standards and Laboratory-Scale Ozonations

For the derivatization and analysis of standards, water was purified by reacting doubly distilled water with excess DNPH and distilling over the purified water. In this way, we were able to remove numerous carbonyl-containing impurities detected in several sources of purified water, including Milli-Q water, distilled water, and reagent water purchased from Fisher Scientific. Munch and coworkers (22) have observed similar carbonyl impurities in purified and bottled reagent water, and reported that purified water rapidly absorbs volatile carbonyl compounds from the air and that a newly opened bottle of reagent water can contain μg/L levels of carbonyl impurities.

## Synthesis and Purification of *Syn-Anti* DNPH Isomers

To isolate and purify the *anti* isomers from the *syn* isomers, the following procedure was used (23). A solution of 2,4-dinitrophenylhydrazine was prepared by dissolving 1.0 g of 2,4-dinitrophenylhydrazine in 5 mL of concentrated sulfuric acid and was added to a mixture of 7 mL of water and 25 mL of ethanol. Derivatives of aldehyde and ketone standards were prepared in the following manner. A solution of the aldehyde or ketone (1.58 mmol dissolved in 5 mL of ethanol) was added to 10 mL of the DNPH solution. The product precipitated immediately as yellow-orange crystalline material. This solution was stirred at room temperature for 5 min, and the precipitate was collected on a fritted glass funnel. About half of the material was removed, dried under vacuum, and recrystallized from ethanol. The remainder was washed with 20 mL of a 5 % aqueous solution of $NaHCO_3$, filtered, and washed again with 20 mL of DNPH-purified water. The crystals thus obtained were dried under vacuum and recrystallized from ethanol. These remaining crystals have been shown to be the *anti* form of the derivatized aldehyde or ketone (24), with the *syn* form being selectively removed.

$^1$H Nuclear magnetic resonance (NMR) spectroscopy was used to confirm that the *anti* form was purified, according to a previously published study (described later) (24). For this work, $^1$H NMR spectra were recorded on a Bruker 300 MHz Fourier transform high resolution spectrometer, using $CDCl_3$ as the solvent.

## Laboratory-Scale Ozonations of Humic Acid

Laboratory-scale batch ozonations were carried out at pH 7 by the addition of ozone (from an approximately 30 mg/L stock solution of ozone in water, at a dose of 2:1 ozone to dissolved organic carbon) to solutions of Suwannee River humic acid in DNPH-purified water (5 mg/L), containing 1 mM sodium bicarbonate buffer. Sodium hydroxide was used to adjust the pH to 7. Ozone stock solutions were prepared by sparging ozone (from an Orec ozone generator) into DNPH-purified water maintained close to 0°C. Ozone concentrations were determined by measuring the UV absorbance (at 254 nm) of the stock solution, and based on the actual concentration, an appropriate dose was applied to the humic acid solution to achieve a dose of 2:1 ozone to dissolved organic carbon. A purified water blank and a raw water blank (containing humic acid and buffer, but not ozonated) were derivatized and analyzed in the same manner as the ozonated samples, so that actual DBPs could be distinguished from impurities.

## Ozonations at Full-Scale Treatment Plants

(1) Valdosta Water Treatment Plant. The Valdosta Water Treatment Plant, located in Valdosta, GA, is a 15 mgd facility that treats ground water (1.0 mg/L total organic carbon [TOC]) obtained from a nearby well field. Raw water is first subjected to air stripping (to remove $H_2S$), after which ozone is applied at a dose of 3.0 mg/L and a contact time of 90 sec. Samples were collected following ozonation. Chlorine is then applied at the plant to achieve a residual of 1.7 mg/L for the finished water; however, those samples are not addressed in this study. Raw water was also collected prior to ozonation as a blank to distinguish DBPs from raw water impurities. (2) Lanier Water Treatment Plant. The Lanier Water Treatment Plant, located in Gwinnett County, GA (in metropolitan Atlanta), is a 150 mgd facility that treats surface water (1.2 mg/L TOC) pumped from Lake Lanier. This water is collected into a 37-million-gallon reservoir, after which it is ozonated at a dose of 0.5 mg/L and a contact time of 4 min. Samples were collected following ozonation and also prior to ozonation to distinguish DBPs from raw water impurities. The ozonated water is then treated at the plant with ferric chloride and dimethylamine-type polymers, followed by filtration through anthracite and granular activated carbon (GAC). Finally, chlorine is added to achieve a residual of 1.6 mg/L for the finished water. Samples treated with post-chlorination are not addressed in this study. Samples were derivatized the same day they were collected.

## Chemical Standards

Standards of 2-hydroxybutanal, 3-hydroxybutanal, 4-hydroxybutanal, 1,4-butanedial, 2-ketobutanal, and 3-ketobutanal were synthesized under contract by CanSyn Corp. (Toronto, ON). Standards of 6-hydroxy-2-hexanone and 5-ketohexanal were synthesized under contract with Majestic Research, Inc. (Athens, GA); these synthesis procedures are reported elsewhere (12). All other chemical standards were purchased at the highest available purity from Aldrich.

# Results

Derivatization with a pre-charged derivatizing agent is often used by LC/MS researchers to enhance the signal of charged ions (usually positive). However, when attempting to identify low μg/L levels of drinking water DBPs in a complex mixture, this approach was not effective. Once formed, pre-charged derivatives are not easily separated from salts, are difficult to concentrate, and are not easily separable by LC. The presence of salts enables the formation of confusing adduct ions (e.g., $[M+Na]^+$, $[M+Ca]^+$) in the mass spectrometer, and without concentration of the sample, MS signals are not of sufficient quality to allow unknown identifications to be made. As a consequence, a different approach was applied, whereby derivatives are not pre-charged, but can be readily charged in the electrospray interface.

Derivatization with DNPH accomplishes this (Figure 1). Similar to the PFBHA reaction, the $NH_2$ group of the DNPH molecule reacts with carbonyl groups, forming a hydrazone. Because the derivative is not pre-charged, it can be easily concentrated onto an ordinary C18 cartridge. This allows for improved detection and for removal of salts (salts are unretained on the C18 phase). However, because the two nitro groups render the NH group acidic, negative ions are formed with abundance with electrospray-MS, allowing for detection of drinking water DBPs at low μg/L levels (for a 1 L sample of water). DNPH derivatization followed by LC/MS analysis enabled the analysis of many carbonyl compounds with polar functional groups that are not effectively analyzed by other methods, in addition to the analysis of traditional unsubstituted aldehydes and ketones that have been previously reported as ozone DBPs. Detection limits achieved with this method were 25 pg, or 0.5 μg/L for a 1L water sample (as determined for octanal and 3-hydroxy-2-butanone). If a larger sample of water is taken, detection limits can be lowered to sub-μg/L levels. Table 1 lists the compounds that were analyzed by this procedure.

Most of the derivatized aldehydes and ketones studied shared common fragments that were observed in the mass spectra obtained using in-source, collisionally induced dissociation (CID) at a cone voltage of 30V. These common fragments included $(NO_2)^-$, $(M-NO)^-$, and $(M-HNO_2)^-$. Because these ions were fairly consistent, neutral loss scans for these ions could be used to selectively detect these DNPH derivatives, eliminating the chemical noise background. Figure 2 shows an example of an electrospray mass spectrum obtained for 2-pentanone at 15V and 30V.

This method was also used to distinguish aldehydes from ketones. With the PFBHA-GC/MS procedure, aldehydes and ketones can be distinguished from their EI spectra through careful comparisons of minor differences in the relative abundances (*11*); however, these determinations are tedious due to the similarity of fragment ions formed. With negative-ion electrospray/MS, using CID at a cone voltage of 30V, however, a unique ion (*m/z* 163) was formed only for aldehydes and not for ketones (as observed for a series of C3 to C10 aldehydes and ketones, and also for hydroxy- and carboxy-substituted aldehydes and ketones). This ion is evident in the CID mass spectrum of pentanal (Figure 3), and the absence of this ion is evident in the CID mass spectrum of a corresponding ketone, 2-pentanone (Figure 2). This unique fragment ion was attributed to fragmentation of the $CH-CH_2$ bond of the aldehyde (Figure 4), along with the loss of a $NO_2$ group, and would not be possible for ketones. To determine whether this fragmentation was sequential (i.e., initial loss of $NO_2$ followed by cleavage of the $CH-CH_2$ bond), or whether it occurred by a concerted mechanism, a parent ion MS/MS analysis was carried out of the *m/z* 163 ion. In this analysis, only one parent ion was evident—the molecular ion. Therefore, the formation of this unique aldehyde fragment ion appears to be occurring through a concerted mechanism. This fragment ion has also been recently reported in APCI-MS/MS studies of short-chain aldehydes in air studies carried out by Kölliker *et al.* (*17*), who attributed the stability of this ion to the formation of a pyrazole ring after loss of the nitro group at the 2-position.

**Table 1. Compounds Analyzed by the DNPH-LC/MS Method**

| Compound | MW | Derivatized MW |
|---|---|---|
| Aldehydes | | |
| *Formaldehyde | 30 | 209 |
| *Acetaldehyde | 44 | 223 |
| *Propanal | 58 | 237 |
| *Butanal | 72 | 251 |
| *Pentanal | 86 | 265 |
| *Hexanal | 100 | 279 |
| *Heptanal | 114 | 293 |
| *Octanal | 128 | 307 |
| *Nonanal | 142 | 321 |
| *Decanal | 156 | 335 |
| Ketones | | |
| *Acetone | 58 | 237 |
| *2-Butanone | 72 | 251 |
| *2-Pentanone | 86 | 265 |
| *2-Hexanone | 100 | 279 |
| 2-Heptanone | 114 | 293 |
| 2-Octanone | 128 | 307 |
| 2-Nonanone | 142 | 321 |
| 2-Decanone | 156 | 335 |
| Di-aldehydes | | |
| *Glyoxal | 58 | 417 |
| Keto-aldehydes | | |
| *Methyl glyoxal (2-ketopropanal) | 72 | 431 |
| *5-Ketohexanal | 114 | 473 |

**Table 1.** *Continued*

| Compound | MW | Derivatized MW |
|---|---|---|
| Hydroxy-aldehydes | | |
| 2-Hydroxybutanal | 88 | 267 |
| 3-Hydroxybutanal | 88 | 267 |
| 4-Hydroxybutanal | 88 | 267 |
| 2,5-Dihydroxybenzaldehyde | 138 | 317 |
| 2,2-Dimethyl-3-hydroxypropionaldehyde | 102 | 281 |
| Hydroxy-ketones | | |
| 1-Hydroxyacetone | 74 | 253 |
| 3-Hydroxy-2-butanone | 88 | 267 |
| 4-Hydroxy-2-butanone | 88 | 267 |
| 5-Hydroxy-2-pentanone | 102 | 281 |
| 4-Hydroxy-3-methyl-2-butanone | 102 | 281 |
| 3-Hydroxy-3-methyl-2-butanone | 102 | 281 |
| *6-Hydroxy-2-hexanone | 116 | 295 |
| 4-Hydroxy-4-methyl-2-pentanone | 116 | 295 |
| 1,3-Dihydroxyacetone | 90 | 269 |
| Aldo-acids | | |
| *Glyoxylic acid | 74 | 253 |
| Keto-acids | | |
| *Pyruvic acid (2-ketopropanoic acid) | 88 | 267 |
| 5-Ketohexanoic acid | 130 | 309 |
| *Ketomalonic acid (2-ketopropanedioic acid) | 118 | 297 |
| *Oxalacetic acid (2-ketobutanedioic acid) | 132 | 311 |

NOTE: Asterisks indicate ozone DBPs (from drinking water plant or laboratory-scale ozonation of humic material). Molecular weights for di-aldehydes and keto-aldehydes represent DNPH derivatives that have been derivatized at both carbonyl groups.

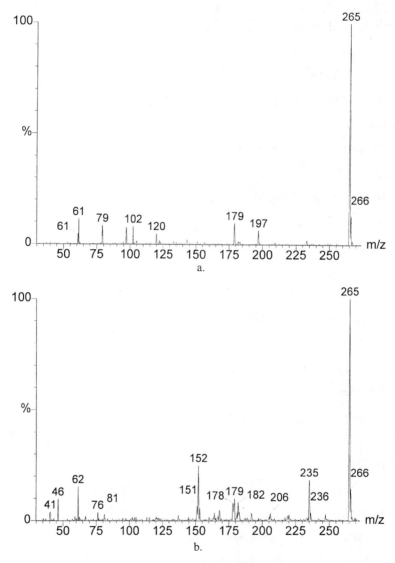

Figure 2. Electrospray mass spectra of 2-pentanone obtained at (a) 15V and (b) 30 V.

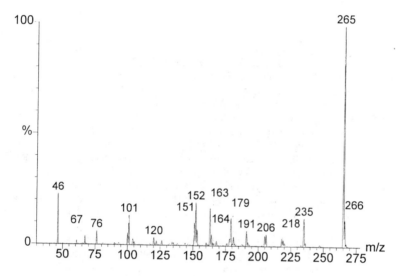

Figure 3. Electrospray mass spectrum of pentanal obtained at 30V.

Pentanal derivative
MW 265

2-Pentanone derivative
MW 265

*Figure 4. Fragmentation of DNPH-derivatized pentanal (left) and 2-pentanone (right).*

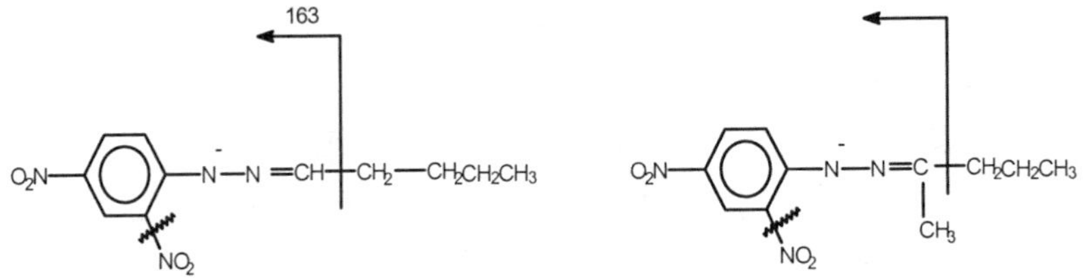

Aldehydes and ketones could also be distinguished by their LC chromatographic peaks. In the DNPH reaction, two isomers are formed (*syn* and *anti*). Using the chromatographic conditions reported here, ketones show 2 chromatographic peaks for the *syn* and *anti* isomers, whereas the aldehydes showed only one peak. Both *syn* and *anti* isomers are formed for the aldehydes, but they co-elute using our chromatographic conditions. We were able to determine this by changing to a more polar elution gradient (25:75 acetonitrile/water to 75:25 acetonitrile/water). Not only do straight chain aldehydes and ketones show this chromatographic behavior, but also hydroxy-branched aldehydes and ketones exhibit it, as well. Using a previously published method (*23*), the *anti* isomer of derivatized aldehyde and ketone standards were selectively purified. The identity of the *anti* isomer was confirmed with $^1$H NMR spectroscopy (*25*), and it was determined that the *anti* isomer was the second of the two isomers in the LC/MS chromatograms.

Diode array UV spectra were collected along with mass spectra to determine if the UV spectra could be used to help identify ozone DBPs. For the standards analyzed (Table 1), the UV spectra of the derivatized aldehydes and ketones were very similar. Small positive shifts of approximately 4-6 nm in the $\lambda_{max}$ were consistently observed when comparing the spectra of ketones to aldehydes with the same number of carbon atoms. Unsubstituted aldehydes (C3 to C10) showed a $\lambda_{max}$ in the range of 359 to 367 nm. The corresponding ketones (C3 to C10) showed a $\lambda_{max}$ in the range of 367 to 373 nm. However, this difference is probably too small to allow a definitive determination for an unknown compound (whether it was an aldehyde or a ketone). Hydroxy-substituted ketones analyzed (Table 1) showed a $\lambda_{max}$ in the same range as for the unsubstituted aldehydes (357 to 361 nm). The two aromatic aldehydes analyzed, 2,5-dihydroxybenzaldehyde and benzaldehyde, did show a substantial shift in the $\lambda_{max}$ to 393 nm and 390 nm, respectively. This suggests that the UV spectrum could possibly indicate the presence of aromaticity (when the carbonyl is in conjugation with the aromatic group). Pyruvic acid (a keto-acid) and ketomalonic acid (a keto-diacid) exhibited a $\lambda_{max}$ of 369 and 375 nm, respectively, which corresponds to the absorbance range observed for the non-substituted ketones.

Once it was established that polar carbonyl-containing compounds could be identified with this method, studies were conducted to determine whether some were present in actual ozonated drinking water samples. Three types of drinking water were analyzed: (1) laboratory-scale ozonations of Suwannee River humic acid, (2) drinking water from a full-scale ozone drinking water treatment plant in Valdosta, GA, and (3) drinking water from a full-scale ozone drinking water plant in Gwinnett County, GA (Lanier Water Treatment Plant). The criteria used for listing an identified compound as a DBP was its presence in the treated samples in quantities at least 2 to 3 times greater than in the untreated, raw water (as judged by comparing chromatographic peak areas). We believe it is important to distinguish a compound as a DBP, even if small amounts of the compound are present in the raw water. Many compounds that are common pollutants, or that are used industrially, have also been proven to be DBPs. Formaldehyde is one example—it is used industrially and is a common air contaminant, but it is also a proven ozone DBP. As a result, we did not want to omit potentially important DBPs, but we recognize, at the same time, that our criteria may screen out a few DBPs that are formed in small quantities. Compounds that were found to be DBPs are indicated with asterisks in Table 1. Polar-substituted aldehydes and ketones that were identified include pyruvic acid, glyoxylic acid, ketomalonic acid, 5-ketohexanal, 6-hydroxy-2-hexanone, and 1,3-dihydroxyacetone (Table 1). The compound identified as 1,3-dihydroxyacetone has not been reported previously and was found in water from one of the full-scale plants (Valdosta). It was judged to be a DBP because it was present in this ozonated sample, but not in the raw water blank. In the ozonated water, this compound showed a derivatized molecular ion of $m/z$ 269 in its electrospray mass spectrum and one chromatographic peak, indicative of an aldehyde or a symmetric ketone. Its 30V CID mass spectrum indicated a ketone structure, with a lack of a strong $m/z$ 163 ion in its mass spectrum (Figure 5). Therefore, from this information, and from the molecular weight, we believe that this compound is likely 1,3-dihydroxyacetone. A standard of 1,3-dihydroxyacetone was then analyzed, and both its mass spectrum and LC retention time matched that of our unknown in the ozonated water sample.

## Conclusions

In conclusion, the qualitative DNPH-LC/MS method described was a useful procedure for analyzing polar carbonyl-containing compounds in drinking water. With this procedure, many ozone DBPs were identified in drinking water at full-scale treatment plants and in drinking water obtained from the laboratory-scale ozonation of Suwannee River humic acid. Because this method can be used for analyzing highly polar DBPs with multiple polar substituents, it offers advantages to the PFBHA derivatization-GC/MS method. It also provides the advantage of allowing aldehydes to be easily distinguished from ketones by their chromatographic behavior and by a unique ion ($m/z$ 163) that is observed for aldehydes and not for ketones. However, because the detection limits for the DNPH-LC/MS method are not as low as for the PFBHA-GC/MS method (LC/MS is typically not as sensitive as GC/MS), it is recommended to use the DNPH method as a supplement to the PFBHA method—not as a replacement.

## Acknowledgments

We would like to thank Sandy Smith and Dieter Franz of the Lanier Water Treatment Plant in Gwinnett County, GA, and Bob Moore of the Valdosta Water

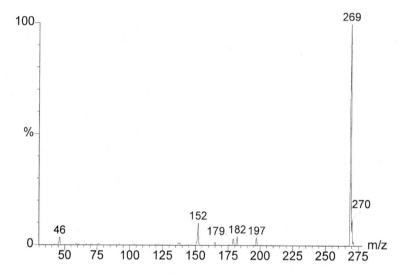

Figure 5. Electrospray mass spectrum of DBP identified as 1,3-dihydroxyacetone (30V).

387

Treatment Plant for graciously supplying ozonated drinking water samples. We would also like to thank Dr. Gary Siuzdak of Scripps Research Institute and Dr. Jeff Keever and Dr. Robert Voyksner of Research Triangle Institute for providing MS/MS spectra; and Dr. Guang-ri Sun for providing NMR analyses.

## References

1. Richardson, S.D. Drinking Water Disinfection By-products. In *The Encyclopedia of Environmental Analysis and Remediation*; Meyers, R.A., Ed.; John Wiley & Sons: New York, 1998; Vol. 3, pp. 1398-1421.
2. Glaze, W.H.; Weinberg, H.S. *Identification and Occurrence of Ozonation By-Products in Drinking Water*, American Water Works Association Research Foundation: Denver, CO, 1993.
3. Glaze, W.H.; Koga, M.; Cancilla, D. *Environ. Sci. Technol.* **1989**, *23*, 838.
4. Haag, W.R.; Hoigne, J. *Environ. Sci. Technol.* **1983**, *17*, 261.
5. Coleman, W.E.; Munch, J.W.; Ringhand, H.P.; Kaylor, W.H.; Mitchell, D.E. *Ozone Sci. Eng.* **1992**, *14*, 51.
6. Glaze, W.H. *Environ. Health Perspec.* **1986**, *69*, 151.
7. LeLacheur, R.M.; Sonnenberg, L.B.; Singer, P.C.; Christman, R.F.; Charles, M.J. *Environ. Sci. Technol.* **1993**, *27*, 2745.
8. Anderson, L.J.; Johnson, J.D.; Christman, R.F. *Org. Geochem.* **1985**, *8*, 65.
9. Lawrence, J.; Tosine, H.; Onuska, R.I.; Comba, M.E. *Ozone Sci. Eng.* **1992**, *14*, 55.
10. Killops, S.D. *Water Res.* **1986**, *20*, 153.
11. Richardson, S.D.; Thruston, Jr., A.D.; Caughran, T.V.; Chen, P.H.; Collette, T.W.; Floyd, T.L.; Schenck, K.M.; Lykins, Jr., B.W.; Sun, G.; Majetich, G. Identification of New Ozone Disinfection By-Products in Drinking Water, *Environ. Sci. Technol.*, in press.
12. Krasner, S.W.; Gramith, J.T.; Means, E.G.; Patania, N.L.; Najm, I.N.; Aieta, E.M. Formation and Control of Brominated Organic Disinfection By-Products. In *Proceedings (Water Quality for the New Decade)*, AWWA Annual Conference, Philadelphia, PA, June 1991. American Water Works Association: Denver, CO, 1991.
13. Xie, Y.; Reckhow, D.A.; Springborg, D.C. *J. Am. Water Works Assoc.* **1998**, *90*, 131.
14. Spaulding, R.S.; Frazey, P.; Rao, X.; Charles, M.J. *Anal. Chem.* **1999**, *71*, 3420.
15. Grosjean, E.; Grosjean, D. *Intern. J. Environ. Anal. Chem.* **1995**, *61*, 47.
16. Schlitt, H. *J. Chromatogr. A.* **1997**, *762*, 187.
17. Kölliker, S.; Oehme, M. *Anal. Chem.* **1998**, *70*, 1979.
18. Grosjean, E.; Green, P.G.; Grosjean, D. *Anal. Chem.* **1999**, *71*, 1851.
19. Riggin, R.M. *Compendium of methods for the determination of toxic organic compounds in ambient air.* EPA-600/4-84-04. U.S. Environmental Protection Agency: Research Triangle Park, NC, 1984.
20. Purdue, L.H.; Dayton, D.P.; Rice, J.; Bursey, J. *Technical assistance document for sampling and analysis of ozone precursors.* EPA-600/4-84-041. Atmospheric Research and Exposure Assessment Laboratory, U.S. Environmental Protection Agency: Research Triangle Park, NC, Oct 1991.

388

21. Intersociety Committee. *Methods of Air Sampling and Analysis*, 3rd Ed.; Lodge, J.P., Jr., Ed.; Lewis Publishers: Chelsea, MI, 1989, pp. 293-295.

22. Munch, J.W.; Munch, D.J.; Winslow, S.D.; Wendelken, S.C.; Pepich, B.V. A User's Guide to Aldehyde Analysis Using PFBHA Derivatization and GC/ECD Detection: Avoiding the Pitfalls. In *Proceedings*, AWWA Water Quality Technology Conference, San Diego, CA, Nov 1998. American Water Works Association: Denver, CO, 1998.

23. Behforouz, M.; Bolan, J.L.; Flynt, M.S. *J. Org. Chem.* **1985**, *50*, 1186.

24. Binding, N.; Muller, W.; Witting, U. *Fresenius J. Anal. Chem.* **1996**, *356*, 315.

25. Karobatos, G.J.; Vane, F.M.; Taller, R.A.; Hsi, N.J. *J. Am. Chem. Soc.* **1964**, *86*, 3351.

Chapter 26

# Identification of New Drinking Water Disinfection By-Products Formed in the Presence of Bromide

Susan D. Richardson[1], Alfred D. Thruston, Jr.[1], Tashia V. Caughran[1],
Paul H. Chen[1], Timothy W. Collette[1], Terrance L. Floyd[1],
Kathleen M. Schenck[2], and Benjamin W. Lykins, Jr.[2]

[1]National Exposure Research Laboratory, U.S. Environmental Protection
Agency, Athens, GA 30605
[2]National Risk Management Research Laboratory, U.S. Environmental
Protection Agency, Cincinnati, OH 45268

Using a combination of mass spectrometry and infrared spectroscopy,
disinfection by-products (DBPs) were identified in ozonated drinking
water containing elevated bromide levels, and in ozonated water
treated with secondary chlorine or chloramine. Only one brominated
by-product—dibromoacetonitrile—was found in the water treated
with only ozone. Many more by-products were identified when
secondary chlorine or chloramine was applied after ozonation. A
number of these by-products have not been reported previously.
When comparing low-bromide water to water with elevated bromide,
a tremendous shift in speciation was observed for samples treated with
secondary chlorine or chloramine. Without high bromide levels,
chlorinated species dominate (e.g., chloroform, trichloroacetaldehyde,
tetrachloro-propanone, dichloroacetonitrile, trichloronitromethane);
with elevated bromide levels (1 mg/L), these shift to brominated
species       (e.g.,      bromoform,            tribromoacetaldehyde,
tetrabromopropanone, dibromo-acetonitrile, tribromonitromethane).
An entire family of bromo- and mixed chlorobromopropanones was
identified that were not present in library databases, and have not been
reported previously. They were observed mainly in the ozone-
chloramine samples, but were also present in ozone-chlorine-treated
water. These brominated by-products were also observed in water
treated with only chloramine or chlorine.

# Introduction

Since the Safe Drinking Water Act (SDWA) was instituted, water treatment plants have had to control the levels of trihalomethanes (THMs) in drinking water. The Stage 1 DBP rule, which was promulgated in December 1998, reduces the maximum contaminant level (MCL) of THMs from 100 µg/L to 80 µg/L, and sets new MCLs for DBPs that have not been previously regulated (haloacetic acids [HAAs], bromate, and chlorite) (*1*). Because of these new regulations and also because of concern over the ability of chlorine to effectively inactivate microorganisms such as *Giardia* and *Cryptosporidium*, alternative disinfectants to chlorine are now seriously being considered by many water municipalities.

Ozone is one of the most popular alternatives to chlorine. Ozone is a strong oxidizer and disinfectant capable of inactivating *Giardia* and *Cryptosporidium*, and it does not usually produce THMs or other chlorinated DBPs. One drawback to the use of ozone, however, is the formation of bromate (BrO$_3^-$). Bromate is of concern because it has been shown to cause cancer in laboratory animals (*2*). Bromate is formed when elevated levels of bromide are present in raw water (*3*). Bromide is present in raw water of coastal U.S. cities and also in the groundwater of some cities in the Western U.S. Specifically in Texas, bromide is introduced into the groundwater through the use of brine for extracting crude oil. The highest bromide levels in the U.S. have approached 2 to 3 mg/L.

In addition to bromate, other brominated DBPs have been identified in waters with elevated bromide that were disinfected with ozone. Bromide can cause the formation of brominated DBPs through its reaction with ozone to form hypobromous acid (HOBr), which then reacts with humic material to form brominated compounds. Some previously reported brominated DBPs include hypobromite, bromoform, bromopicrin, dibromoacetonitrile, bromoacetone, cyanogen bromide, bromoacetic acids, bromo-ketones, bromonitriles, bromoalkanes, and bromohydrins (*3-9*). Health effects data is not available on a large number of these compounds, but there is concern because brominated compounds are generally more toxic than their chlorinated analogs.

There is much uncertainty over the by-products formed in waters high in bromide when ozone is the chosen disinfectant and when secondary treatments such as chlorine or chloramine are applied following treatment with ozone. Few studies from drinking water plants have addressed the effect of bromide, and it has been a goal of the U.S. Environmental Protection Agency to determine what by-products are formed under these conditions. It is important to determine if other brominated by-products, besides bromate, exist that pose health risks to consumers. Our objective was to identify all organic compounds in sample extracts obtained from ozone treatments of bromide-enriched water. Often newly identified DBPs are not found in mass spectral libraries; therefore, a combination of mass spectrometry techniques, as well as infrared spectroscopy, was used to identify the unknowns.

## Experimental

Ozonations were carried out at a pilot plant in Jefferson Parish, LA, which uses Mississippi River water as the raw water source. Because the raw water does not contain elevated levels of bromide (0.054 mg/L yearly average), sodium bromide was added to the raw water (prior to ozonation) at a concentration of 1 mg/L. Three rounds of treatment were performed, and all rounds included secondary treatment with chlorine or chloramine (to achieve a residual of 2 to 3 mg/L). Ozone doses were approximately 4.3, 3.0, and 4.3 mg/L for the three rounds studied. Samples were collected following filtration through dual media (sand and anthracite), which was not biologically active. Ozonated raw water without added bromide was also studied; those results can be found elsewhere (*10*). For each round, the untreated, bromide-enhanced raw water was collected to enable distinction between the actual DBPs of ozone and chemicals present in the raw water.

All water samples, except water used in pentafluorobenzylhydroxylamine (PFBHA) derivatizations and water analyzed for THMs and HAAs (described below), were concentrated by adsorption on Amberlite XAD resins. Details about the preparation of these resins can be found elsewhere (*11*). Water samples were acidified to pH 2 by the in-line addition of HCl, prior to passage through columns containing a combination of XAD-8 resin over XAD-2 resin. The columns were eluted with ethyl acetate. Residual water was removed from the ethyl acetate eluents by using separatory funnels to drain off the water layers, followed by the addition of sodium sulfate. After removal of an aliquot for mutagenicity testing, the ethyl acetate eluents (equivalent to approximately 75 L of treated water) were shipped on cold packs to the National Exposure Research Laboratory in Athens, GA. At the laboratory, the samples were concentrated to 1 mL by rotary evaporation.

In addition to the raw water controls, four blanks were also analyzed: (1) ethyl acetate passed through the XAD resins and concentrated in the same manner as the treated samples; (2) deionized, distilled water passed through the XAD resins and concentrated; (3) deionized, distilled water treated with chlorine and concentrated; and (4) deionized, distilled water treated with chloramine and concentrated. The latter two blanks were done to determine whether there were any artifacts due to reaction of secondary disinfectants with the ethyl acetate or with resin material.

For PFBHA derivatizations, 750 mL of treated water (and raw water as a control) were derivatized according to a procedure published by Sclimenti et al. (*12*). Derivatized aldehydes and ketones were then extracted with hexane and concentrated to 1 mL by rotary evaporation. Methylations were performed using $BF_3$/methanol (*13*); THMs were measured using EPA Method 551 (*14*); and haloacetic acids were measured using EPA Method 552.0 (*14*). Samples analyzed for THMs and HAAs were quenched with sodium sulfite prior to extraction. The method detection limit (MDL) for THMs was 0.1 µg/L; the MDLs for the HAAs were 0.60, 0.20, 0.24, 0.20, 0.25, and 0.20 µg/L for chloro-, bromo-, dichloro-, trichloro-, bromochloro-, and dibromoacetic acid, respectively.

High-resolution gas chromatography/electron ionization-mass spectrometry (GC/EI-MS) and gas chromatography/chemical ionization-mass spectrometry (GC/CI-MS) analyses were performed on a VG 70-SEQ high-resolution, hybrid mass spectrometer, equipped with a Hewlett Packard model 5890A gas chromatograph. The high-resolution mass spectrometer was operated at an accelerating voltage of 8 kV and a resolution of 10,000. Low-resolution GC/EI-MS experiments were carried out on the VG 70-SEQ mass spectrometer, and low-resolution GC/CI-MS experiments were carried out on a Finnigan TSQ 7000. Positive chemical ionization experiments were accomplished by using methane or 2% ammonia in methane gas. Injections of 1 to 2 μL of the extract were introduced via a split/splitless injector onto a J&W Scientific DB-5 chromatographic column (30-m, 0.25 mm i.d., 0.25 μm film thickness). The GC temperature program consisted of an initial temperature of 35°C, which was held for 4 min, followed by an increase at a rate of 9°C/min to 285°C, which was held for 30 min. Transfer lines were held at 280°C, and the injection port was controlled at 250°C.

Gas chromatography/infrared spectroscopy (GC/IR) analyses were performed on a Hewlett Packard Model 5890 Series II GC interfaced to a Hewlett Packard Model 5965B infrared detector (IRD). Spectra were generated at 8 cm$^{-1}$ resolution with a useful range of 4000 to 700 cm$^{-1}$. Injections of 2 μL of the extracts were introduced onto a Restek Rtx-5 column (30-m, 0.32 mm i.d., 0.5 μm film thickness) with a heated on-column injector (280°C). The GC temperature program consisted of an initial temperature of 35°C, which was held for 4 min, followed by an increase at a rate of 9°C/min to 280°C, which was held for 30 min. Transfer lines were held at 280°C, and the light pipe was controlled at 280°C.

A standard of tribromonitromethane (bromopicrin) was graciously provided by Sylvia Barrett and Stuart Krasner of the Metropolitan Water District of Southern California. A standard of dibromonitromethane was synthesized by Majestic Research, Inc. (Athens, GA) (15). All other chemicals and chemicals used in synthesis procedures were either purchased at the highest level of purity from Aldrich, Chem Service, or TCI America, or were previously synthesized and reported elsewhere (10).

## Results

The criteria we used for listing an identified compound as a DBP was its presence in the treated samples in quantities at least 2 to 3 times greater than in the untreated, raw water (as judged from comparing chromatographic peak areas). Although we did not want to omit potentially important DBPs, we recognize that our criteria may miss a few DBPs that are formed in small quantities.

Overall, many DBPs were identified, several of which have not been reported previously. The halogenated DBPs identified are listed in Table 1. Many of the compounds were not present in any spectral library (NIST or Wiley), and many of the ones that were in the libraries did not give conclusive library matches. CI-MS was used to determine molecular weights when molecular ions were not present in the mass spectra (and to confirm molecular ions present); high-resolution MS was used to determine the molecular formulas for the molecular ions and for fragments. GC/IR provided functional group information when needed to solve an unknown structure. Many of the identified

**Table 1. Halogenated DBPs Identified in High-Bromide Waters**

| DBP | Ozone | Ozone-Chlorine | Ozone-Chloramine |
|---|---|---|---|
| **Halo-alkanes/alkenes** | | | |
| Bromochloromethane | | X | |
| Dibromomethane | | X | X |
| Bromodichloromethane | | X | X |
| Dibromochloromethane | | X | X |
| Chloroform | | X | X |
| Bromoform | | X | X |
| Bromochloroiodomethane | | X | X |
| Tribromochloromethane | | X | X |
| 1,1-Dibromopropane | | X | X |
| 2,4-Dibromo-1-butene | | | X |
| 1-Bromohexane | | | X |
| **Halo-aldehydes** | | | |
| Bromochloroacetaldehyde | | X | X |
| Tribromoacetaldehyde (bromal hydrate) | | X | X |
| 2-Bromo-2-methylpropanal | | | X |
| 3-Bromo-4-hydroxy-5-methoxy benzaldehyde | | | X |
| **Halo-ketones** | | | |
| 1-Bromopropanone | | X | X |
| 1,1-Dibromopropanone | | X | X |
| 1,1,1-Trichloropropanone | | | X |
| 1,1-Dibromo-3-chloropropanone | | | X |

*Continued on next page*

**Table 1.** *Continued*

| DBP | Ozone | Ozone-Chlorine | Ozone-Chloramine |
|---|---|---|---|
| 1,1,1-Tribromopropanone | | X | X |
| 1,1,3-Tribromopropanone | | X | X |
| 1,1,3,3-Tetrachloropropanone | | | X |
| 1,1-Dibromo-3,3-dichloropropanone | | | X |
| 1,3-Dibromo-1,3-dichloropropanone | | | X |
| 1,1,3-Tribromo-3-chloropropanone | | | X |
| <u>1,1,3,3-Tetrabromopropanone</u> | | | X |
| 1,1,1,3,3-Pentachloropropanone | | X | |
| 1,1,1,3-Tetrabromo-3-chloropropanone | | X | |
| 1,1,1,3,3-Pentabromo-3-chloropropanone | | X | |
| 5-Bromo-2-pentanone | | X | |
| **Halo-acids** | | | |
| <u>Chloroacetic acid</u> | | X | |
| <u>Bromoacetic acid</u> | | X | X |
| <u>Dichloroacetic acid</u> | | X | X |
| <u>Bromochloroacetic acid</u> | | X | X |
| <u>Dibromoacetic acid</u> | | X | X |
| <u>Trichloroacetic acid</u> | | X | X |
| Dibromochloroacetic acid | | X | |
| <u>Tribromoacetic acid</u> | | X | |
| **Halo-acetonitriles** | | | |
| <u>Dichloroacetonitrile</u> | | | X |
| <u>Bromochloroacetonitrile</u> | | X | X |

**Table 1.** *Continued*

| DBP | Ozone | Ozone-Chlorine | Ozone-Chloramine |
|---|---|---|---|
| Dibromoacetonitrile | X | X | X |
| Tribromoacetonitrile | | X | |
| **Halo-alcohols** | | | |
| 2-Bromoethanol | | | X |
| 3-Chloro-2-butanol | | | X |
| **Halo-nitro-methanes** | | | |
| Bromonitromethane | | X | X |
| Dibromonitromethane | | X | X |
| Tribromonitromethane (bromopicrin) | | X | |
| **Halo-acetates** | | | |
| Methyl bromochloroacetate | | X | |
| Methyl dibromoacetate | | X | |
| **Halo-aromatics** | | | |
| Chloromethyl benzene | | X | |
| 2,4-Dibromophenol | | | X |
| 2,6-Dibromophenol | | | X |
| 2,6-Dibromo-4-methyl phenol | | | X |
| 2,4,6-Tribromophenol | | X | X |
| 4-Chloro-2,6-di-*tert*-butyl phenol | | X | |
| 4-Bromo-2,6-di-*tert*-butyl phenol | | X | |
| **Other Halogenated DBPs** | | | |
| 2,2,2-Trichloroacetamide | | X | |
| 2,4,6-Tribromobenzeneamine | | X | X |

NOTE: Underlined DBPs were confirmed by the analysis of authentic standards; other DBPs listed are tentative identifications. Haloaldehydes are likely present in hydrated form (as for bromal hydrate).

DBPs presented in this paper were confirmed with purchased or synthesized standards. For completeness, THMs and haloacetic acids (HAAs), which are common DBPs of chlorine and chloramine, were analyzed for and quantified in these drinking water samples.

## DBPs from Treatment with Ozone

As mentioned earlier, other researchers have reported a few brominated organics found in ozonated water containing bromide. These include bromoform, dibromoacetonitrile, bromoacetic acid, dibromoacetic acid, cyanogen bromide, bromopicrin, 1,1-dibromoacetone, bromoalkanes, and bromohydrins (3-9). As a result, not only were all GC/MS peaks examined, but reconstructed ion chromatograms were also created for the specific mass spectral ions of all brominated compounds that have been previously reported. This was done in case co-eluting compounds might obscure their detection. After careful examination, the only brominated compound that was found as an actual DBP was dibromoacetonitrile (Table 1). Dibromoacetonitrile was also present in the raw, untreated water, but at levels approximately 20 times lower than in the ozonated sample. Bromoform was found in all ozonated samples (resin extracts), but it was also present at the same levels in the raw, untreated water (with and without the addition of sodium bromide). Therefore, based on our criteria, bromoform was not recognized as a DBP in these samples. It is interesting that in the quantitative analyses of THMs (using EPA Method 551), bromoform was not found above the detection limit in the ozone-bromide samples. This is probably because much less water was concentrated using this method than was concentrated for qualitative analyses (75 L).

Many non-halogenated DBPs were observed, however, these were also seen in ozonated water without added bromide (10). Many of these compounds have been previously observed in other studies (8-9, 16); however, many others have never been reported. This is probably due to the lack of spectral library information on several of these DBPs and to the difficulty in identifying PFBHA derivatives. Mass spectra of the PFBHA derivatives are not present in the spectral libraries, and, in many cases, these PFBHA derivatives exhibit very weak or no molecular ions in their mass spectra. Through the use of high-resolution MS, CI-MS, and IR, we identified compounds not present in the library databases. Through the synthesis of several aldehyde and ketone standards, we were able to identify PFBHA-derivatized aldehydes and ketones. Details on the identification of these DBPs can be found elsewhere (10).

## DBPs from Treatment with Ozone-Chlorine and Ozone-Chloramine

When chlorine or chloramine was used as a secondary disinfectant, following ozonation of the bromide-enriched water, a shift to more brominated species was observed. This effect has been observed by others for THMs and HAAs (17-20) and has been attributed to the formation of HOBr, which is a more effective halogen-substituting agent than HOCl (18-20).

Without elevated levels of bromide present in the water, secondary chlorine and chloramine tend to form predominantly chlorinated compounds, such as the more highly chlorinated species of THMs (chloroform and bromodichloromethane), chlorinated species of HAAs (di- and trichloroacetic acid), chloropropanones, and chloroacetonitriles (*10*). However, with elevated bromide ion present, secondary chlorine and chloramine form more highly brominated THMs (bromoform and dibromochloromethane), brominated species of HAAs (bromoacetic acid, and dibromoacetic acid), brominated propanones, and bromoacetonitriles. Quantitatively, this is evidenced by increased bromoform levels (e.g., from 0.9 µg/L without added bromide to 37.2 µg/L with bromide for ozone-chlorine treatment) and corresponding decreased chloroform levels (e.g., from 19.4 µg/L without added bromide to 1.7 µg/L with bromide). We also saw an increase in dibromoacetic acid levels (e.g., from 2.2 µg/L without added bromide to 12.9 µg/L with bromide) with a corresponding decrease in dichloroacetic acid levels (e.g., from 8.8 µg/L without added bromide to 1.7 µg/L with bromide). This shift is also evident qualitatively, as judged by the presence of several brominated species in the bromide-enriched water that were either not observed in the corresponding low-bromide samples, or were observed at extremely low levels. These compounds include: brominated alkanes, such as dibromoiodomethane; brominated propanones, such as tribromopropanone and tetrabromopropanone; bromoacetonitriles, such as dibromoacetonitrile and tribromoacetonitrile; and bromo-nitro-methanes, such as tribromonitromethane. Several of these DBPs have not been previously reported. These shifts from chlorinated species to brominated species are noted in Table 2.

Because chlorine and chloramine form the chlorinated analogs of these compounds without ozone (*10*), it is likely that most of these brominated DBPs are formed strictly from the reaction of chlorine or chloramine with the natural organic matter (NOM) and elevated bromide present in the water, and not by the combination of ozone-bromide and chlorine or chloramine. The only DBP that appeared to be formed only by the combination of ozone-bromide and chlorine or chloramine was the compound tentatively identified as 3-bromo-4-hydroxy-5-methoxy-benzaldehyde. It was observed in a sample treated with ozone and chloramine. The chlorinated analog of this compound was not observed in a chlorinated water sample from a separate study with the same raw water (*10*). It is possible that 3-bromo-4-hydroxy-5-methoxy-benzaldehyde was formed by the reaction of chloramine-bromide with a pre-formed ozone by-product.

Many of the compounds identified in the ozone-bromide/chlorine and ozone-bromide/chloramine samples have never been reported. This may be due to the fact that many of these compounds, including some with relatively simple structures, were not present in the mass spectral library databases (NIST or Wiley). The family of brominated propanones identified herein are examples of DBPs that were not present in the library databases. These include 1,1-dibromopropanone; 1,1-dibromo-3-chloropropanone; 1,1,1-tribromopropanone; 1,1,3-tribromopropanone; 1,1-dibromo-3,3-dichloropropanone;
1,3-dibromo-1,3-dichloropropanone; 1,1,3,3-tetrabromopropanone; 1,1,3-tribromo-3-chloropropanone; 1,1,1,3-tetrabromo-3-chloropropanone; and 1,1,1,3,3-pentabromo-3-chloropropanone.

The identification of dibromonitromethane illustrates the process used to identify the unknown DBPs. This chemical was found in ozone-bromide treatments involving

**Table 2. Shift in Speciation from Chlorinated to Brominated DBPs (Examples) in Bromide-Enriched Water**

| Low Bromide (0.05 mg/L) Ozone-Chlorine or Ozone-Chloramine | High Bromide (1.0 mg/L) Ozone-Chlorine or Ozone-Chloramine |
|---|---|
| **Chlorinated THM species** | **Brominated THM species** |
| Chloroform, Bromodichloromethane, Dichloroiodomethane | Bromoform, Dibromochloromethane, Dibromoiodomethane |
| **Chlorinated HAA species** | **Brominated HAA species** |
| Chloro-, Dichloro-, and Trichloro-acetic acid | Bromo-, Bromochloro-, Dibromo-acetic acid |
| **Chloro-aldehydes** | **Bromo-aldehydes** |
| Trichloroacetaldehyde (chloral hydrate) | Tribromoacetaldehyde (bromal hydrate) |
| **Chloro-propanones** | **Bromo-propanones** |
| 1,1-Dichloropropanone 1,1,1-Trichloropropanone 1,1,3,3-Tetrachloropropanone 1,1,1,3,3-Pentachloropropanone | 1,1-Dibromopropanone 1,1,1-Tribromopropanone 1,1,3,3-Tetrabromopropanone 1,1,1,3-Tetrabromo-3-chloropropanone |
| **Chloro-acetonitriles** | **Bromo-acetonitriles** |
| Dichloroacetonitrile | Dibromoacetonitrile, Tribromoacetonitrile |
| **Chloro-nitro-methanes** | **Bromo-nitro-methanes** |
| Trichloronitromethane (chloropicrin) | Dibromonitromethane, Tribromonitromethane (bromopicrin) |

secondary chlorine or chloramine. It was also found in low-bromide waters treated with ozone-chlorine, ozone-chloramine, and chlorine or chloramine only (*10*). However, concentrations increased 10-fold in the bromide-enriched water. Figure 1a shows the low-resolution GC/EI-MS spectrum, and Figure 1b shows the GC/IR spectrum obtained. The mass spectrum was not present in either the NIST or the Wiley library databases; nor was the infrared spectrum present in the infrared library database (NIH). Immediately evident from the spectrum is the presence of at least two bromine atoms in the structure, as evidenced by the two-bromine isotopic distribution shown for the ions at $m/z$ 171/173/175. However, this cluster at $m/z$ 171 did not represent the molecular ion; hence, molecular weight information and the overall composition of the structure was missing from the mass spectrum. Attempts to obtain a molecular ion by positive ion

399

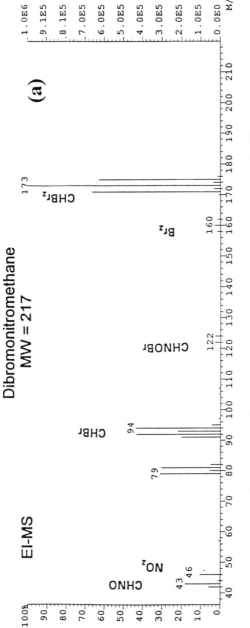

Figure 1. (a) EI mass spectrum and (b) IR spectrum of compound identified as dibromonitromethane.

*Continued on next page.*

400

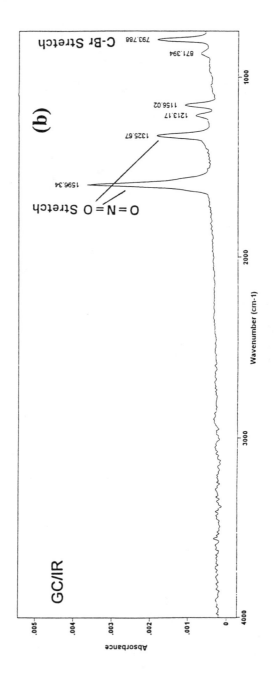

Figure 1. *Continued.*

methane or 2% ammonia in methane failed, probably due to the strong electronegativity of this molecule. However, when high-resolution EI mass spectrometry (10,000 resolution) was applied to this compound, the missing half of the molecule was evident in the lower part of the mass spectrum. The ion at $m/z$ 43, which most often is $C_3H_7$, was clearly the more unusual CHNO fragment, as indicated by its exact mass of 43.006 Da. The ion at $m/z$ 46 was determined to be $NO_2$, with an exact mass of 45.993 Da. Therefore, from the remaining ions identified as CHNOBr ($m/z$ 122/124), $Br_2$ ($m/z$ 158/160/162), and $CHBr_2$ (171/173/175) (Figure 1a), the complete structure was postulated to be dibromonitromethane. GC/IR confirmed this assignment, with definitive O=N=O stretches (anti-symmetric and symmetric) evident in the IR spectrum at 1596 and 1326 cm⁻¹. The C-Br stretch was also evident at 794 cm⁻¹. From this spectral information alone, the degree of confidence in this identification was quite high. To confirm its identity, a standard was synthesized, and the mass spectrum, IR spectrum, and GC retention time of the standard matched that of the tentatively identified compound, thus confirming its identity.

## Conclusions

In conclusion, the effect of elevated bromide levels in the ozone-treated water resulted in only one brominated DBP that was detected—dibromoacetonitrile. It is quite possible that bromate was also present, but inorganic DBPs were not measured in this study. The effect of elevated bromide showed a much more pronouced effect on the DBPs formed when secondary chlorine or chloramine treatment was applied following ozonation. For those samples, many halogenated DBPs were found, many of which have never been reported. Of these compounds, an entire family of halopropanones was identified. It is not known at this time whether any of these brominated DBPs pose a health risk. Preliminary health effects studies are underway to determine the potential risks of some of these selected bromine-containing DBPs.

## Acknowledgments

We wish to acknowledge the assistance of Wayne Koffskey, Chief Chemist, Jefferson Parish Department of Public Utilities, Jefferson Parish, LA. We would also like to acknowledge Dave Cmehil, John Glass Sr., Robert Miller, Paul Ringhand, Brad Smith, Ray Hauck, Mark Domino, and Lucy Garner for their contributions to this study; Yingbo Guo for her assistance with mass spectrometry analyses; and George Yager for assistance with GC/IR analyses.

## References

1. *Fed. Reg.* **1994**, *59* (145), 38668.
2. Kurokawa, Y.; Aoki, S.; Matsushima, Y.; Takamura, N.; Imazawa, T.; Hayashi, Y. *J. Nat. Cancer Inst.* **1986**, *77*, 977.
3. Haag, W.R.; Hoigne, J. *Environ. Sci. Technol.* **1983**, *17*, 261.
4. Krasner, S.W.; Gramith, J.T.; Means, E.G.; Patania, N.L.; Najm, I.N.; Aieta, E.M. Formation and Control of Brominated Organic Disinfection By-Products. In

*Proceedings (Water Quality for the New Decade)*, AWWA Annual Conference, Philadelphia, PA, June 1991. American Water Works Association: Denver, CO, 1991.

5. Amy, G.L.; Siddiqui, M.S. Ozone-Bromide Interactions in Water Treatment. In *Proceedings (Water Research for the New Decade)*, AWWA Annual Conference, Philadelphia, PA, June 1991. American Water Works Association: Denver, CO, 1991.

6. Collette, T.W.; Richardson, S.D.; Thruston, Jr., A.D. *Appl. Spectrosc.* **1994**, *48* (10), 1181.

7. Cavanaugh, J.E.; Weinberg, H.S.; Gold, A.; Sangaiah, R.; Marbury, D.; Glaze, W.H.; Collette, T.W., Richardson, S.D.; Thruston, Jr., A.D. *Environ. Sci. Technol.* **1992**, *26*, 1658.

8. Glaze, W.H.; Weinberg, H.S. *Identification and Occurrence of Ozonation By-Products in Drinking Water*, American Water Works Association Research Foundation: Denver, CO, 1993; pp 13-18.

9. Richardson, S.D. Drinking Water Disinfection By-Products. In *The Encyclopedia of Environmental Analysis and Remediation*; Meyers, R.A., Ed. John Wiley & Sons: New York, 1998; Vol 3, p 1398.

10. Richardson, S.D.; Thruston, Jr., A.D.; Caughran, T.V.; Chen, P.H.; Floyd, T.L.; Collette, T.W. Identification of New Ozone Disinfection By-Products in Drinking Water. *Environ. Sci. Technol.*; in press, 1999.

11. Richardson, S.D.; Thruston, Jr., A.D.; Collette, T.W.; Patterson, K.S.; Lykins, Jr., B.W.; Majetich, G.; Zhang, Y. *Environ. Sci. Technol.* **1994**, *28*, 592.

12. Sclimenti, M.J.; Krasner, S.W.; Glaze, W.H.; Weinberg, H.S. In *Proceedings*, AWWA Water Quality Technology Conference, San Diego, CA, Nov 1990. American Water Works Association: Denver, CO, 1991.

13. Kanniganti, R.; Johnson, J.D.; Ball, L.M.; Charles, M.J. *Environ. Sci. Technol.* **1992**, *26*, 1998.

14. *Methods for the Determination of Organic Compounds in Drinking Water, Supplement 1*. EPA/600/4-90-020. Environmental Monitoring Systems Laboratory, Office of Research and Development, U.S. Environmental Protection Agency: Cincinnati, OH, July 1990.

15. Richardson, S.D.; Thruston, Jr., A.D.; Caughran, T.V.; Chen, P.H.; Collette, T.W.; Floyd, T.L. Identification of New Drinking Water Disinfection By-Products Formed in the Presence of Bromide. *Environ. Sci. Technol.*, in press, 1999.

16. Coleman, W.E.; Munch, J.W., Ringhand, H.P.; Kaylor, W.H.; Mitchell, D.E. *Ozone Sci. Eng.* **1992**, *14*, 51.

17. Shukairy, H.M.; Miltner, R.J.; Summers, R.S. *J. Am. Water Works Assoc.* **1994**, *86* (6), 72.

18. Rook, J.J. *Water Treat. Exam.* **1974**, *23*, 234.

19. Rook, J.J.; Gras, A.A.; van der Heijden, B.G.; de Wee, J. *J. Environ. Sci. Health* **1978**, *A13*, 91.

20. Cooper, W.J.; Zika, R.G.; Steinhauer, M.S. *J. Am. Water Works Assoc.* **1985**, *77* (4), 116.

# INDEXES

# Author Index

404

# Subject Index

# Highlights from ACS Books

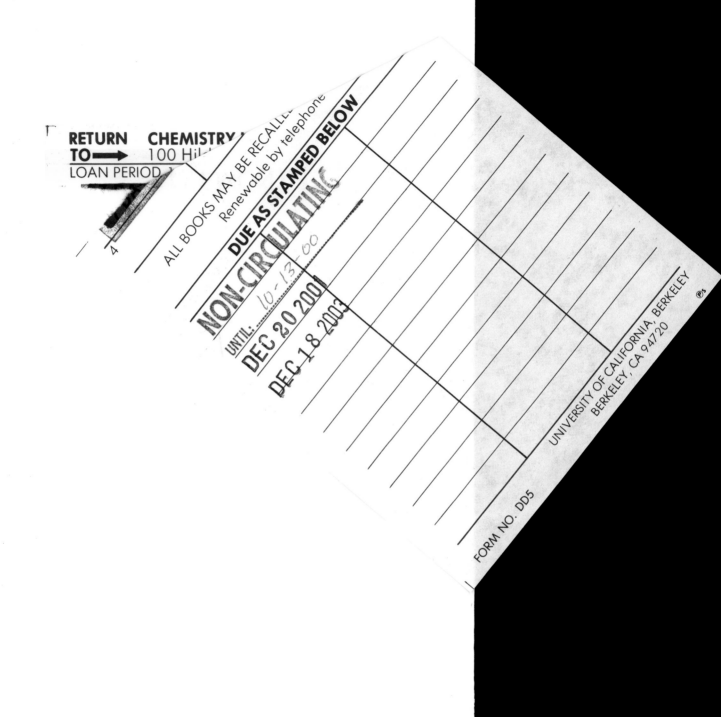